Topics in the Theory of Solid Materials

T0204137

Series in Materials Science and Engineering

Series Editors: **B Cantor**, University of York, UK
M J Goringe, School of Mechanical and Materials Engineering, University of Surrey, UK

Other titles in the series

Microelectronic Materials
C R M Grovenor
Department of Materials, University of Oxford, UK

Physical Methods for Materials Characterisation
P E J Flewitt
Magnox Electric, Berkeley, UK
and
R K Wild
University of Bristol, UK

Aerospace Materials
B Cantor, H Assender and P Grant
Department of Materials, University of Oxford, UK

Fundamentals of Ceramics
M Barsoum
Department of Materials Engineering, Drexel University, USA

Solidification and Casting
B Cantor and K O'Reilly
Department of Materials, University of Oxford, UK

Forthcoming titles in the series

Computer Modelling of Heat, Fluid Flow and Mass Transfer in Materials Processing
C-P Hong
Yonsei University, Korea

Fundamentals of Fibre Reinforced Composite Materials
A R Bunsell and J Renard
Centre des Matériaux, Pierre-Marie Fourt, France

Metal and Ceramic Composites
B Cantor, F P E Dunne and I C Stone
Department of Materials, University of Oxford, UK

High Pressure Surface Science
Y Gogotsi and V Domnich
Department of Materials Engineering, Drexel University, USA

Series in Materials Science and Engineering

Topics in the Theory of Solid Materials

John M Vail

*Department of Physics and Astronomy,
University of Manitoba, Canada*

Institute of Physics Publishing
Bristol and Philadelphia

British Library Cataloguing-in-Publication Data

A catalogue record for this book is available from the British Library.

ISBN 0 7503 0729 3

Library of Congress Cataloging-in-Publication Data are available

Series Editors: **B Cantor and M J Goringe**

Commissioning Editor: Tom Spicer
Production Editor: Simon Laurenson
Production Control: Sarah Plenty
Cover Design: Victoria Le Billon
Marketing: Nicola Newey and Verity Cooke

Published by Institute of Physics Publishing, wholly owned by The Institute of Physics, London

Institute of Physics Publishing, Dirac House, Temple Back, Bristol BS1 6BE, UK

US Office: Institute of Physics Publishing, The Public Ledger Building, Suite 929, 150 South Independence Mall West, Philadelphia, PA 19106, USA

Typeset by Academic + Technical, Bristol
Printed in the UK by MPG Books Ltd, Bodmin, Cornwall

Dedication

To the memory of my parents
Oneita Weaver Vail (née Cannon)
and
Walter Moncrieff Vail

Contents

Preface   **xiii**

1   Strain and stress in continuous media   **1**
  1.1   Introduction   1
  1.2   Deformation: strain and rotation   2
      1.2.1   The strain tensor   4
      1.2.2   The rotation tensor   6
  1.3   Forces and stress   7
  1.4   Linear elasticity   10
      1.4.1   Hooke's law   12
      1.4.2   Isotropic media   13
      1.4.3   Elastic moduli   16
      1.4.4   Stability conditions   18
  1.5   Equilibrium   20

2   Wave propagation in continuous media   **22**
  2.1   Introduction   22
  2.2   Vector fields   22
  2.3   Equation of motion   24
  2.4   Wave propagation   27
      2.4.1   Shear and rotational waves 27
      2.4.2   Dilatational or irrotational waves   28
      2.4.3   General discussion   30
  Appendix to Chapter 2   31

3   Thermal properties of continuous media   **34**
  3.1   Introduction   34
  3.2   Classical thermodynamics   34
      3.2.1   The Maxwell relations   34
      3.2.2   Elastic constants, bulk moduli and specific heats   35
  3.3   Thermal conduction and wave motion   38
  3.4   Wave attenuation by thermal conduction   41

4   Surface waves   **48**
   4.1   Introduction   48
   4.2   Rayleigh waves   49
   4.3   Boundary conditions   50
   4.4   Dispersion relation   51
   4.5   Character of the wave motion   54

5   Dislocations   **57**
   5.1   Introduction   57
   5.2   Description of dislocations   57
   5.3   Deformation fields of dislocations   62
     5.3.1   Screw dislocation   63
     5.3.2   Edge dislocation   65
   5.4   Uniform dislocation motion   69
   5.5   Further study of dislocations   71

6   Classical theory of the polaron   **73**
   6.1   Introduction   73
   6.2   Equations of motion   74
   6.3   The constant-velocity polaron   78
   6.4   Polaron in a magnetic field: quantization   86

7   Atomistic quantum theory of solids   **90**
   7.1   Introduction   90
   7.2   The hamiltonian of a solid   90
   7.3   Nuclear dynamics: the adiabatic approximation   91
   7.4   The harmonic approximation   93
   7.5   Phonons   94
     7.5.1   Periodic boundary conditions for bulk properties   94
     7.5.2   The dynamical matrix of the crystal   97
     7.5.3   The normal modes of crystal vibration 100
     7.5.4   Electrons and phonons: total energy   102
   7.6   Statistical thermodynamics of a solid   103
     7.6.1   Partition function of the crystal   104
     7.6.2   Equation of state of the crystal   105
     7.6.3   Thermodynamic internal energy of the crystal; phonons as bosons   107
   7.7   Summary   108

8   Phonons   **109**
   8.1   Introduction   109
   8.2   Monatomic linear chain   110
   8.3   Diatomic linear chain   116
   8.4   Localized mode of a point defect   121

9   Classical atomistic modelling of crystals   **126**
   9.1   Introduction   126
   9.2   The shell model for insulating crystals   126
   9.3   Cohesive energy of a crystal   129
   9.4   Elastic constants   131
   9.5   Dielectric and piezoelectric constants   135

10 Classical atomic diffusion in solids   **140**
   10.1   Introduction   140
   10.2   The diffusion equation   141
          10.2.1   Derivation   141
          10.2.2   Planar source problem   144
   10.3   Diffusion as a random walk   147
   10.4   Equilibrium concentration of point defects   151
   10.5   Temperature dependence of diffusion: the Vineyard relation   154
   Appendix to Chapter 10: Stirling's formula   161

11 Point defects in crystals   **163**
   11.1   Introduction   163
          11.1.1   Crystals and defects   163
          11.1.2   Modelling of point defects in ionic crystals   165
   11.2   Classical diffusion   168
          11.2.1   Copper and silver diffusion in alkali halides   168
          11.2.2   Dissociation of the oxygen-vacancy defect complex
                   in BaF_2 171
   11.3   Defect complex stability   173
   11.4   Impurity charge-state stability   176
          11.4.1   Nickel in MgO   176
          11.4.2   Oxygen in BaF_2 177
   11.5   Optical excitation   177
          11.5.1   Frenkel exciton and impurity absorption in MgO   178
          11.5.2 Cu^+ in NaF   179
          11.5.3 O^- in BaF_2 179
   11.6   Spin densities   181
          11.6.1   F center in NaF   181
          11.6.2 F_2^+ center in NaF   182
          11.6.3 $(F_2^+)^*$ center in NaF   182
   11.7   Local band-edge modification   183
          11.7.1   Valence band edge in NiO:Li   183
          11.7.2   Conduction band edge in $BaF_2:O^-$ 184
   11.8   Electronic localization   185
   11.9   Quantum diffusion   187
   11.10 Effective force constants for local modes   189

11.11 Summary   190
Appendix to Chapter 11: the ICECAP method   193

12 Theoretical foundations of molecular cluster computations   196
  12.1 Introduction   196
  12.2 Hartree–Fock approximation   197
    12.2.1   The approximation   197
    12.2.2   Normalization   201
    12.2.3   Total energy   202
    12.2.4   Charge density and exchange charge   204
    12.2.5   The single-particle density functional   207
  12.3 The Fock equation   208
    12.3.1   The variational derivation   208
    12.3.2   Total energy algorithm   214
    12.3.3   Solution of the Fock equation   214
  12.4 Localizing potentials   218
  12.5 Embedding in a crystal   221
    12.5.1   Introduction   221
    12.5.2   Approximate partitioning with a localizing potential   223
    12.5.3   Summary   227
  12.6 Correlation   229
  12.7 One-, two- and N-particle density functionals   238
    12.7.1   Introduction   238
    12.7.2   Density functional of Hohenberg and Kohn   239
    12.7.3   Reduced density matrices   241
    12.7.4   The many-fermion system   245
    12.7.5   The density functional and the two-particle density
        operator   248

13 Paramagnetism and diamagnetism in the electron gas   250
  13.1 Introduction   250
  13.2 Paramagnetism of the electron gas   251
    13.2.1   The total energy   251
    13.2.2   The magnetic susceptibility   254
    13.2.3   Solution at low temperature   257
    13.2.4   Solution at high temperature   266
  13.3 Diamagnetism of the electron gas   269
    13.3.1   Introduction   269
    13.3.2   The Landau levels   270
    13.3.3   The Fermi distribution   273
    13.3.4   Energy considerations   281
    13.3.5   Magnetization: the de Haas–van Alphen effect 283
    13.3.6   Diamagnetism at $T = 0$ 291
Appendix to Chapter 13   294

14 Charge density waves in solids **298**

14.1 Introduction 298

14.2 Effective electron–electron interaction 299

14.3 The Hartree equation: uniform and periodic cases 303

    14.3.1 The Hartree approximation 303

    14.3.2 The uniform solution 306

    14.3.3 The periodic solution 309

14.4 Charge density waves: the Mathieu equation 313

    14.4.1 The Mathieu equation 313

    14.4.2 Solution away from the band gap 316

    14.4.3 Solution near the band gap 318

    14.4.4 The self-consistency condition 322

    14.4.5 The total energy 328

14.5 Discussion 331

References **334**

Exercises **339**

Answers **346**

Author index **353**

Subject index **357**

Preface

The title of this work has been carefully chosen to reflect its contents. 'Topics' connotes the fact that the work presents a collection of subjects that, while they give valuable insights, do not represent a comprehensive account of the theory of solid materials. 'Theory' means that mathematical modelling prevails, and discussion of experimental methods and results is largely absent. 'Solid Materials', rather than 'Solid State Physics', is meant to signal that although the author is a physicist by profession, this work presents aspects of solid state science that are often omitted from, or unemphasized in, university courses in solid state physics.

All of the material in this work has been taught by me either in third or fourth year undergraduate university courses, or in a beginning graduate level course in solid state physics, with two exceptions: the polaron in Chapter 6, and the one- two- and N-particle density functionals in section 12.7. In my opinion, neither of these topics is above the level of the other chapters. In academia, a distinction is made between solid state physics, taught by physics departments, and materials science, often taught in engineering departments, with a third somewhat distinct viewpoint often represented in chemistry departments. While I do not claim to have bridged these distinctions, I have introduced material that has come to my attention as a result of my research having overlapped areas outside of traditional solid state physics. Part of my motivation has been to present material that will tend to nudge the various sub-disciplines to increase their overlap. In particular, and somewhat simplistically put, I want to encourage solid state physicists to learn more about topics other than quantum mechanical features of electronic structure, and other materials scientists to learn more about such features. Another motivation has been to focus attention on basic theoretical concepts in the science of solid materials, to provide a clear understanding of fundamental properties. I hope that, at least in places, the mathematical approach reflects the connection with computational simulation which, at its best, can quantitatively complement experiment in the investigation of solid state properties. That at least is the intention of Chapter 9 on classical atomistic modelling of crystals, and of Chapter 11, with its appendix, on point defects in crystals.

Computational simulation is a highly refined and fast developing technique in the science of materials while, it seems to me, the study of the analytical mathematical methods upon which much of the relevant software is based is diminishing. I hope that this work will encourage the student to believe that the mathematical theory is accessible and, by following through the derivations, to develop more facility with it. To that end, many of the derivations are more detailed than is usually found in advanced textbooks. To keep all this manageable, the physical models upon which the mathematical formulations are based are, to adapt a saying attributed to Einstein, as simple as possible, but no simpler. A collection of questions, with answers provided for most of them, is given at the end of the work. Working these problems will greatly enhance the student's grasp of the material.

I can claim that each chapter addresses properties and phenomena of solids that are of basic scientific importance. Some, such as Chapters 1 to 3 on elastic media, and Chapter 8 on phonons, are more traditional than others. Many other topics are equally basic, and the lecturer who uses this book as a text may well replace some of its topics with others that he/she favors. Nevertheless, I believe that the student will be well served by learning any of the material herein. In order to maintain the clarity of the theoretical approach, I have largely omitted discussion of ancillary examples and of experimental methods and results. Chapter 11 is an exception, where the detailed comparison between a wide range of computed and experimental results is given to illustrate the value and promise of computational modelling, and the essential nature of experiment to that field in providing an anchor for it. A balanced and comprehensive view of solids requires further reading and study, and in many chapters I have given references to works that can provide this. The student's view needs to be broadened beyond that of the present work both in terms of the range of basic phenomena, and particularly in terms of experimental methods and phenomenology.

I can only claim that a few chapters are wholly or largely my original work. These include Chapters 6, 11 and 14 on the polaron, point defects and charge density waves respectively, and to a lesser degree Chapter 12 on molecular clusters. Otherwise, I have relied largely on 'golden oldies', works of established pedagogical excellence by leaders, or in some cases giants, in their fields. Chapters 1 to 4 largely follow Landau and Lifshitz's *Theory of Elasticity* for the continuum theory of solids, with important input from Bhatia and Singh's *Mechanics of Deformable Media*. The introduction to dislocations, Chapter 5, follows closely the classic by the Weertmans. Chapter 6 includes my own previously unpublished work on the classical polaron, with an added section on Schafroth's discussion of quantization for effective potentials that are velocity-dependent. Chapter 7 on the general quantum theory of solids closely follows an article by Maradudin in 1962. The simple one-dimensional illustration of phonon properties in Chapter 8 is modelled on a presentation by Kittel. In

Chapter 9 I reconstruct an example of classical atomistic modelling by my former Harwell colleague, John Harding. Classical atomic diffusion in Chapter 10 closely follows the book by Borg and Dienes. Chapter 11 summarizes most of my research with many collaborators during the past fifteen years on the computation of point defect properties in insulating crystals. Chapter 12, on the theoretical foundations of molecular cluster computations, reflects what I have learned in the course of the computations of Chapter 11, and incorporates particularly Pryce's elegant derivation of the Fock equation, and Thouless' treatment of many-body perturbation theory, for correlation. In Chapter 13, the discussion of electron paramagnetism follows Huang's *Statistical Mechanics*, and diamagnetism is discussed according to Pippard's brilliant exposition. Chapter 14 on charge density waves follows a partly pedagogical work of my own from long ago (1964).

Of the fourteen chapters, nine of the first ten are based on classical physics, the exception being Chapter 7, which lays out the atomistic basis of materials science in quantum mechanical terms. The preceding six chapters are based on continuum mechanics, although dislocations in Chapter 5 require some acknowledgment of atomic structure, and the polaron in Chapter 6 is a discrete particle, or rather, quasiparticle. These nine classical chapters might be considered to fall within the field of traditional materials science, with some overlap into the realm of the traditional physics of solids. I think that materials physics pedagogy needs to move more into this area. Chapter 7, although it is quantum mechanical, uses only the ideas of Schrödinger's equation and the simple harmonic oscillator. It should therefore be readily accessible to the materials science student, with appropriate leadership from the lecturer.

Chapters 11 to 14 are unabashedly quantum mechanical. In my view they are legitimate fare for students of materials science, but they will remain closed to such students unless they have training in quantum mechanics at the level of senior undergraduate physics courses. The possible exception is parts of Chapter 12, which remain of specialist interest to quantum chemists and solid state physicists. I believe that the current generation of materials scientists need to have training in quantum mechanics because so much of modern materials science is based on quantum mechanical features and properties, and increasingly so. These chapters represent essentially standard fare for physics students.

The whole fourteen chapters are appropriate for study by physics students, and probably also for students of materials chemistry. Chapters 1 to 4 were part of a third year half course on the continuum physics of materials (about 35 hours of lecture in all) that included a sizable component of fluid mechanics. The first half of Chapter 13, on paramagnetism of the electron gas, was given as an example of the application of statistical thermodynamics at the end of a third year full course on thermodynamics. Both these third year courses (in the honors program in our department, slightly above the level of some physics major programs in the US) were also available for fourth year credit, and for graduate credit. Chapter 8, on phonons, is fairly standard in fourth year solid state physics courses, and was given as such. Chapters 5, 7, 9, 10, 11 and part of Chapter 12 were given in a fourth year half course on the atomistic physics of materials, also available for graduate credit. Most of this material was also given at one time or another in a beginning graduate course in solid state physics. Diamagnetism of the electron gas (in Chapter 13) and charge density waves (Chapter 14) were also given in this graduate course. Those who consult the original sources will realize that such material can only be presented effectively to such students through a major pedagogical effort, which I hope is reflected in the present work.

The present generation of students in the science of materials have a huge new opportunity from the developments taking place in the interdisciplinary area that overlaps molecular biology and related topics on one side, and traditional solid state science on the other. To take advantage, students trained in physics will need to have a special kind of introduction to biology. Students of biology will need to be given the mathematics and physics to enable them to understand molecular structure and processes at the quantum cluster level. Although I have not addressed this issue in the text, I mention it here because I feel that it is of central importance that all students of the physical and life sciences now upgrade their training to include significant new holdings in this overlap area. The responsibility for this will lie mainly with their teachers. I believe that much of the material of Chapter 12 will come to be needed by a much wider range of science students than has been the case in the past.

I am grateful to Miguel A Blanco of the University of Oviedo for providing the cover illustration, from a computational study of interstitial diffusion channels in β-Ga_2O_3. It is with pleasure and humility that I acknowledge influences, assistance and support without which this work could not have been done. I thank my teachers, long ago now, at the University of Manitoba and at Brandeis University; my research colleagues in the United Kingdom and the United States, as well as my research students in Canada. All these are a numerous band, too many to name here, and their much-appreciated roles have been complex, worthy of an academic autobiography, but here is not the place for it. I am grateful to the University of Manitoba for having put before me a steady stream of students who inspired my pedagogical efforts, and for having supported me as a Senior Scholar during the past two years; my editor, Tom Spicer, at Institute of Physics Publishing for his patience; Wanda Klassen of the University of Manitoba for having typed the entire manuscript and its revisions with enthusiasm and uncanny accuracy; and my beloved wife, Audrey, for her unflagging support, patience and encouragement.

John M Vail
Winnipeg

Chapter 1

Strain and stress in continuous media

1.1 Introduction

When no external force is applied to a solid object, and it is in equilibrium, it has some particular shape. When forces are applied, its shape changes. This effect, referred to as strain, will be defined precisely in section 1.2. For a segment of material of the object, forces acting upon it may originate outside the object, or may come from adjacent segments of the material. Those internal forces that act on the surface of the segment constitute the so-called stress, which will be precisely defined in section 1.3. Stresses produce strain; the relationship between the two will be discussed in section 1.4. Equilibrium will be discussed in section 1.5.

Although we know that a solid is made up of atoms, which in turn consist of nuclei and electrons, our human senses are incapable of detecting this fine-scale structure. We can only detect variations in material properties over a spatial scale that is referred to as macroscopic, let us say over a scale of the order of a millimetre, or perhaps even as small as 0.01 mm. There are many circumstances where, from the viewpoint of our human senses, the material is a continuum. Chapters 1 to 6 describe the properties of solids in terms of this continuum model. Nevertheless, we shall find that it is helpful to remember that there is a limiting smallness of scale for all materials that exist under conditions like those at the earth's surface ('terrestrial' conditions), and this scale corresponds to atomic dimensions, represented by the angstrom, namely 10^{-10} m.

For the purpose of describing the state of a solid throughout its volume and at its surface, we shall conceptually divide the material up into a very large number of small volume segments. For the continuum model, these segments must still be large compared with atomic dimensions (10^{-10} m), but small compared with macroscopic dimensions (10^{-5} m, let us say). Between the two regimes there are five orders of magnitude, corresponding to a linear distance of $\sim 10^5$ atomic diameters. Thus, for example, the small continuum volume segments might be a thousand atoms in diameter, but yet also be one hundred thousandth of the diameter of a 1 cc sample of the material.

In Chapter 9, we shall derive detailed relationships between an atomic-scale model of a solid, and the continuum model properties.

The content of Chapters 1 to 4 are mainly based on the textbooks by Landau and Lifshitz (1970) and by Bhatia and Singh (1986).

1.2   Deformation: strain and rotation

Suppose that a sample of solid material has been deformed. In general it means that the material (the set of atoms) within a small volume at arbitrary position $\vec{r}$ has been displaced to a new position $(\vec{r} + \vec{u}(\vec{r}))$. We therefore describe the deformation by the displacement field or deformation vector field $\vec{u}(\vec{r})$. More generally, $\vec{u}$ may also depend on time: see Chapter 2.

The nature of the deformation is not best described by $\vec{u}(\vec{r})$, but by the effect of the deformation on small lengths, areas, and volumes within the sample. We therefore consider a small vector segment $d\vec{r}$ at $\vec{r}$ in the sample. More precisely, we consider the material (the set of atoms) that coincides with $d\vec{r}$. Throughout this work, we want to limit the discussion to small deformations. We shall specify more fully what 'small deformations' means quantitatively as we go along. One aspect of it is, however, that under deformation, $d\vec{r}$ is deformed mainly into another vector $d\vec{r}'$: that is, the atoms that originally coincided with the straight-line segment $d\vec{r}$ before deformation, approximately coincide with the straight-line segment $d\vec{r}'$ after deformation. We comment that we use the notation $d\vec{r}$, which properly refers to an infinitesimal, but which here refers to an entity that is infinitesimal only in the context of a macroscopic scale, but not down to an atomic scale. The methods of calculus will nevertheless be applied to such entities.

In figure 1.1, the extremities of $d\vec{r}$, at $\vec{r}$ and $(\vec{r} + d\vec{r})$ respectively, are shown displaced to $(\vec{r} + \vec{u}(\vec{r}))$ and to $(\vec{r} + d\vec{r} + \vec{u}(\vec{r} + d\vec{r}))$ respectively. Then

$$d\vec{r}' = [\vec{r} + d\vec{r} + \vec{u}(\vec{r} + d\vec{r})] - [\vec{r} + \vec{u}(\vec{r})]$$

$$= d\vec{r} + [\vec{u}(\vec{r} + d\vec{r}) - \vec{u}(\vec{r})]$$

$$= d\vec{r} + (d\vec{r} \cdot \vec{\nabla})\vec{u}(\vec{r}). \tag{1.1}$$

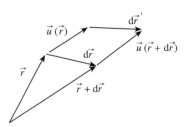

Figure 1.1. The effect of the deformation field $\vec{u}(\vec{r})$ on a segment of material coinciding with a small vector $d\vec{r}$: see equation (1.1).

The final expression in equation (1.1) comes from the Taylor expansion of $\vec{u}(\vec{r} + d\vec{r})$ to first order small quantities, $d\vec{r}$. This first-order approximation corresponds to *linear elasticity*, which will govern our entire discussion of continuous media.

In equation (1.1) we have used vector notation, the first of three notations that will be introduced. The second is index notation:

$$dx_i' = (dx_i + dx_j\,\partial_j u_i).\qquad(1.2)$$

Equation (1.2) comes from identifying the components of a vector such as $d\vec{r}$ by dx_i where $i = 1, 2,$ or 3 corresponds to $x_1, x_2,$ or x_3 cartesian components. Also in equation (1.2), $\partial_j \equiv \partial/\partial x_j$, and $dx_j\,\partial_j \equiv \sum_{j=1}^{3} dx_j\,\partial_j$; that is, the Einstein summation convention is applied, in which the summation symbol is omitted, but implied by a repeated index such as j, in a sum such as that arising from the dot product $(d\vec{r}\cdot\vec{\nabla})$. Equation (1.2) can be put in a more suggestive form,

$$dx_i' = (\delta_{ij} + \partial_j u_i)\,dx_j,\qquad(1.3)$$

where δ_{ij} is the Kronecker delta (unity if $i = j$, zero otherwise); note the Einstein summation in both terms. In equation (1.3), the quantity $\partial_j u_i$ is a cartesian tensor of second rank which describes the *deformation tensor*. We can then think of $(\delta_{ij} + \partial_j u_i)$ along with the summation process as an operator that transforms $d\vec{r}$ into $d\vec{r}'$.

We now introduce the third notation for this work. The elements of a second-rank cartesian tensor may be displayed in an obvious way as a three-by-three matrix. Thus we may represent the deformation tensor $\partial_j u_i$ by $\underline{\underline{u}}$, a matrix whose element in row j and column i is $\partial_j u_i$. Quite generally, we may write

$$\underline{\underline{u}} = (\underline{\underline{\varepsilon}} + \underline{\underline{t}}),\qquad(1.4)$$

where $\underline{\underline{\varepsilon}}$ is the symmetric tensor,

$$\varepsilon_{ij} = \tfrac{1}{2}(\partial_i u_j + \partial_j u_i),\qquad(1.5)$$

called the *strain* tensor, and $\underline{\underline{t}}$ is the antisymmetric tensor,

$$t_{ij} = \tfrac{1}{2}(\partial_i u_j - \partial_j u_i),\qquad(1.6)$$

called the *rotation* tensor, for reasons that will become clear shortly. In the next two subsections we shall see that decomposing $\underline{\underline{u}}$ into $\underline{\underline{\varepsilon}}$ and $\underline{\underline{t}}$ corresponds to separately identifying the two distinct characteristics of deformation, namely compression (or dilatation) and shear on one hand, and rotation on the other, both of which, in general, are position-dependent in the material. Thus, although our notation will not always show it, $\underline{\underline{\varepsilon}} = \underline{\underline{\varepsilon}}(\vec{r})$ and $\underline{\underline{t}} = \underline{\underline{t}}(\vec{r})$.

1.2.1   The strain tensor

In this section, consider a deformation such that $\underline{t} = 0$. In matrix notation equation (1.2) becomes

$$\underline{dr'} = (\underline{\underline{I}} + \underline{\underline{u}}) \cdot \underline{dr} \qquad (1.7)$$

where $\underline{\underline{I}}$ is the identity matrix (diagonal, with unit elements), and $\underline{dr}$ and $\underline{dr'}$ are column matrices with elements dx_i $(i = 1, 2, 3)$, etc. With $\underline{t} = 0$, this becomes

$$\underline{dr'} = (\underline{dr} + \underline{\underline{\varepsilon}} \cdot \underline{dr}) \qquad (1.8)$$

Let the coordinate system be the principal axes of ε_{ij} at position $\vec{r}$, i.e. the coordinate system in which $\underline{\underline{\varepsilon}}$ is diagonal, with diagonal elements $\varepsilon^{(\alpha)}$, $\alpha = 1, 2, 3$. In this case, at $\vec{r}$, we have from equation (1.8)

$$dx'_\alpha = (1 + \varepsilon^{(\alpha)}) \, dx_\alpha. \qquad (1.9)$$

(There is no Einstein summation in equation (1.9).) Equation (1.9) has a simple interpretation: $\varepsilon^{(\alpha)}$ is the fractional elongation (positive or negative) in the principal-axis direction α:

$$\varepsilon^{(\alpha)} = \frac{(dx'_\alpha - dx_\alpha)}{dx_\alpha}. \qquad (1.10)$$

Now throughout our presentation of continuum mechanics, we shall be concerned with the deformation and dynamics of the material in a small volume (ΔV) at arbitrary position $\vec{r}$ in the material. Consider here (ΔV) which is a small rectangular parallelopiped before deformation,

$$(\Delta V) = dx_1 \, dx_2 \, dx_3, \qquad (1.11)$$

with edges oriented in the principal-axis directions. From equation (1.9), this volume becomes $(\Delta V)'$, where

$$(\Delta V)' = (1 + \varepsilon^{(1)})(1 + \varepsilon^{(2)})(1 + \varepsilon^{(3)})(\Delta V). \qquad (1.12)$$

To first-order in the small quantities ε_{ij}, i.e. in $\varepsilon^{(\alpha)}$, we then have

$$(\Delta V)' = [1 + \varepsilon^{(1)} + \varepsilon^{(2)} + \varepsilon^{(3)}](\Delta V). \qquad (1.13)$$

But $(\varepsilon^{(1)} + \varepsilon^{(2)} + \varepsilon^{(3)})$ is the sum of diagonal elements of $\underline{\underline{\varepsilon}}$, defined as the trace of $\underline{\underline{\varepsilon}}$, denoted $\mathrm{Tr}(\underline{\underline{\varepsilon}})$. From equation (1.13) we conclude that the fractional volume change at $\vec{r}$ is given by $\mathrm{Tr}(\underline{\underline{\varepsilon}})$:

$$\frac{[(\Delta V)' - (\Delta V)]}{(\Delta V)} = \mathrm{Tr}(\underline{\underline{\varepsilon}}). \qquad (1.14)$$

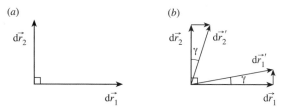

Figure 1.2. The shear effect of strain element ε_{12} on two orthogonal small vectors $d\vec{r}_1$ and $d\vec{r}_2$: see equations (1.16)–(1.19).

We note that this result is independent of coordinate system, and is valid at all points $\vec{r}$. This is because $\mathrm{Tr}(\underline{\varepsilon})$ is a scalar, as is obvious from equation (1.15),

$$\mathrm{Tr}(\underline{\varepsilon}) = \varepsilon_{ii} = \partial_i u_i = \vec{\nabla} \cdot \vec{u}, \tag{1.15}$$

where we note Einstein summation in equation (1.15).

Equations (1.10) and (1.14) above summarize the information about deformation which is contained in the diagonal elements of $\underline{\varepsilon}$. Now consider the off-diagonal elements. For simplicity consider the case where $\underline{\varepsilon}$ has the form

$$\underline{\varepsilon} = \begin{pmatrix} 0 & \varepsilon_{12} & 0 \\ \varepsilon_{12} & 0 & 0 \\ 0 & 0 & 0 \end{pmatrix}. \tag{1.16}$$

Let us examine what such a strain does to two small vectors, perpendicular in the x_1–x_2 plane (see figure 1.2a). Denote these vectors as

$$d\vec{r}_1 = dx_1\,\hat{k}_1, \qquad d\vec{r}_2 = dx_2\,\hat{k}_2, \tag{1.17}$$

where $\hat{k}_1$ and $\hat{k}_2$ are unit vectors in directions x_1 and x_2 respectively. Then from equation (1.8)

$$\underline{dr}_1' = \begin{pmatrix} 1 & \varepsilon_{12} & 0 \\ \varepsilon_{12} & 1 & 0 \\ 0 & 0 & 0 \end{pmatrix} \begin{pmatrix} dx_1 \\ 0 \\ 0 \end{pmatrix} = \begin{pmatrix} dx_1 \\ \varepsilon_{12}\,dx_1 \\ 0 \end{pmatrix}$$

and similarly

$$\underline{dr}_2' = \begin{pmatrix} \varepsilon_{12}\,dx_2 \\ dx_2 \\ 0 \end{pmatrix}.$$

These last two equations are

$$d\vec{r}_1' = (\hat{k}_1 + \varepsilon_{12}\hat{k}_2)\,dx_1, \qquad d\vec{r}_2' = (\varepsilon_{12}\hat{k}_1 + \hat{k}_2)\,dx_2. \tag{1.18}$$

The vectors $\mathrm{d}\vec{r}_1'$ and $\mathrm{d}\vec{r}_2'$ are shown along with $\mathrm{d}\vec{r}_1$ and $\mathrm{d}\vec{r}_2$ in figure 1.2*b*. It is clear from equations (1.18) that the shear angle γ is given, for small ε_{12}, in radian measure:

$$\gamma \approx \varepsilon_{12}. \qquad (1.19)$$

From equation (1.16), with $\mathrm{Tr}(\underline{\varepsilon}) = 0$, we see that there is no fractional volume change from ε_{12}. The reader should show directly that the area subtended by $\mathrm{d}\vec{r}_1'$ and $\mathrm{d}\vec{r}_2'$ in the x_1–x_2 plane is the same as that subtended by $\mathrm{d}\vec{r}_1$ and $\mathrm{d}\vec{r}$, but only to first order. Explicitly,

$$(\mathrm{d}\vec{r}_1' \times \mathrm{d}\vec{r}_2') = (1 - \varepsilon_{12}^2)(\hat{k}_1 \times \hat{k}_2)\,\mathrm{d}x_1\,\mathrm{d}x_2,$$

$$(\mathrm{d}\vec{r}_1 \times \mathrm{d}\vec{r}_2) = (\hat{k}_1 \times \hat{k}_2)\,\mathrm{d}x_1\,\mathrm{d}x_2.$$

1.2.2 The rotation tensor

From the antisymmetry of the rotation tensor t_{ij}, equation (1.6), we see that it has the matrix form

$$\underline{t} = \begin{pmatrix} 0 & -R_3 & R_2 \\ R_3 & 0 & -R_1 \\ -R_2 & R_1 & 0 \end{pmatrix} \qquad (1.20)$$

where

$$R_1 = -t_{23} = -\tfrac{1}{2}(\partial_2 u_3 - \partial_3 u_2)$$

is in fact the x_1 component of the vector:

$$\vec{R} = -\tfrac{1}{2}(\vec{\nabla} \times \vec{u}). \qquad (1.21)$$

Similarly t_{31} and t_{12} give the x_2 and x_3 components of the vector $\vec{R}$, equation (1.21).

Consider now the special case where $\underline{\varepsilon} = 0$. Then, from equations (1.7) and (1.4),

$$\underline{\mathrm{d}r}' = (\underline{\mathrm{d}r} + \underline{t} \cdot \underline{\mathrm{d}r}). \qquad (1.22)$$

But from equation (1.20), $\underline{t} \cdot \underline{\mathrm{d}r}$ is a column vector with elements equal to the components of $(\vec{R} \times \mathrm{d}\vec{r})$, i.e.

$$\mathrm{d}\vec{r}' = [\mathrm{d}\vec{r} + (\vec{R} \times \mathrm{d}\vec{r})]. \qquad (1.23)$$

Then, since $(\mathrm{d}\vec{r} \cdot \vec{R} \times \mathrm{d}\vec{r}) = 0$, it follows, to first order in the small quantities R_i, that

$$|\mathrm{d}\vec{r}'|^2 = |\mathrm{d}\vec{r}|^2, \qquad (1.24)$$

from equation (1.23). This means that a deformation $\underline{t}$, with $\underline{\varepsilon} = 0$, causes no compression or dilatation in any direction, and therefore no volume change. We conclude that, for the most general deformation, to first order, all volume change is contained in $\mathrm{Tr}(\underline{\varepsilon})$, as in equation (1.14).

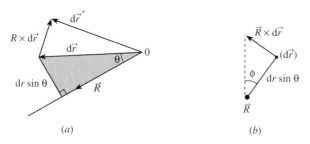

Figure 1.3. (*a*) The deformation effect of rotation vector $\vec{R}$ on small vector $d\vec{r}$; (*b*) the plane orthogonal to $\vec{R}$ containing ($\vec{R} \times d\vec{r}$): see equations (1.23)–(1.25).

We now show that equation (1.23) represents a rotation of $d\vec{r}$ through an angle of $|\vec{R}|$ radians about the direction of $\vec{R}$. It is for this reason that $\vec{R}$ is called the rotation vector: through its magnitude and direction it gives the rotation which deformation $\underline{t}$, equation (1.20), produces for arbitrary small vector $d\vec{r}$, at any point in the material. In figure 1.3*a*, we show the vectors $d\vec{r}$, $\vec{R}$, $\vec{R} \times d\vec{r}$, and $d\vec{r}'$ in relation with each other. Now consider the plane perpendicular to $\vec{R}$, looking in the direction of $(-\vec{R})$, figure 1.3*b*, which contains the line denoted ($dr \sin \theta$), that projects $d\vec{r}$ onto the direction of $\vec{R}$. The point denoted ($\vec{R}$) is the intersection of the line of $\vec{R}$ with this plane; the points denoted ($d\vec{r}$) and ($d\vec{r}'$) are the positions $d\vec{r}$ and $d\vec{r}'$ relative to the origin 0 (figure 1.3*a*): they lie in the plane of figure 1.3*b*. The vector $\vec{R} \times d\vec{r}$ also lies in this plane. It is perpendicular to the plane of $\vec{R}$ and $d\vec{r}$, shown shaded in figure 1.3*a*. Since this plane is seen edge-on in figure 1.3*b*, $\vec{R} \times d\vec{r}$ is perpendicular to the line denoted $dr \sin \theta$, which is part of this plane edge. Now since $|\vec{R}|$ is a first-order small quantity compared with $|d\vec{r}| = dr$, the angle φ will be small, in general. Then ($\vec{R} \times d\vec{r}$) nearly coincides with a segment of a circle of radius $dr \sin \theta$ centered on the point ($\vec{R}$) in figure 1.3*b*. Thus in radian measure,

$$\varphi \approx \frac{|\vec{R} \times d\vec{r}|}{dr \sin \theta} = |\vec{R}|. \tag{1.25}$$

But this angle φ is the angle by which $d\vec{r}$ must be rotated about the direction of $\vec{R}$ (perpendicular to the plane of figure 1.3*b*), to come into coincidence with vector $d\vec{r}'$. It follows that the deformation $\underline{t}$, equation (1.20), see also equations (1.4) and (1.6), accounts only for rotation at a particular point in the material.

1.3   Forces and stress

When no forces act on a sample of material, it will come to an equilibrium configuration. Applied forces will then produce deformations, discussed in section 1.2. These forces are of two types: forces applied to the surface of the sample by contact with other matter or radiation, and so-called body forces arising from external fields, principally electric, magnetic and gravitational. Thus an object at rest and in equilibrium on the earth's surface will be deformed relative to its configuration in a gravity-free environment, such as a space station. Dielectric and magnetic materials will be deformed in electric and magnetic fields respectively, relative to their equilibrium field-free configurations. Electrostriction and piezoelectricity, discussed in section 9.5, are examples of such electrical effects. Very often, but not always, these body-force effects are negligible. In any case, we shall see how to incorporate both surface and body forces in our analysis of deformation.

Consider now a small segment of material internal to the sample. Suppose this material is deformed by surface and body forces, or as part of a non-equilibrium process (for example as part of a free-body oscillation: see Chapter 2). Then this internal segment is subject to forces arising from contact with the adjacent material in the sample. These forces are basically interatomic in origin: we therefore refer to them as short-range forces. They are discussed in more detail in section 9.2. Some internal forces are not so short-ranged, for example in dielectric media, where the atoms are actually charged ions, and so-called Madelung forces are present, along with induced electric dipole effects. In this chapter we consider only truly short-range internal effects.

Let us consider the short-range internal force acting on a segment of volume dV of the material. We express it by

$$d\vec{F} = \vec{f}(\vec{r})\,dV$$

where $\vec{f}(\vec{r})$ is force per unit volume at $\vec{r}$. The total internal force on an arbitrary volume V of the material arises from surface effects. The necessary relationship between volume and surface integrals of this sort is given by Gauss's theorem, which for a vector field $\vec{f}$ as here is

$$\int_V dV f_i = \int_{S(V)} ds\, n_j \sigma_{ji}, \tag{1.26}$$

where

$$f_i = \partial_j \sigma_{ji}. \tag{1.27}$$

The second-rank cartesian tensor σ_{ji} is called the *stress tensor*. In equation (1.26), $S(V)$ is the surface that bounds volume V, and n_j is the unit outward normal vector to $S(V)$, which we shall denote $\hat{n}$ as part of the vector notation that includes $\vec{f}$.

We now show that the stress tensor σ_{ji} is symmetrical, provided the resultant torque $\vec{\tau}$ on arbitrary volume V is due to internal forces only. In that case,

$$\vec{\tau} = \int_V dV (\vec{r} \times \vec{f}). \tag{1.28}$$

From equation (1.27), this becomes, in index notation,

$$\tau_i = \int_V dV \, \varepsilon_{ijk} x_j f_k = \int_V dV \, \varepsilon_{ijk} x_j \partial_l \sigma_{lk}. \tag{1.29}$$

In equation (1.29), we have used the fact that

$$(\vec{r} \times \vec{f})_i = \varepsilon_{ijk} x_j f_k$$

with Einstein summation convention throughout, where ε_{ijk} is the Levi-Cività symbol whose value is zero unless i, j and k are all different, and ± 1 when (i, j, k) is a cyclic or non-cyclic permutation of $(1, 2, 3)$ respectively. Integration by parts of the right-hand side of equation (1.29) gives

$$\tau_i = \int_{S(V)} ds \, \varepsilon_{ijk} x_j \sigma_{lk} n_l - \int_V dV \, \varepsilon_{ijk} (\partial_l x_j) \sigma_{lk}. \tag{1.30}$$

Now since $\partial_l x_j = \delta_{lj}$, the volume term in equation (1.30) becomes

$$\int_V dV \, \varepsilon_{ijk} \sigma_{jk} \tag{1.31}$$

whose component for $i = 1$ is

$$\int_V dV (\sigma_{32} - \sigma_{23}). \tag{1.32}$$

Similarly for components $i = 2, 3$. But since τ_i, equation (1.30), is the torque on volume V due to internal forces only, and these forces are purely of surface origin, the volume integral in equations (1.30)–(1.32) must be zero. It follows that, since V is arbitrary,

$$\sigma_{ij} = \sigma_{ji}. \tag{1.33}$$

Since σ_{ij} expresses only internal short-range forces, it does not contribute to any non-zero torque of external origin, and therefore the symmetry of the stress tensor, equation (1.33), is not invalidated by such torques.

Just as in section 1.2 we obtained simple physical interpretations for the elements ε_{ij} of the strain tensor and t_{ij} of the rotation tensor, here we can get a simple picture of the elements σ_{ij} of the stress tensor. Consider a small segment ds of surface area within the material. Without loss of generality it may be assumed to be approximately planar and normal to the x_1 axis, defined by the unit vector $\hat{n}_1$. This surface $d\vec{s} = dx_2 \, dx_3 \, \hat{n}_1$ is illustrated in figure 1.4. Now refer to equation (1.26). At some arbitrary interior point in the material, let V be a very small volume segment, ΔV, of which $d\vec{s}$ is

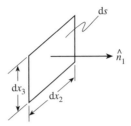

Figure 1.4. A segment of surface $\mathrm{d}\vec{s} = \mathrm{d}x_2\,\mathrm{d}x_3\,\hat{n}_1$, part of the boundary of a small volume ΔV: see equations (1.34).

part of the boundary $S(V)$. Then approximately

$$(\Delta V)f_i = \int_{S(\Delta V)} \mathrm{d}s\,n_j\sigma_{ji}.$$

But here $\hat{n} = \hat{n}_1 = (1,0,0)$ is the unit outward normal. Thus the contribution from $\mathrm{d}\vec{s} = \mathrm{d}x_2\,\mathrm{d}x_3\,\hat{n}_1$ to the net force $(\Delta V)\vec{f}$ is as follows:

$$\text{to } (\Delta V)f_1 : \quad \mathrm{d}s\,n_j\sigma_{ji} = \mathrm{d}s\,n_1\sigma_{11} = \mathrm{d}s\,\sigma_{11} \qquad (1.34a)$$

$$\text{to } (\Delta V)f_2 : \quad \mathrm{d}s\,n_1\sigma_{12} = \mathrm{d}s\,\sigma_{12} \qquad\qquad\quad (1.34b)$$

$$\text{to } (\Delta V)f_3 : \quad \mathrm{d}s\,n_1\sigma_{13} = \mathrm{d}s\,\sigma_{13} \qquad\qquad\quad (1.34c)$$

where $\mathrm{d}s = \mathrm{d}x_2\,\mathrm{d}x_3$. From equations (1.34a)–(1.34c) we conclude that $\sigma_{11}(\vec{r})$ is the force per unit area due to internal forces at $\vec{r}$ in the $\hat{n}_1$ direction on an area segment that lies perpendicular to $\hat{n}_1$, and σ_{12} and σ_{13} are mutually orthogonal tangential forces per unit area on such an area segment. Similar definitions of all the other elements of σ_{ij} follow from this discussion. Clearly the diagonal elements of $\underline{\underline{\sigma}}$ refer to compressive or dilatational forces, and the off-diagonal elements refer to shear forces.

1.4   Linear elasticity

We observe that when forces are applied to a material object, it becomes deformed. In previous sections we have described this deformation throughout the material in terms of strain and rotation, and we have described internal forces of interatomic short-range nature in terms of stress. We have mentioned body forces, and we now introduce forces of external origin acting on the physical surface of the material object. Specifically, we denote the force per unit volume due to body forces of external origin by $\vec{b}$. For example, in the case of gravity,

$$\vec{b}(\vec{r}) = \rho(\vec{r}) \cdot \vec{g} \qquad (1.35)$$

where $\vec{g}$ is the acceleration due to gravity, and $\rho(\vec{r})$ is the mass density of the material. Similarly, we define a position-dependent pressure, i.e. force per unit area, by $\vec{P}(\vec{r})$. Since by definition the force per unit area internally is given by $\underline{\sigma}$ according to equations (1.34a)–(1.34c), namely that σ_{ji} is the force per unit area in direction i on a segment of material surface oriented normal to direction j, we conclude that

$$P_i = n_j \sigma_{ji}, \qquad (1.36)$$

where the unit vector n_j is outward normal to the physical surface.

Let us now consider the work done on a material object by body and surface forces $\vec{b}(\vec{r})$ and $\vec{P}(\vec{r})$ in a reversible, quasi-static process, as required for classical equilibrium thermodynamics (see for example Pippard, 1964). Denote it ΔW:

$$\Delta W = \left\{ \int_V dV\, b_i(\Delta u_i) + \int_{S(V)} ds\, P_i(\Delta u_i) \right\}. \qquad (1.37)$$

From equation (1.36) and Gauss's theorem,

$$\int_{S(V)} ds\, P_i(\Delta u_i) = \int_{S(V)} ds\, n_j \sigma_{ji}(\Delta u_i)$$

$$= \int_V dV\, \partial_j [\sigma_{ji}(\Delta u_i)]$$

$$= \int_V dV [(\partial_j \sigma_{ji})(\Delta u_i) + \sigma_{ij} \partial_j (\Delta u_i)]. \qquad (1.38)$$

But $\partial_j \sigma_{ji} = f_i$, the internal force per unit volume (see equations (1.26)–(1.27)). Thus, from equations (1.37) and (1.38),

$$\Delta W = \int_V dV \{(b_i + f_i)(\Delta u_i) + \sigma_{ji} \partial_j (\Delta u_i)\}. \qquad (1.39)$$

In equation (1.39), the first term is the total work done on the material due to the combination of external body forces $\vec{b}$ and internal forces $\vec{f}$. But in a quasi-static process, these forces are in equilibrium throughout, so their sum must be zero for the particular deformation (Δu_i) discussed here. We therefore conclude that

$$\Delta W = \int_V dV\, \sigma_{ji} \partial_j (\Delta u_i). \qquad (1.40)$$

We can express equation (1.40) in terms of strain, as follows:

$$\sigma_{ji} \partial_j (\Delta u_i) = \sigma_{ji} \Delta (\partial_j u_i)$$

$$= \tfrac{1}{2} \{\sigma_{ij} \Delta (\partial_j u_i) + \sigma_{ji} \Delta (\partial_j u_i)\}$$

$$= \tfrac{1}{2} \{\sigma_{ij} [\Delta (\partial_j u_i) + \Delta (\partial_i u_j)]\}$$

$$= \sigma_{ij} \Delta \{\tfrac{1}{2} (\partial_i u_j + \partial_j u_i)\} = \sigma_{ij} \Delta (\varepsilon_{ij}). \qquad (1.41)$$

In equation (1.41) we have used the symmetry of σ_{ji}, equation (1.33), and the definition of ε_{ij}, equation (1.5), and have interchanged summation indices i and j in one term. Thus, for a homogeneous system, or for a small, approximately homogeneous segment of a system, the combined first and second laws of thermodynamics takes the form

$$\Delta U = \{T(\Delta S) + \Delta W\}$$
$$= \{T(\Delta S) + \sigma_{ij}(\Delta \varepsilon_{ij})\}, \qquad (1.42)$$

where in equation (1.42) the extensive quantities U, W and S are here taken to be per unit volume: see equations (1.40) and (1.41). In equation (1.42), U is the internal energy, T is Kelvin temperature, and S is entropy.

1.4.1 Hooke's law

Now, for the moment, ignore thermodynamic conditions, and consider strain $\underline{\varepsilon}$ produced by stress $\underline{\sigma}$. For a stable system, a non-zero stress is required to produce a non-zero strain. In general, $\underline{\sigma}$ is a material-dependent function of $\underline{\varepsilon}$, i.e. a function of six variables, namely the six independent elements of $\underline{\varepsilon}$. For small strain $\underline{\varepsilon}$, this relationship is approximately linear:

$$\sigma_{ij} = c_{ijkl} \varepsilon_{kl}. \qquad (1.43)$$

The material-dependent fourth rank cartesian tensor c_{ijkl} is the *elastic constant tensor*. The approximate linear relationship, equation (1.43), is known as *Hooke's law*. For a homogeneous system, c_{ijkl} is independent of position $\vec{r}$: we shall assume homogeneity in what follows. The tensor c_{ijkl} has $3^4 = 81$ elements, but the symmetry of $\underline{\sigma}$ and of $\underline{\varepsilon}$ means that a much smaller number are independent, as we shall see.

A mathematically convenient reformulation of stress, strain, and Hooke's law takes advantage of the symmetry of $\underline{\sigma}$ and $\underline{\varepsilon}$. It is based on the *Voigt notation*, as follows. Define σ_n and ε_n ($n = 1, \ldots, 6$):

$$
\begin{aligned}
\sigma_1 &= \sigma_{11} & \varepsilon_1 &= \varepsilon_{11} \\
\sigma_2 &= \sigma_{22} & \varepsilon_2 &= \varepsilon_{22} \\
\sigma_3 &= \sigma_{33} & \varepsilon_3 &= \varepsilon_{33} \\
\sigma_4 &= \sigma_{23} & \varepsilon_4 &= 2\varepsilon_{23} \\
\sigma_5 &= \sigma_{31} & \varepsilon_5 &= 2\varepsilon_{31} \\
\sigma_6 &= \sigma_{12} & \varepsilon_6 &= 2\varepsilon_{12}
\end{aligned}
\qquad (1.44)
$$

In equations (1.44) note particularly the factors of 2 in ε_4, ε_5 and ε_6. From equation (1.42), this enables us to write the energy density increment:

$$dW = \sigma_{ij}\, d\varepsilon_{ij} = \sum_{n=1}^{6} \sigma_n\, d\varepsilon_n. \qquad (1.45)$$

From equation (1.45) in Voigt notation, we conclude that

$$\frac{\partial W}{\partial \varepsilon_n} = \sigma_n \tag{1.46}$$

and

$$\frac{\partial^2 W}{\partial \varepsilon_m \partial \varepsilon_n} = \frac{\partial \sigma_n}{\partial \varepsilon_m} = \frac{\partial \sigma_m}{\partial \varepsilon_n} = \frac{\partial^2 W}{\partial \varepsilon_n \partial \varepsilon_m}. \tag{1.47}$$

But from Hooke's law we can write

$$\sigma_n = \sum_n c_{nm}\varepsilon_m, \tag{1.48}$$

where c_{nm} are elastic constants in Voigt notation. The matrix c_{nm} is 6×6; it has 36 elements. Furthermore, from equations (1.48) and (1.47),

$$c_{nm} = \frac{\partial \sigma_n}{\partial \varepsilon_m} = \frac{\partial \sigma_m}{\partial \varepsilon_n} = c_{mn}. \tag{1.49}$$

Equation (1.49) shows that c_{nm} is symmetrical. It therefore has at most $\sum_{n=1}^{6} = 21$ independent elements, a large reduction from the 81 elements of c_{ijkl}, equation (1.43).

In terms of Hooke's law, equation (1.48), we now have the work per unit volume associated with an infinitesimal strain, from equation (1.45):

$$dW = \sum_{nm} c_{nm}\varepsilon_m \, d\varepsilon_n. \tag{1.50}$$

Clearly, in this approximation, the energy density is a homogeneous expression of second-order in the elements of ε_n. From Euler's theorem on homogeneous functions,

$$\sum_n \varepsilon_n \frac{\partial W}{\partial \varepsilon_n} = 2W. \tag{1.51}$$

From equations (1.46) and (1.48) this becomes

$$W = \frac{1}{2}\sum_n \varepsilon_n \sigma_n = \frac{1}{2}\sum_{nm} c_{nm}\varepsilon_m\varepsilon_n. \tag{1.52}$$

Such a second-order approximation to the energy density is called the *harmonic approximation*, equivalent to the approximation of *linear elasticity* embodied in Hooke's law.

1.4.2 Isotropic media

We now consider an isotropic material, and refer to the elastic constant tensor c_{ijkl} defined in equation (1.43). For an isotropic solid, the measured elastic constants must be independent of coordinate system: they must be scalars. The same kind of derivation that led to equation (1.52) for the energy density in Voigt notation leads to

$$W = \tfrac{1}{2} c_{ijkl} \varepsilon_{ij} \varepsilon_{kl}. \tag{1.53}$$

But if the elastic constants are scalar, then they must be multiplied by those sets of second-order terms in elements of $\underline{\underline{\varepsilon}}$ that are scalars. Thus only the first- and second-order scalar invariants of $\underline{\underline{\varepsilon}}$ must be involved, the first squared and the second standing alone.

The first three orders of scalar invariants of a second-rank tensor can be obtained as follows. Consider

$$\underline{\underline{\beta}} = (\underline{\underline{\varepsilon}} + \gamma \underline{\underline{I}}), \tag{1.54}$$

where $\underline{\underline{I}}$ is the identity and γ is a scalar. If we evaluate the determinant, we obtain

$$\det(\underline{\underline{\beta}}) = \{\det(\underline{\underline{\varepsilon}}) + \gamma \theta + \gamma^2 \Delta + \gamma^3\} \tag{1.55}$$

where

$$\theta = \{(\varepsilon_{11}\varepsilon_{22} + \varepsilon_{22}\varepsilon_{33} + \varepsilon_{33}\varepsilon_{11}) - (\varepsilon_{12}^2 + \varepsilon_{23}^2 + \varepsilon_{31}^2)\}, \tag{1.56}$$

and $\Delta = \mathrm{Tr}(\underline{\underline{\varepsilon}})$, see equations (1.14) and (1.15). In equation (1.55), since γ^3 is scalar, so must all other terms be, namely Δ, θ and $\det(\underline{\underline{\varepsilon}})$. We conclude that, with only scalar elastic constants in equation (1.53) for isotropic media,

$$W = (a\Delta^2 + b\theta) \tag{1.57}$$

where a and b are scalar elastic constants, i.e. there are only two of them which are independently measurable.

It is conventional to write equation (1.57) in terms of a more appealing form than θ, equation (1.56): this leads us to the so-called *Lamé elastic constants* λ and μ, in place of a and b. It is based on the recognition that Δ^2 and θ can be combined as follows:

$$(\Delta^2 - 2\theta) = \varepsilon_{ij}\varepsilon_{ij}. \tag{1.58}$$

The proof is straightforward:

$$\Delta^2 = \varepsilon_{ii}\varepsilon_{jj}$$
$$= \{(\varepsilon_{11}^2 + \varepsilon_{22}^2 + \varepsilon_{33}^2) + 2(\varepsilon_{11}\varepsilon_{22} + \varepsilon_{22}\varepsilon_{33} + \varepsilon_{33}\varepsilon_{11})\}.$$

Combining this with equation (1.56) for θ, using the symmetry $\varepsilon_{ij} = \varepsilon_{ji}$ then leads to equation (1.58). Thus equation (1.57) becomes

$$W = \{a\Delta^2 + \tfrac{1}{2}b(\Delta^2 - \varepsilon_{ij}\varepsilon_{ij})\}$$

or

$$W = \{ \tfrac{1}{2}\lambda\varepsilon_{ii}\varepsilon_{jj} + \mu\varepsilon_{ij}\varepsilon_{ij} \}$$

where

$$\tfrac{1}{2}\lambda = (a + \tfrac{1}{2}b) \qquad \text{and} \qquad \mu = -\tfrac{1}{2}b. \tag{1.59}$$

We next use the Voigt notation, equation (1.44), to obtain the elastic constants as in equation (1.49) with equation (1.46):

$$\sigma_n = \frac{\partial W}{\partial\varepsilon_n} \tag{1.46}$$

$$c_{nm} = \frac{\partial\sigma_n}{\partial\varepsilon_m}. \tag{1.49}$$

From equation (1.59)

$$\sigma_1 = \sigma_{11} = \frac{\partial W}{\partial\varepsilon_{11}} = \lambda\Delta + 2\mu\varepsilon_{11}$$

$$= \lambda(\varepsilon_{11} + \varepsilon_{22} + \varepsilon_{33}) + 2\mu\varepsilon_{11}$$

or $\qquad \sigma_1 = (\lambda + 2\mu)\varepsilon_1 + \lambda(\varepsilon_2 + \varepsilon_3).$ $\qquad$ (1.60)

Similarly

$$\sigma_2 = (\lambda + 2\mu)\varepsilon_2 + \lambda(\varepsilon_3 + \varepsilon_1). \tag{1.61}$$

$$\sigma_3 = (\lambda + 2\mu)\varepsilon_3 + \lambda(\varepsilon_1 + \varepsilon_2). \tag{1.62}$$

Also

$$\sigma_4 = \sigma_{23} = \frac{\partial W}{\partial\varepsilon_{23}} = 2\mu\varepsilon_{23} = \mu\varepsilon_4,$$

and similarly $\sigma_5 = \mu\varepsilon_5$ and $\sigma_6 = \mu\varepsilon_6$. Returning to equation (1.49) above, with equation (1.60),

$$c_{11} = \frac{\partial\sigma_1}{\partial\varepsilon_1} = (\lambda + 2\mu)$$

and similarly for c_{22} and c_{33}, from equations (1.61) and (1.62). Also

$$c_{44} = \frac{\partial\sigma_4}{\partial\varepsilon_4} = \mu,$$

and similarly for c_{55} and c_{66}. Furthermore,

$$c_{12} = \frac{\partial\sigma_1}{\partial\varepsilon_2} = \lambda,$$

and similarly for c_{23} and c_{31}. Also

$$c_{4n} = 0 \qquad \text{except for } n = 4,$$

and similarly for c_{5n} and c_{6n}. In matrix form we therefore have

$$
\underline{\underline{c}} = \begin{pmatrix}
(\lambda + 2\mu) & \lambda & \lambda & 0 & 0 & 0 \\
\lambda & (\lambda + 2\mu) & \lambda & 0 & 0 & 0 \\
\lambda & \lambda & (\lambda + 2\mu) & 0 & 0 & 0 \\
0 & 0 & 0 & \mu & 0 & 0 \\
0 & 0 & 0 & 0 & \mu & 0 \\
0 & 0 & 0 & 0 & 0 & \mu
\end{pmatrix}.
\tag{1.63}
$$

In summary, for an isotropic material,

$$c_{11} = c_{22} = c_{33} = (\lambda + 2\mu),$$

$$c_{12} = c_{23} = c_{31} = \lambda,$$

$$c_{44} = c_{55} = c_{66} = \mu.$$

From these we note the so-called *Cauchy relation*:

$$c_{11} = (c_{12} + 2c_{44}).
\tag{1.64}$$

This general result can be used as a test for detailed calculations based on approximate models and computational methods, as for an atomistic model (see Chapter 9, for example), when applied to a cubic crystal. Landau and Liftshitz (1970) discuss the criterion for equation (1.64) to be applicable to a cubic crystalline material (p. 42).

1.4.3 Elastic moduli

In this section we discuss elastic moduli that are commonly used to describe isotropic bulk solids. The first of these is *Poisson's ratio*: the magnitude of the ratio of lateral to longitudinal strain. Suppose that the material is simply stretched in the x_1 direction: $\sigma_1 > 0$, and $\sigma_k = 0$ ($k \neq 1$). From isotropy, $\varepsilon_2 = \varepsilon_3$, and applying this to equation (1.48) with the isotropic Voigt elastic constants from equation (1.63), we obtain

$$\sigma_1 = (\lambda + 2\mu)\varepsilon_1 + 2\lambda\varepsilon_2,
\tag{1.65}$$

$$0 = \lambda\varepsilon_1 + 2(\lambda + \mu)\varepsilon_2.
\tag{1.66}$$

If we denote Poisson's ratio by ν, then from equation (1.66) we get

$$\nu \equiv -\frac{\varepsilon_2}{\varepsilon_1} = \frac{\lambda}{2(\lambda + \mu)}.
\tag{1.67}$$

From equations (1.65) and (1.66) together we get so-called *Young's modulus*, denoted E,

$$E \equiv \frac{\sigma_1}{\varepsilon_1} = \frac{\mu(2\mu + 3\lambda)}{(\lambda + \mu)}.
\tag{1.68}$$

We next discuss *compressibility K* or *bulk modulus* $B = K^{-1}$. By definition, the compressibility is the magnitude of fractional volume change per unit pressure change. Since increasing pressure produces decreasing volume (at least for hydrostatic, or isotropic pressure),

$$K = -\frac{1}{V}\left(\frac{\partial V}{\partial p}\right). \tag{1.69}$$

Now in terms of strain, equation (1.14) gives,

$$\frac{\delta V}{V} = Tr(\delta\underline{\underline{\varepsilon}}). \tag{1.70}$$

where δV is a small change of volume V. For bulk properties, the volume segment (ΔV) in equation (1.14) becomes the total volume V here. Now, for hydrostatic pressure, from the discussion following equations (1.34) we conclude that

$$\sigma_{ij} = -p\delta_{ij}, \tag{1.71}$$

or, in Voigt notation,

$$\sigma_k = -p \qquad (k = 1, 2, 3). \tag{1.72}$$

Where we consider an infinitesimal volume change δV as in equation (1.70) above, let it be induced by an infinitesimal pressure change δp:

$$\delta\sigma_k = -\delta p \qquad (k = 1, 2, 3) \tag{1.73}$$

$$\delta\sigma_k = 0 \qquad (k = 4, 5, 6) \tag{1.74}$$

If we now apply the relationship

$$\delta\sigma_n = \sum_{nm} c_{nm}\delta\varepsilon_m \tag{1.75}$$

and use the isotropy results, equations (1.63), (1.73) and (1.75), we obtain simply

$$\delta\sigma_1 = -\delta p = (3\lambda + 2\mu)\delta\varepsilon_1. \tag{1.76}$$

We conclude that the compressibility K, equation (1.69), is given by

$$K = -\frac{1}{\delta p}\left(\frac{\delta V}{V}\right) = \frac{1}{(3\lambda + 2\mu)\delta\varepsilon_1}3\delta\varepsilon_1$$

or

$$K = (\lambda + \tfrac{2}{3}\mu)^{-1} = B^{-1}. \tag{1.77}$$

In arriving at equation (1.77) we have used equation (1.70) with equation (1.76).

Finally we introduce the *shear modulus G*. It is defined by the shear stress per unit angular deformation. The shear deformation is illustrated in figure

1.2*b*; the angular deformation is (2γ) or, from equation (1.19), approximately $(2\varepsilon_{12}) = \varepsilon_4$ in Voigt notation (see equations (1.44)). The shear stress is given by σ_{ij} $(i \neq j)$: see equations (1.34b), (1.34c). In Voigt notation, with isotropy, $\sigma_{ij} = \sigma_k$ $(k = 4, 5, 6)$ for $i \neq j$. Consider the case $\sigma_k = 0$ $(k = 1, 2, 3)$ with $\sigma_{k'} = \sigma_4 \neq 0$ $(k' = 4, 5, 6)$. Then equations (1.48) and (1.63), the strain–stress relations with isotropy, give

$$\sigma_4 = \mu\varepsilon_4. \tag{1.78}$$

Thus for the shear modulus G we have

$$G = \frac{\sigma_{12}}{(2\gamma)} = \frac{\sigma_4}{(2\gamma)} = \frac{(\mu\varepsilon_4)}{\varepsilon_4}. \tag{1.79}$$

From equation (1.79) we see that the Lamé constant μ is simply the shear modulus.

1.4.4 Stability conditions

Let us consider stability of equilibrium at constant temperature. The combined first and second laws of thermodynamics can then be examined in terms of the Helmholtz free energy F. Specifically,

$$dF = (-S\,dT + dW). \tag{1.80}$$

The stability condition is then that F increases with any possible change of state. At constant temperature, this becomes that W must increase with any deformation. Now for isotropic media we had

$$W = (\tfrac{1}{2}\lambda\varepsilon_{ii}\varepsilon_{jj} + \mu\varepsilon_{ij}\varepsilon_{ij}). \tag{1.59}$$

This is clearly positive definite for $\underline{\underline{\varepsilon}} \neq 0$, if $\lambda > 0$ and $\mu > 0$, for it is of the form

$$W = \left\{ \tfrac{1}{2}\lambda(\varepsilon_{ii})^2 + \mu \sum_{\alpha,\beta}(\varepsilon_{\alpha\beta})^2 \right\}. \tag{1.81}$$

It is not clear that λ and μ are both positive, however, because $(\varepsilon_{ii})^2$ and $\sum_{\alpha,\beta}(\varepsilon_{\alpha\beta})^2$ are not independent. Certainly, if the diagonal elements of $\underline{\underline{\varepsilon}}$ are all zero, the deformation is a pure shear, and

$$W = \mu \sum_{\alpha \neq \beta} \varepsilon_{\alpha\beta}^2, \tag{1.82}$$

which for $W > 0$ requires $\mu > 0$. For a hydrostatic compression without shear, we note that, in principal axes, the off-diagonal elements $\underline{\underline{\varepsilon}}$ will be zero, and the diagonal elements $\varepsilon^{(\alpha)}$ will be equal. In that case we have

$$W = \{\tfrac{1}{2}\lambda(3\varepsilon^{(\alpha)})^2 + \mu[3(\varepsilon^{(\alpha)})^2]\}$$

$$= 3\{(\tfrac{3}{2}\lambda + \mu)(\varepsilon^{(\alpha)})^2\}, \tag{1.83}$$

which requires

$$(\lambda + \tfrac{2}{3}\mu) = B > 0. \tag{1.84}$$

In equation (1.84) we have used equation (1.77). We therefore have

$$B > 0, \qquad \mu > 0, \tag{1.85}$$

i.e. bulk and shear moduli are both positive.

It remains to determine the constraints on Poisson's ratio ν, equation (1.67), and Young's modulus E, equation (1.68). The reader can show, though it may be tedious, that from equations (1.67), (1.68) and (1.77)

$$B = \frac{E}{3(1 - 2\nu)} \qquad \text{and} \qquad \mu = \frac{E}{2(1 + \nu)}. \tag{1.86}$$

Now, in equation (1.68) for E, we note first that $(2\mu + 3\lambda) > 0$, equation (1.84), and then again from equation (1.84) with $B > 0$ and $\mu > 0$, equation (1.85), that $(\lambda + \mu) = (B + \tfrac{1}{3}\mu) > 0$. Thus

$$E > 0. \tag{1.87}$$

From the definition of Young's modulus, equation (1.68), this simply says that for isotropic elastic media, tension always results in stretching, never contraction. Then from B in equation (1.86) with $B > 0$, $E > 0$ [equations (1.85) and (1.87)] we have

$$(1 - 2\nu) > 0 \qquad \text{or} \qquad \nu < \tfrac{1}{2}, \tag{1.88}$$

and also from μ in equation (1.86), with $\mu > 0$, $E > 0$,

$$(1 + \nu) > 0 \qquad \text{or} \qquad \nu > -1. \tag{1.89}$$

From equation (1.86) we see that with $\nu \lesssim \tfrac{1}{2}$, we have the compressibility $K = B^{-1} \approx 0$. That is, for a nearly incompressible medium, Poisson's ratio is nearly $\tfrac{1}{2}$.

While negative values of ν, indicating lateral *swelling* under longitudinal tension, used to be thought not to exist in practice (see Landau and Lifshitz, 1970, p. 14, first footnote), this is no longer the case. Materials with $\nu < 0$ are referred to as *auxetic*, and are now the subject of much experimental and theoretical analysis. A representative reference is Baughman *et al.* (1998), which analyses a wide range of cubic crystalline solids, while giving references to works on many types of more exotic materials. For anisotropic materials, it is also possible for ν to be greater than 0.5, the theoretical limit for cubic crystals being 2.0, with values at least up to 1.68 having been deduced. Of most interest for the case discussed in this section, namely *isotropic* materials, is the fact that at least one such material has been found to have $\nu < 0$. Lakes (1987) has produced open-celled polymer foams, which are isotropic, with Poisson's ratios ranging down to -0.7. Both of the papers cited in this paragraph are highly recommended to the reader.

1.5   Equilibrium

We now consider what the condition for equilibrium requires of the deformation field $\vec{u}(\vec{r})$. The condition is simply that, in the absence of body forces, the force per unit volume, $\vec{f}(\vec{r})$, equation (1.27), should be zero

$$\partial_j \sigma_{ji} = 0. \tag{1.90}$$

For linear elasticity in an isotropic material, we begin with equation (1.59):

$$W = \tfrac{1}{2}\lambda \varepsilon_{ii}\varepsilon_{jj} + \mu \varepsilon_{ij}\varepsilon_{ij}. \tag{1.91}$$

From equation (1.45) we deduce the analogue of equation (1.46)

$$\sigma_{ij} = \frac{\partial W}{\partial \varepsilon_{ij}}. \tag{1.92}$$

From equations (1.91) and (1.92):

$$\sigma_{ij} = (\lambda \varepsilon_{kk}\delta_{ij} + 2\mu \varepsilon_{ij}). \tag{1.93}$$

To apply this in equation (1.90) we note that

$$\partial_j(\varepsilon_{kk}\delta_{ij}) = \delta_{ij}\partial_j\varepsilon_{kk} = \partial_i\varepsilon_{kk}.$$

We now introduce $\vec{u}(\vec{r})$ through equation (1.5), and obtain:

$$\partial_i\varepsilon_{kk} = \partial_i\partial_k u_k$$

and

$$\partial_j\varepsilon_{ij} = \tfrac{1}{2}(\partial_j\partial_i u_j + \partial_j\partial_j u_i).$$

Thus the equilibrium condition, equation (1.90), is

$$\lambda\partial_i\partial_k u_k + \mu(\partial_j\partial_i u_j + \partial_j\partial_j u_i) = 0$$

or

$$(\lambda + \mu)\partial_i(\partial_k u_k) + \mu\partial_j\partial_j u_i = 0. \tag{1.94}$$

In vector notation, this is simply

$$(\lambda + \mu)\vec{\nabla}(\vec{\nabla}\cdot\vec{u}) + \mu\nabla^2\vec{u} = 0. \tag{1.95}$$

This second-order, linear, homogeneous partial differential equation, with the appropriate boundary condition, equation (1.36), gives us the deformation of a solid material in the absence of external body forces. For future reference, we write equation (1.95) in an alternative form, using the vector relationship

$$\vec{\nabla}\times(\vec{\nabla}\times\vec{u}) = \vec{\nabla}(\vec{\nabla}\cdot\vec{u}) - \nabla^2\vec{u}. \tag{1.96}$$

We then obtain

$$(\lambda + 2\mu)\vec{\nabla}(\vec{\nabla}\cdot\vec{u}) - \mu\vec{\nabla}\times(\vec{\nabla}\times\vec{u}) = 0. \tag{1.97}$$

Solution of equation (1.97), to determine the deformation field $\vec{u}(\vec{r})$, requires specific boundary conditions, typified by equation (1.36). Body forces may also be added to equations (1.90), (1.95) and (1.97). Such solutions can be carried out analytically for many systems that have high symmetry, in terms of the special functions of mathematical physics. Many older text-books contain much material of this sort. Excellent representative references are Green and Zerna (1968) and Sokolnikoff (1956). Currently, detailed analyses by computational methods are carried out using programs that incorporate the special functions for high-symmetry cases, and that use quite different methods for more general cases. Such work can be found in the periodical *Computer Methods in Applied Mechanics and Engineering* (North-Holland, Amsterdam). In section 5.3 we apply and discuss the equilibrium condition in relation to edge and screw dislocations.

Chapter 2

Wave propagation in continuous media

2.1 Introduction

In Chapter 1 we introduced time-independent deformation of a continuous solid. In this chapter we shall see the fundamental time-dependent deformations that can occur in an infinite continuous medium. In Chapter 4 we shall see the characteristic effect of an infinite planar surface upon the deformation dynamics. The fundamental dynamical entities are, of course, waves, whose combinations and interactions are as rich and significant in the field of elastic properties as they are in the field of electromagnetism, and in all other field theories. Amongst the results that we shall obtain is the existence of two distinct speeds of wave propagation. Before any of this, we must discuss in section 2.2 the fact that any vector field in an infinite medium, dropping faster than (distance)$^{-1}$ at large distance, consists of irrotational and shear components. In section 2.3 we derive the equation of motion in the absence of thermal effects. In section 2.4 we determine the propagation speeds of irrotational and shear waves.

As we mentioned in Chapter 1, the contents of this chapter are largely based on the textbooks by Landau and Lifshitz (1970) and Bhatia and Singh (1986).

2.2 Vector fields

For an arbitrary vector field $\vec{f}(\vec{r})$, recall the relationship of equation (1.96) whence

$$\nabla^2 \vec{f} = \vec{\nabla}(\vec{\nabla} \cdot \vec{f}) - \vec{\nabla} \times (\vec{\nabla} \times \vec{f}). \tag{2.1}$$

Recall also the representation of a Dirac delta function in the form

$$\delta(\vec{r} - \vec{r}') = \frac{-1}{4\pi} \nabla^2 \frac{1}{|\vec{r} - \vec{r}'|}. \tag{2.2}$$

This result is most easily seen from electrostatics. Consider Maxwell's equation,

$$\vec{\nabla} \cdot \vec{E} = \frac{\rho(\vec{r})}{\varepsilon_0}, \tag{2.3}$$

where $\vec{E}$ is the electric field due to a charge distribution $\rho(\vec{r})$, and ε_0 is the permittivity of free space. For a unit point charge, $\rho(\vec{r}) = \delta(\vec{r})$, the application of Gauss's theorem easily leads to Coulomb's law,

$$\vec{E}(\vec{r}) = \frac{\vec{r}}{4\pi\varepsilon_0 r^3}. \tag{2.4}$$

The above generalizes to the case of the electric field at $\vec{r}'$ due to a unit point charge at $\vec{r}$,

$$\vec{\nabla}' \cdot \vec{E}(\vec{r}') = \frac{1}{4\pi\varepsilon_0} \vec{\nabla}' \cdot \left(\frac{\vec{r}' - \vec{r}}{|\vec{r}' - \vec{r}|^3} \right) = \frac{\delta(\vec{r}' - \vec{r})}{\varepsilon_0}, \tag{2.5}$$

where $(\vec{\nabla}' \cdot)$ is divergence with respect to components of $\vec{r}'$, and for a function of $(\vec{r}' - \vec{r})$, $(\vec{\nabla}' \cdot) = (-\vec{\nabla} \cdot)$. It follows from equation (2.5) that

$$\delta(\vec{r} - \vec{r}') = \frac{1}{4\pi} \vec{\nabla} \cdot \frac{(\vec{r} - \vec{r}')}{|\vec{r} - \vec{r}'|^3}, \tag{2.6}$$

and since

$$\frac{(\vec{r} - \vec{r}')}{|\vec{r} - \vec{r}'|^3} = -\vec{\nabla} \frac{1}{|\vec{r} - \vec{r}'|}, \tag{2.7}$$

we obtain equation (2.2) from equations (2.6) and (2.7). Note that while this derivation is motivated by electrostatics, it does not depend on electrostatics.

We now consider an arbitrary vector field $\vec{F}(\vec{r})$, and write

$$\vec{F}(\vec{r}) = \int d^3r' \vec{F}(\vec{r}')\delta(\vec{r} - \vec{r}')$$

and use equations (2.2) and (2.1) to obtain

$$\vec{F}(\vec{r}) = -\int d^3r' \vec{F}(\vec{r}') \frac{1}{4\pi} \nabla^2 \frac{1}{|\vec{r}' - \vec{r}|}$$

$$= -\frac{1}{4\pi} \nabla^2 \int d^3r' \frac{\vec{F}(\vec{r}')}{|\vec{r}' - \vec{r}|}$$

$$= (\vec{\nabla} \times \vec{A}) + \vec{\nabla}\varphi \tag{2.8}$$

where

$$\vec{A}(\vec{r}) = \vec{\nabla} \times \int \frac{d^3r'}{4\pi} \frac{\vec{F}(\vec{r}')}{|\vec{r}' - \vec{r}|}, \qquad \varphi(\vec{r}) = -\vec{\nabla} \cdot \int \frac{d^3r'}{4\pi} \frac{\vec{F}(\vec{r}')}{|\vec{r}' - \vec{r}|}. \tag{2.9}$$

Now in general $\vec{\nabla} \times (\vec{\nabla}\varphi) = 0$ and $\vec{\nabla} \cdot (\vec{\nabla} \times \vec{A}) = 0$. It then follows from equation (2.8) that

$$\vec{F}(\vec{r}) = \{\vec{F}_L(\vec{r}) + \vec{F}_T(\vec{r})\}, \tag{2.10}$$

where

$$\vec{F}_L = \vec{\nabla}\varphi \rightarrow \vec{\nabla} \times \vec{F}_L = 0 \tag{2.11}$$

and

$$\vec{F}_T = \vec{\nabla} \times \vec{A} \rightarrow \vec{\nabla} \cdot \vec{F}_T = 0. \tag{2.12}$$

Equations (2.10)–(2.12) are the important result, that a vector field $\vec{F}(\vec{r})$ is the sum of a curl-less part and a div-less part, given respectively by $\vec{\nabla}\varphi$ and $\vec{\nabla} \times \vec{A}$. Furthermore, φ and $\vec{A}$ are determined explicitly, for given $\vec{F}$, by equations (2.9), and the integrals in equations (2.9) exist if $\vec{F}(\vec{r})$ drops off faster than r^{-1} for large r. In the context of continuous media, equations (2.10)–(2.12) given us the result that the time-dependent distortion field $\vec{u}(\vec{r}, t)$ has the form

$$\vec{u}(\vec{r}, t) = [\vec{u}_L(\vec{r}, t) + \vec{u}_T(\vec{r}, t)], \tag{2.13}$$

where

$$\vec{\nabla} \times \vec{u}_L = 0, \qquad \vec{\nabla} \cdot \vec{u}_T = 0. \tag{2.14}$$

The notations L (longitudinal) and T (transverse) will be understandable after section 2.3. More details of the above derivation can be found in the work by Morse and Feshbach (1953), section 1.5, pp. 52–54. In particular, for $\vec{F}(\vec{r})$ as defined, it is not hard to show that equations (2.9) reduce to

$$\vec{A}(\vec{r}) = \int \frac{d^3r'}{4\pi} \frac{[\vec{\nabla}' \times \vec{F}(\vec{r}')]}{|\vec{r} - \vec{r}'|}, \qquad \varphi(\vec{r}) = -\int \frac{d^3r'}{4\pi} \frac{[\vec{\nabla}' \cdot \vec{F}(\vec{r}')]}{|\vec{r} - \vec{r}'|} \tag{2.15}$$

(see Appendix to this chapter). From equations (2.15) it follows that, if $\vec{\nabla} \times \vec{F} = 0$ and $\vec{\nabla} \cdot \vec{F} = 0$ everywhere, then $\vec{A}$ and φ are also, as are $\vec{F}_L$ and $\vec{F}_T$, equations (2.11) and (2.12), whence from equation (2.10) $\vec{F}$ itself is identically zero.

2.3 Equation of motion

We limit our discussion to isotropic elastic solids. Recall from Chapter 1 that the force per unit volume $\vec{f}$ is given, from equations (1.27) and (1.95), as

$$\vec{f} = [(\lambda + \mu)\vec{\nabla}(\vec{\nabla} \cdot \vec{u}) + \mu\nabla^2\vec{u}], \tag{2.16}$$

and this will not be zero for a time-dependent system where $\vec{u} = \vec{u}(\vec{r}, t)$. For future use, we rewrite equation (2.16), using equation (1.96), as

$$\vec{f} = [(\lambda + \mu)\vec{\nabla} \times (\vec{\nabla} \times \vec{u}) + (\lambda + 2\mu)\nabla^2\vec{u}]. \tag{2.17}$$

In equations (2.16) and (2.17), λ and μ are the Lamé elastic constants: see the discussion of equations (1.58) and (1.59).

The equation of motion for the medium is now obtained from Newton's second law of mechanics in the form

$$\vec{f} = \rho\vec{a}, \tag{2.18}$$

where ρ is the mass density and $\vec{a}$ is the acceleration. We must be careful in applying equation (2.18) to the continuous medium, because we must distinguish between the position of a segment of mass, whose motion is governed by equation (2.18), and the general position vector $\vec{r}$ in the medium. We shall be dealing with the force per unit volume $\vec{f}$ at a point $\vec{r}$ in the medium: from equation (2.17) or (2.18), since $\vec{u} = \vec{u}(\vec{r}, t)$, we have $\vec{f} = \vec{f}(\vec{r}, t)$. Thus $\vec{f}(\vec{r}, t)$ accelerates the material at $\vec{r}$. Denote the position of this material at time t by $\vec{r}_m(t)$. What is the change of $r_m(t)$ in time $\Delta t \gtrsim 0$? There are two contributions, in first order. One comes from the explicit time dependence of $\vec{u}(\vec{r}, t)$:

$$(\Delta\vec{r}_m)_1 = [\vec{u}(\vec{r}_m(t), t + \Delta t) - \vec{u}(\vec{r}_m(t), t)]. \tag{2.19}$$

The other comes from the fact that at time t the material is in motion:

$$(\Delta\vec{r}_m)_2 = [\vec{u}(\vec{r}_m(t + \Delta t), t) - \vec{u}(\vec{r}_m(t), t]. \tag{2.20}$$

We now evaluate the total displacement $\Delta\vec{r}_m$ from equations (2.19) and (2.20):

$$\Delta\vec{r}_m = [(\Delta\vec{r}_m)_1 + (\Delta\vec{r}_m)_2]. \tag{2.21}$$

In this way we arrive at the velocity field $\vec{v}(\vec{r}, t)$ of the material:

$$\vec{v}(\vec{r}_m(t), t) = \lim_{\Delta t \to 0} \left(\frac{\Delta\vec{r}_m}{\Delta t} \right)$$

$$= \left\{ \lim_{\Delta t \to 0} \frac{[\vec{u}(\vec{r}_m(t), t + \Delta t) - \vec{u}(\vec{r}_m(t), t)]}{\Delta t} \right.$$

$$\left. + \lim_{\Delta t \to 0} \frac{[\vec{u}(\vec{r}_m(t + \Delta t), t) - \vec{u}(\vec{r}_m(t), t)]}{\Delta t} \right\}. \tag{2.22}$$

Since

$$\Delta\vec{r}_m = [\vec{r}_m(t + \Delta t) - \vec{r}_m(t)],$$

we identify equation (2.22) as

$$\vec{v}(\vec{r}_m(t), t) = \frac{\partial}{\partial t}\vec{u}(\vec{r}_m(t), t) + (\vec{v} \cdot \vec{\nabla}_m)\vec{u}(\vec{r}_m(t), t), \tag{2.23}$$

where $\vec{\nabla}_m$ is grad with respect to $\vec{r}_m$. But $\vec{r}_m$ is the position of the material being acted on by $\vec{f}(\vec{r}, t)$. Thus we must now identify $\vec{r}$ and $\vec{r}_m$. From equation

(2.23) we conclude that

$$\vec{v}(\vec{r}, t) = \left(\frac{\partial}{\partial t} + \vec{v} \cdot \vec{\nabla} \right) \vec{u}(\vec{r}, t). \tag{2.24}$$

The operator

$$\left(\frac{\partial}{\partial t} + \vec{v} \cdot \vec{\nabla} \right) \equiv \frac{D}{Dt} \tag{2.25}$$

is called the *material* time derivative: it gives the time rate of change of a property (in this case the displacement) of a segment of material at point $\vec{r}$ in the medium, at time t. We therefore conclude that the acceleration $\vec{a}(\vec{r}, t)$ of the segment of material is

$$\vec{a} = \frac{D\vec{v}}{Dt} = \left(\frac{\partial}{\partial t} + \vec{v} \cdot \vec{\nabla} \right)^2 \vec{u}, \tag{2.26}$$

from equations (2.24) and (2.25). Accordingly, the equation of motion, equation (2.18), becomes

$$\vec{f} = \rho \left(\frac{\partial}{\partial t} + \vec{v} \cdot \vec{\nabla} \right)^2 \vec{u}, \tag{2.27}$$

with $\vec{f}$ given by equations (2.16) or (2.17) for an elastic solid. In equation (2.27), $\vec{f}$, ρ, $\vec{v}$ and $\vec{u}$ are all functions of $(\vec{r}, t)$.

For an elastic solid, we are assuming that deformation $\vec{u}$ is a first-order small quantity. We now further assume that the velocity field is first-order small. This will be valid, given small displacements, if elastic constants are not too strong, in the absence of external forces. In that case, the equation of motion, equation (2.27), reduces to

$$\vec{f} = \rho \frac{\partial^2}{\partial t^2} \vec{u}. \tag{2.28}$$

The linear approximation has the effect of neglecting any contribution to the dynamics from the terms in $(\vec{v} \cdot \vec{\nabla})$ in equation (2.27). These are the terms that represent a contribution to time rates of change due to motion of the material, as discussed following equation (2.20).

Finally, we note that deviations of the density $\rho(\vec{r}, t)$ from the mean density ρ_0 are first-order small, and therefore are to be neglected in equation (2.28). We see this as follows. Consider a small quantity of material which when undeformed has volume V_0. Its total mass is $(\rho_0 V_0)$. When it is deformed, it has volume V and density $\rho(\vec{r}, t)$, if it is located at position $\vec{r}$ at time t. Thus,

$$\rho_0 V_0 = \rho V. \tag{2.29}$$

Now from our discussion of strain, in Chapter 1, we saw, equations (1.14) and (1.15),

$$\frac{(V - V_0)}{V} = \vec{\nabla} \cdot \vec{u}. \tag{2.30}$$

Thus, from equations (2.29) and (2.30),

$$\rho V = \rho(V_0 + V\vec{\nabla} \cdot \vec{u}) = \rho_0 V_0. \tag{2.31}$$

But to zeroth order, neglecting the term in $\vec{\nabla} \cdot \vec{u}$, we have from equation (2.31)

$$\rho \approx \rho_0, \tag{2.32}$$

and this approximation is adequate to maintain the right-hand side of equation (2.28) at first order. Thus, combining equations (2.16), (2.28) and (2.32), we have the equation of motion,

$$\left\{ \mu\nabla^2\vec{u} - \rho_0 \frac{\partial^2\vec{u}}{\partial t^2} + (\lambda + \mu)\vec{\nabla}(\vec{\nabla} \cdot \vec{u}) \right\} = 0 \tag{2.33}$$

and similarly, from equation (2.17),

$$\left\{ (\lambda + 2\mu)\nabla^2\vec{u} - \rho_0 \frac{\partial^2\vec{u}}{\partial t^2} + (\lambda + \mu)\vec{\nabla} \times (\vec{\nabla} \times \vec{u}) \right\} = 0. \tag{2.34}$$

Equations (2.33) and (2.34) are clearly wave-like equations for the dynamical behavior of the deformation $\vec{u}(\vec{r}, t)$ of an isotropic elastic solid continuum in the absence of external forces.

2.4   Wave propagation

2.4.1   Shear and rotational waves

Equation (2.33) for the deformation dynamics reduces to the simple wave equation if $\vec{\nabla} \cdot \vec{u} = 0$. From equations (1.14) and (1.15), this is the condition for a wave that involves no volume change anywhere in the medium. Such deformations were shown in Chapter 1 to be characteristic of local shears and rotations (end of section 1.2.1 and equation (1.24) and discussion there). Such waves are therefore referred to as *shear* or *rotational waves*. In that case we have, from equation (2.33) with $\vec{\nabla} \cdot \vec{u} = 0$,

$$\left(\nabla^2 - \frac{1}{v_T^2}\frac{\partial^2}{\partial t^2}\right)\vec{u} = 0 \tag{2.35}$$

where

$$v_T = (\mu/\rho_0)^{1/2} \tag{2.36}$$

is the propagation speed of the wave. If we operate from the left on equation (2.35) with $(\frac{1}{2}\text{curl}) = (\frac{1}{2}\vec{\nabla}\times)$, we obtain:

$$\left(\nabla^2 - \frac{1}{v_T^2}\frac{\partial^2}{\partial t^2}\right)\vec{R} = 0 \qquad (2.37)$$

where $\vec{R} = -\frac{1}{2}\vec{\nabla}\times\vec{u}$ is the rotation defined and discussed in section 1.2.2. We therefore see that both shear $\vec{u}$ (with $\vec{\nabla}\cdot\vec{u} = 0$) and rotation $\vec{R}$ propagate with speed v_T, equation (2.36).

The wave equation (2.35) has plane wave solutions. Linear combinations of plane waves are also solutions, where the linear coefficients are determined by initial and boundary conditions. Consider

$$\vec{u}_T(\vec{r}, t) = \hat{e}_T \exp[i(\vec{k}\cdot\vec{r} - \omega t)], \qquad (2.38)$$

where $\hat{e}_T$ is a unit polarization vector, $\vec{k}$ is the propagation vector giving both direction of propagation and wavelength $\lambda = 2\pi/k$, and ω is the angular frequency of the wave ($2\pi \times$ normal frequency). Substituting $\vec{u}$ from equation (2.38) into the wave equation (2.35) gives simply

$$\left(-k^2 + \frac{\omega^2}{v_T^2}\right) = 0,$$

whence the so-called dispersion relation for shear and rotational waves:

$$\omega_T(k) = v_T k, \qquad (2.39)$$

giving the specific relationship between frequency and wavelength for such a wave in such a medium. The condition $\vec{\nabla}\cdot\vec{u} = 0$ becomes, from equation (2.38),

$$\vec{k}\cdot\hat{e}_T = 0, \qquad (2.40)$$

which shows that the polarization $\hat{e}_T$ (i.e. direction of displacement) of such a wave is perpendicular, or transverse, to the direction of propagation $\vec{k}$. This is why the subscript T, for transverse, has been used here, and in equations (2.10) and (2.12). The interpretation that the field with $\vec{\nabla}\cdot\vec{u} = 0$ is transverse applies only to the present case, for a wave in an infinite medium. We shall see in Chapter 4 that the div-less part of a surface wave is not purely transverse.

2.4.2   Dilatational or irrotational waves

Consider now a deformation field for which $\vec{\nabla}\times\vec{u} = 0$: it is *irrotational*, since then $\vec{R} = 0$; see equation (1.21). In this case the form of equation (2.34) for the motion is convenient. It reduces to

$$\left(\nabla^2 - \frac{1}{v_L^2}\frac{\partial^2}{\partial t^2}\right)\vec{u} = 0 \qquad (2.41)$$

where

$$v_L = [(\lambda + 2\mu)/\rho_0]^{1/2} \tag{2.42}$$

is the propagation speed. Note, from equations (2.36) and (2.42),

$$v_L > v_T. \tag{2.43}$$

Now, if we operate on equation (2.41) from the left with div $= (\vec{\nabla} \cdot)$, we obtain

$$\left(\nabla^2 - \frac{1}{v_L^2}\frac{\partial^2}{\partial t^2}\right)(\vec{\nabla} \cdot \vec{u}) = 0. \tag{2.44}$$

From equation (2.30), and in Chapter 1, we saw that $(\vec{\nabla} \cdot \vec{u})$ is the fractional volume change, which we refer to as the *dilatation*, allowing it to be positive or negative where, in ordinary language, negative dilatation is simply compression. Thus we see from equations (2.41) and (2.44) that dilatational waves $\vec{u}$ with $\vec{\nabla} \times \vec{u} = 0$, and dilatation $(\vec{\nabla} \cdot \vec{u})$ itself, propagate with speed $v_L > v_T$, equation (2.43).

Again consider plane-wave solutions of equation (2.41), where now

$$\vec{u}(\vec{r}, t) = \hat{e}_L \exp[i(\vec{k} \cdot \vec{r} - \omega t)]. \tag{2.45}$$

We obtain the dispersion relation,

$$\omega_L(k) = v_L k. \tag{2.46}$$

Regarding equations (2.39) and (2.46), we see that there are two dispersion relations, one for rotational and another for irrotational waves, both linear, the latter having a greater slope, due to the condition $v_L > v_T$, equation (2.43). Because of the continuum approximation, the range of k-values is continuous from zero to infinity, corresponding to wavelengths from infinity to zero. These dispersion relations are illustrated in figure 2.1. For a real solid, since the material has atomic structure, there is characteristically a shortest wavelength, and thus a largest propagation vector $\vec{k}$. Furthermore, the dispersion relation for a material consisting of discrete atoms is not linear. These matters are discussed further in Chapter 8, and illustrated in figure 8.1.

Applied to plane waves, equation (2.45), the irrotational condition $(\vec{\nabla} \times \vec{u}) = 0$ gives

$$(\vec{k} \times \hat{e}_L) = 0. \tag{2.47}$$

This shows that the deformation is polarized in a direction $\hat{e}_L$ which is parallel to the propagation direction given by $\vec{k}$. Thus, for an infinite medium, irrotational waves are longitudinal, whence the notation L in this section, and in equations (2.10) and (2.11). The interpretation that the curl-less component of a wave is longitudinal does not carry over to other cases, however, as will be seen in Chapter 4 on surface waves.

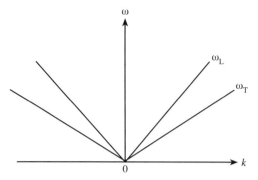

Figure 2.1. Dispersion relations $\omega_T(k)$ and $\omega_L(k)$ for angular frequencies of transverse (T) and longitudinal (L) components of travelling waves in an infinite, continuous homogeneous, isotropic, elastic medium: see equations (2.36) and (2.39), and (2.42) and (2.46) respectively, with equation (2.43).

2.4.3 General discussion

So far, we have considered only waves for which either $(\vec{\nabla} \cdot \vec{u}) = 0$ or $(\vec{\nabla} \times \vec{u}) = 0$. Now consider the general case, where neither of these conditions is satisfied. From the discussion of section 2.2, it nevertheless follows that

$$\vec{u}(\vec{r}, t) = [\vec{u}_L(\vec{r}, t) + \vec{u}_T(\vec{r}, t)] \tag{2.48}$$

where

$$(\vec{\nabla} \cdot \vec{u}_T) = 0, \qquad (\vec{\nabla} \times \vec{u}_L) = 0. \tag{2.49}$$

We shall show that, even when neither $\vec{u}_L$ or $\vec{u}_T$ is zero, the two components of $\vec{u}$ still propagate independently with the characteristic speeds v_L and v_T respectively.

The proof depends on the fact that if $(\vec{\nabla} \times \vec{F}) = 0$ and $(\vec{\nabla} \cdot \vec{F}) = 0$ then $\vec{F} = 0$ everywhere. First, substitute equation (2.48) into equation (2.33). Then we have

$$\{(\mu\nabla^2 - \rho_0\partial_t^2)(\vec{u}_L + \vec{u}_T) + (\lambda + \mu)\vec{\nabla}(\vec{\nabla} \cdot \vec{u}_L)\} = 0, \tag{2.50}$$

having used $(\vec{\nabla} \cdot \vec{u}_T) = 0$, and having introduced the notation $\partial_t = \partial/\partial t$. Operating on equation (2.50) from the left with $(\vec{\nabla}\times)$, we get

$$\vec{\nabla} \times \{(\mu\nabla^2 - \rho_0\partial_t^2)\vec{u}_T\} = 0, \tag{2.51}$$

having used $(\vec{\nabla} \times \vec{u}_L) = 0$ and $\vec{\nabla} \times \vec{\nabla}\Phi = 0$ for arbitrary Φ, where here $\Phi = (\vec{\nabla} \cdot \vec{u}_L)$. Furthermore,

$$\vec{\nabla} \cdot \{(\mu\nabla^2 - \rho_0\partial_t^2)\vec{u}_T\} = (\mu\nabla^2 - \rho_0\partial_t^2)(\vec{\nabla} \cdot \vec{u}_T) = 0, \tag{2.52}$$

since $(\vec{\nabla} \cdot \vec{u}_T) = 0$. It follows that the quantity in $\{\}$ brackets in equations (2.51) and (2.52) is zero everywhere, since its curl and its div are both zero.

Thus

$$\left(\nabla^2 - \frac{1}{v_T^2}\,\partial_t^2\right)\vec{u}_T = 0, \tag{2.53}$$

i.e. $\vec{u}_T$ in equation (2.48) propagates with speed v_T, independent of $\vec{u}_L$. Similarly, substitute equation (2.48) into equation (2.34) and operate from the left with div $= (\vec{\nabla}\cdot)$. Then

$$\vec{\nabla}\cdot\{[(\lambda + 2\mu)\nabla^2 - \rho_0\partial_t^2]\vec{u}_L\} = 0, \tag{2.54}$$

having used $(\vec{\nabla}\times\vec{u}_L) = 0$ and $\vec{\nabla}\cdot(\vec{\nabla}\times\vec{a}) = 0$ for arbitrary $\vec{a}$, where here $\vec{a} = (\vec{\nabla}\times\vec{u}_T)$. Furthermore,

$$\vec{\nabla}\times\{[(\lambda + 2\mu)\nabla^2 - \rho_0\partial_t^2]\vec{u}_L\} = [(\lambda + 2\mu)\nabla^2 - \rho_0\partial_t^2](\vec{\nabla}\times\vec{u}_L) = 0. \tag{2.55}$$

It follows from equations (2.54) and (2.55) that the quantity in { } brackets is zero everywhere, since its div and curl are both zero:

$$\left(\nabla^2 - \frac{1}{v_L^2}\,\partial_t^2\right)\vec{u}_L = 0, \tag{2.56}$$

so that $\vec{u}_L$ in equation (2.48) propagates with speed v_L, independent of $\vec{u}_T$. To summarize, an arbitrary wave motion $\vec{u}(\vec{r}, t)$ has components $\vec{u}_L$ and $\vec{u}_T$ that propagate at different speeds v_L and v_T respectively, independent of each other. In an infinite medium it is the longitudinal and transverse components of the wave that are propagating with these particular speeds. More generally, it is the curl-less and div-less parts of the wave that propagate this way.

In this chapter, it is the wave motion of an infinite, isotropic, continuous elastic medium that we have analysed. This is essential underpinning for all kinds of properties and processes of systems that may be realistically related to such a medium. Thus these waves will in practice be studied in connection with equilibrium and non-equilibrium thermal properties (Chapter 3), with surfaces (Chapter 4) and interfaces, with extended defects (Chapter 5), and with electronic properties (Chapter 6). Their counterparts in atomically modelled systems (Chapters 7 and 8) are equally important in all the above contexts, as well as for point defects (Chapter 11), for charge density waves (Chapter 14) and superconductivity, and for interactions of condensed matter with quantum-mechanical probes such as photons and neutrons.

Appendix to Chapter 2

Consider the first integral in equation (2.9):

$$\vec{A}(\vec{r}) = \vec{\nabla}\times\int\frac{\mathrm{d}^3r'}{4\pi}\,\frac{\vec{F}(\vec{r}')}{|\vec{r} - \vec{r}'|}. \tag{A2.1}$$

The curl operator ($\vec{\nabla}\times$) acts only on $\vec{r}$ in ($|\vec{r} - \vec{r}'|^{-1}$), so far as differentiation is concerned. In fact, in general,

$$\vec{\nabla} \times (g(\vec{r})\vec{F}(\vec{r})) = \{(\vec{\nabla}g \times \vec{F}) + g(\vec{\nabla} \times \vec{F})\}, \qquad (A2.2)$$

whence

$$\vec{\nabla} \times (g(\vec{r})\vec{F}(\vec{r}')) = \vec{\nabla}g(\vec{r}) \times \vec{F}(\vec{r}'). \qquad (A2.3)$$

Thus, equation (A2.1) is

$$\vec{A}(\vec{r}) = \int \frac{d^3r'}{4\pi} \, \vec{\nabla}\left(\frac{1}{|\vec{r} - \vec{r}'|}\right) \times \vec{F}(\vec{r}') \qquad (A2.4)$$

$$= -\int \frac{d^3r'}{4\pi} \, \vec{\nabla}'\left(\frac{1}{|\vec{r} - \vec{r}'|}\right) \times \vec{F}(\vec{r}'). \qquad (A2.5)$$

We can apply equation (A2.2) to equation (A2.5):

$$\vec{A}(\vec{r}) = -\int \frac{d^3r'}{4\pi} \, \vec{\nabla}' \times \left(\frac{\vec{F}(\vec{r}')}{|\vec{r} - \vec{r}'|}\right) + \int \frac{d^3r'}{4\pi} \left(\frac{\vec{\nabla}' \times \vec{F}(\vec{r}')}{|\vec{r} - \vec{r}'|}\right). \qquad (A2.6)$$

The first integral in equation (A2.6) can be transformed into a surface integral, by Gauss's theorem,

$$\int_V d^3r' \, \vec{\nabla}' \times \left(\frac{\vec{F}(\vec{r}')}{|\vec{r} - \vec{r}'|}\right) = \oint_S ds' \, \hat{n}' \times \left(\frac{\vec{F}(\vec{r}')}{|\vec{r} - \vec{r}'|}\right), \qquad (A2.7)$$

where $\hat{n}'$ is the unit outward normal vector to S. For an infinite volume V, the surface S is at infinity. The integrand will be zero there, provided

$$\lim_{\vec{r}' \to \infty} \left(\frac{\vec{F}(\vec{r}')}{|\vec{r} - \vec{r}'|}\right) = 0, \qquad (A2.8)$$

i.e. provided that for large $\vec{r}$

$$|\vec{F}(\vec{r})| \lesssim r^{-(1+\alpha)}, \qquad (A2.9)$$

for $\alpha > 0$. In that case, equation (A2.6) reduces to

$$\vec{A}(\vec{r}) = \int \frac{d^3r'}{4\pi} \frac{[\vec{\nabla}' \times \vec{F}(\vec{r}')]}{|\vec{r} - \vec{r}'|}. \qquad (A2.10)$$

A similar derivation applies to the second integral in equation (2.9):

$$\varphi(\vec{r}) = -\vec{\nabla} \cdot \int \frac{d^3r'}{4\pi} \frac{\vec{F}(\vec{r}')}{|\vec{r} - \vec{r}'|}. \qquad (A2.11)$$

We note that

$$\vec{\nabla} \cdot (g(\vec{r})\vec{F}(\vec{r})) = \{\vec{\nabla}g \cdot \vec{F} + g\vec{\nabla} \cdot \vec{F}\}, \qquad (A2.12)$$

whence

$$\vec{\nabla} \cdot (g(\vec{r})\vec{F}(\vec{r}')) = \vec{\nabla}g(\vec{r}) \cdot \vec{F}(\vec{r}'). \tag{A2.13}$$

Thus equation (A2.11) is

$$\varphi(\vec{r}) = -\int \frac{d^3r'}{4\pi} \, \vec{\nabla}\left(\frac{1}{|\vec{r}-\vec{r}'|}\right) \cdot \vec{F}(\vec{r}') \tag{A2.14}$$

$$= \int \frac{d^3r'}{4\pi} \, \vec{\nabla}'\left(\frac{1}{|\vec{r}-\vec{r}'|}\right) \cdot \vec{F}(\vec{r}') \tag{A2.15}$$

$$= \left\{\int \frac{d^3r'}{4\pi} \, \vec{\nabla}' \cdot \left(\frac{\vec{F}(\vec{r}')}{|\vec{r}-\vec{r}'|}\right) - \int \frac{d^3r'}{4\pi} \frac{[\vec{\nabla}' \cdot \vec{F}(\vec{r}')]}{|\vec{r}-\vec{r}'|}\right\}. \tag{A2.16}$$

The first integral in equation (A2.16) transforms as follows:

$$\int_V d^3r' \, \vec{\nabla}' \cdot \left(\frac{\vec{F}(\vec{r}')}{|\vec{r}-\vec{r}'|}\right) = \oint_S ds\,\hat{n}' \cdot \left(\frac{\vec{F}(\vec{r})}{|\vec{r}-\vec{r}'|}\right), \tag{A2.17}$$

which is zero for infinite volume under the same conditions, equations (A2.8) and (A2.9), as before. Thus

$$\varphi(\vec{r}) = -\int \frac{d^3r'}{4\pi} \frac{[\vec{\nabla}' \cdot \vec{F}(\vec{r}')]}{|\vec{r}-\vec{r}'|}. \tag{A2.18}$$

This completes the proof of equations (2.15), i.e. equations (A2.10) and (A2.18).

Chapter 3

Thermal properties of continuous media

3.1 Introduction

In previous chapters we have discussed the mechanical properties of continuous solid media at constant temperature. In this chapter we shall first discuss the classical, reversible, equilibrium thermodynamics of such a system, in section 3.2. Then in section 3.3 we shall introduce thermal conduction. This can be expected to affect wave motion, on the basis that local compression of the material in the wave may produce heat, some of which may be conducted away from the compressed region irreversibly. The result is that the mechanical energy of the wave may be reduced as the wave propagates through the medium. Thus in section 3.4 we analyse the damping of waves by thermal conduction. While the phenomenon is of basic importance in itself, it is also a good illustration of a wider class of dissipative processes. Our presentation in this chapter largely follows the textbook of Bhatia and Singh (1986). However, as always, further valuable insight can be gained from Landau and Lifshitz (1970), especially section 35, for this topic.

3.2 Classical thermodynamics

3.2.1 The Maxwell relations

In Chapter 1, we briefly introduced the combined first and second laws of classical thermodynamics, equation (1.42),

$$\mathrm{d}U = (T\,\mathrm{d}S + \sigma_{ij}\,\mathrm{d}\varepsilon_{ij}), \tag{3.1}$$

in terms of stress σ_{ij} and strain ε_{ij}. In Voigt notation this takes the form

$$\mathrm{d}U = \left(T\,\mathrm{d}S + \sum_{n=1}^{6} \sigma_n\,\mathrm{d}\varepsilon_n\right); \tag{3.2}$$

see equation (1.45). We introduce Helmholtz and Gibbs free energies, F and G respectively,

$$F = (U - TS),$$ (3.3)

$$G = \left(F - \sum_n \sigma_n \varepsilon_n \right),$$ (3.4)

whence from equations (3.2)–(3.4):

$$dF = -S \, dT + \sum_n \sigma_n \, d\varepsilon_n,$$ (3.5)

$$dG = -S \, dT - \sum_n \varepsilon_n \, d\sigma_n.$$ (3.6)

The point here is that the combined first and second laws in terms of U, F, and G are appropriate for determining thermodynamic relations in terms of different pairs of variables: entropy and strain, temperature and strain, and temperature and stress, respectively. We note here, as in Chapter 1, that extensive quantities such as U, F, G and S are taken to be *per unit volume*.

From equations (3.5) and (3.6) we obtain two of the Maxwell relations, as follows. From equation (3.5),

$$\left(\frac{\partial F}{\partial T} \right)_\varepsilon = -S, \qquad \left(\frac{\partial F}{\partial \varepsilon_n} \right)_T = \sigma_n,$$

whence

$$\left(\frac{\partial^2 F}{\partial T \, \partial \varepsilon_n} \right) = -\left(\frac{\partial S}{\partial \varepsilon_n} \right)_T = \left(\frac{\partial \sigma_n}{\partial T} \right)_\varepsilon \equiv \beta_n,$$ (3.7)

defining β_n. Equation (3.7) is the Maxwell relation expressing a calorimetric coefficient (involving ΔS or $T \, \Delta S = \Delta Q$: heat increment) in terms of a thermometric measurement, involving $(\partial / \partial T)_\varepsilon$. Since thermometric measurements are usually much easier to perform accurately than are calorimetric measurements, one practical value of the Maxwell relations is to obviate the latter type of measurement. Similarly from equation (3.6) we obtain

$$\left(\frac{\partial S}{\partial \sigma_n} \right)_T = \left(\frac{\partial \varepsilon_n}{\partial T} \right)_\sigma \equiv \alpha_n$$ (3.8)

defining α_n. In the above equations, $\underline{\varepsilon}$ and $\underline{\sigma}$ stand for the six elements of Voigt strains or stresses respectively.

3.2.2 Elastic constants, bulk moduli and specific heats

We begin with isothermally measured elastic constants: see equations (1.48) and (1.49). Thus, in terms of independent variables $(T, \underline{\varepsilon})$:

$$d\sigma_n = \left\{ \left(\frac{\partial \sigma_n}{\partial T} \right)_{\varepsilon} dT + \sum_m \left(\frac{\partial \sigma_n}{\partial \varepsilon_m} \right)_T d\varepsilon_m \right\}$$

$$= \left(\beta_n \, dT + \sum_m c_{nm}^T \, d\varepsilon_m \right), \tag{3.9}$$

where we have used equations (3.7) and (1.49) for β_n and for c_{nm} respectively, and c_{nm}^T is the isothermal Voigt elastic constant. Similarly,

$$dS = \left(\frac{\partial S}{\partial T} \right)_{\varepsilon} dT + \sum_m \left(\frac{\partial S}{\partial \varepsilon_m} \right)_T d\varepsilon_m$$

$$= \left\{ (T^{-1} c_{\varepsilon}) \, dT - \sum_m \beta_m \, d\varepsilon_m \right\}, \tag{3.10}$$

where we have again used equation (3.7) for β_m, and the definition of the specific heat capacity at constant strain, c_{ε}:

$$c_{\varepsilon} = T \left(\frac{\partial S}{\partial T} \right)_{\varepsilon}. \tag{3.11}$$

We can now obtain a relationship between adiabatic and isothermal elastic constants by combining equations (3.9) and (3.10), as follows. Solve equation (3.10) for (dT):

$$dT = \frac{T}{c_{\varepsilon}} \left\{ dS + \sum_m \beta_m \, d\varepsilon_m \right\}. \tag{3.12}$$

Substitute for dT from equation (3.12) into equation (3.9):

$$d\sigma_n = \left\{ \left(\frac{\beta_n T}{c_{\varepsilon}} \right) dS + \sum_m \left[\left(\frac{T\beta_n \beta_m}{c_{\varepsilon}} \right) + c_{nm}^T \right] d\varepsilon_m \right\}. \tag{3.13}$$

Thus, adiabatically, with S held constant, and $(dS) = 0$, equation (3.13) becomes

$$\left(\frac{\partial \sigma_n}{\partial \varepsilon_m} \right)_S \equiv c_{nm}^S = \left(\frac{T\beta_n \beta_m}{c_{\varepsilon}} + c_{nm}^T \right)$$

or

$$\left(c_{nm}^S - c_{nm}^T \right) = \left(\frac{T\beta_n \beta_m}{c_{\varepsilon}} \right), \tag{3.14}$$

where c_{nm}^S are adiabatic Voigt elastic constants.

Returning to equation (3.9), we can obtain an interesting and useful result for the coefficients β_n, equation (3.7). Keeping $\underline{\sigma}$ constant, we obtain

$$\beta_n = -\sum_m c_{nm}^T \left(\frac{\partial \varepsilon_m}{\partial T} \right)_{\underline{\sigma}} = -\sum_m c_{nm}^T \alpha_m, \tag{3.15}$$

from equation (3.8). We now consider the case of isotropy. First note that in equation (3.8), α_n are the coefficients of thermal expansion at constant stress. For an isotropic solid, the diagonal elements in the principal-axis coordinate system of the tensor,

$$\alpha_{ij} = \left(\frac{\partial \varepsilon_{ij}}{\partial T} \right)_{\underline{\sigma}}, \tag{3.16}$$

are equal, and the off-diagonal elements are zero. The reason is simply that, for such a medium, thermal expansion is the same in all directions, and cannot induce shear: see also Bhatia and Singh (1986, section 3.7, pp. 40–41). It follows that in equation (3.15), for an isotropic solid, $\alpha_m = \alpha_1 \equiv \alpha$, defining α, for $m = 1, 2, 3$, and zero otherwise. Then using equation (3.15) with the Voigt elastic constant matrix equation (1.63), we obtain

$$\beta_n = \beta_1 \equiv \beta = (3\lambda^T + 2\mu^T)\alpha \tag{3.17}$$

for $n = 1, 2, 3$, defining β, and zero otherwise, where superscripts T on λ and μ indicate isothermal Lamé constants. When this result is applied to equation (3.14), we see

$$(c_{nm}^S - c_{nm}^T) = \frac{T\beta^2}{c_v} \tag{3.18}$$

for n and m both <4, zero otherwise, where $c_v = c_\varepsilon$ is the specific heat capacity at constant volume, for an isotropic solid. One result of equation (3.18) is

$$c_{44}^S - c_{44}^T = 0 \rightarrow \mu^S = \mu^T, \tag{3.19}$$

see equation (1.63). Thus the isothermal and adiabatic shear moduli (equation (1.79)) are equal. We therefore omit superscripts on it henceforth. The other result from equation (3.18) is that

$$(c_{11}^S - c_{11}^T) = (c_{12}^S - c_{12}^T) = \frac{T\beta^2}{c_v}. \tag{3.20}$$

From equation (1.64) and (1.63), c_{11} and c_{12} can be taken to be the two independent elastic constants for an isotropic solid.

We now recall the expression for the bulk modulus of such a solid, equation (1.77), which becomes

$$B_S = (\lambda^S + \tfrac{2}{3}\mu), \tag{3.21}$$

$$B_T = (\lambda^T + \tfrac{2}{3}\mu). \tag{3.22}$$

But, again from equation (1.63),

$$c_{11} = (\lambda + 2\mu), \qquad c_{12} = \lambda, \tag{3.23}$$

whence

$$B_S = \tfrac{1}{3}(c_{11}^S + 2c_{12}^S), \tag{3.24}$$

$$B_T = \tfrac{1}{3}(c_{11}^T + 2c_{12}^T). \tag{3.25}$$

From equations (3.24) and (3.25), along with equation (3.20),

$$(B_S - B_T) = (c_{11}^S - s_{11}^T) = (c_{12}^S - c_{12}^T) = \frac{T\beta^2}{c_v}. \tag{3.26}$$

If we think of relationships such as those in equation (3.26) in terms of measurement, as we should, we encounter once again the wonder to be found in the science of thermodynamics. The results are also useful, as we shall now see.

3.3   Thermal conduction and wave motion

We now re-examine the dynamics of an isotropic elastic solid, as discussed in Chapter 2. There we said that, in linear approximation,

$$\rho_0 \partial_t^2 u_j = \partial_i \sigma_{ji}; \tag{3.27}$$

see equations (2.28), (2.32) and (1.27). In Chapter 2, thermodynamic conditions were not considered, as we indicated they must be, in section 3.1. In analysing the spatial variation of $\underset{=}{\sigma}$, equation (3.27), we must consider increments associated with temperature, assuming non-uniform temperature distribution, as well as with strain as in Chapter 2. Thus

$$\partial_i \sigma_{ji} = \left(\frac{\partial \sigma_{ji}}{\partial T}\right)_{\underset{=}{\varepsilon}} (\partial_i T) + \left(\frac{\partial \sigma_{ji}}{\partial \varepsilon_{kl}}\right)_T (\partial_i \varepsilon_{kl}). \tag{3.28}$$

The second term in equation (3.28) has been evaluated in Chapter 1, section 1.5, equations (1.90)–(1.97). In vector notation it is

$$\{(\lambda^T + \mu)\vec{\nabla}(\vec{\nabla} \cdot \vec{u}) + \mu \nabla^2 \vec{u}\} = \{(\lambda^T + 2\mu)\vec{\nabla}(\vec{\nabla} \cdot \vec{u}) - \mu \vec{\nabla} \times (\vec{\nabla} \times \vec{u})\}, \tag{3.29}$$

in terms of the deformation field $\vec{u}$, where we have given two alternative forms, and we have used equation (3.19) to write $\mu^T = \mu$. Now from equations (3.7), which is in Voigt notation, and equation (3.17), we see that

$$\left(\frac{\partial \sigma_{ij}}{\partial T}\right)_{\underset{=}{\varepsilon}} = \beta \delta_{ij}. \tag{3.30}$$

It follows that in equation (3.28)

$$\left(\frac{\partial \sigma_{ij}}{\partial T}\right)_{\underline{\underline{\varepsilon}}} \partial_i T = \beta \partial_j T. \tag{3.31}$$

Thus combining equations (3.28), (3.29) and (3.31), equation (3.27) becomes

$$\{\mu \nabla^2 \vec{u} - \rho_0 \partial_t^2 \vec{u} + (\lambda^{\mathrm{T}} + \mu) \vec{\nabla}(\vec{\nabla} \cdot \vec{u})\} = \beta \vec{\nabla} T. \tag{3.32}$$

We see, by comparing equation (3.32) with equation (2.33) that a thermal gradient ($\vec{\nabla} T$) introduces a source term to the wave equation. A thermal gradient is in turn associated with thermal conduction, as we shall now see. Thermal conduction is described in terms of *heat flux density* $\vec{J}(\vec{r}, t)$,

$$\vec{J}(\vec{r}, t) = -\chi \vec{\nabla} T(\vec{r}, t), \tag{3.33}$$

where χ is the *coefficient of thermal conductivity*, a material constant. Equation (3.33) is taken to be empirical: that heat flows as a result of a thermal gradient, from hot to cold. In the absence of sources or sinks of heat in the material sample, we have energy conservation within an arbitrary fixed volume V of the material,

$$\int_{S(V)} \mathrm{d}s \, \hat{n} \cdot \vec{J}(\vec{r}, t) = -\frac{\mathrm{d}}{\mathrm{d}t} \int_V \mathrm{d}V \, Q(\vec{r}, t), \tag{3.34}$$

where $\hat{n}$ is the unit *outward* normal vector to surface $S(V)$, expressing the fact that the net normal *inward* heat flux through the surface $S(V)$ bounding the volume V accounts entirely for the rate of change of the total amount of heat (thermal energy) within volume V. In equation (3.34) $Q(\vec{r}, t)$ is the thermal energy density within volume V. Since the volume V is fixed,

$$\frac{\mathrm{d}}{\mathrm{d}t} \int_V \mathrm{d}V \, Q = \int_V \mathrm{d}V \, \frac{\mathrm{D}}{\mathrm{D}t} Q, \tag{3.35}$$

where $\mathrm{D}/\mathrm{D}t$ is the material derivative discussed throughout section 2.3 and defined in equation (2.25). The term $(\vec{v} \cdot \vec{\nabla})Q$ that comes from equation (2.25) expresses the phenomenon of *convection*, that is heat transfer due to material flow $\vec{v}$. In a solid, this effect is absent. Equation (3.34), with equation (3.35), then reduces to

$$-\int_V \mathrm{d}V \, \frac{\partial Q}{\partial t} = \int_{S(V)} \mathrm{d}s \, \hat{n} \cdot \vec{J} = \int_V \mathrm{d}V \, \vec{\nabla} \cdot \vec{J}. \tag{3.36}$$

In equation (3.36) we have used Gauss's theorem. Because the volume V is arbitrary, we conclude from equation (3.36) that

$$-\frac{\partial Q}{\partial t} = \vec{\nabla} \cdot \vec{J}. \tag{3.37}$$

Now an increment of heat density ΔQ is given in terms of entropy density (entropy per unit volume) by

$$\Delta Q = T(\Delta S). \tag{3.38}$$

Thus equation (3.37) becomes

$$-T\left(\frac{\partial S}{\partial t}\right)_{\vec{r}} = \vec{\nabla} \cdot \vec{J}. \tag{3.39}$$

Our notation in equation (3.39) emphasizes that the partial derivative is not a thermodynamic coefficient, but a derivative at a fixed position $\vec{r}$, which henceforth we denote ∂_t. Introducing thermal conductivity, assumed to be independent of position $\vec{r}$, from equation (3.33) into equation (3.39), we obtain the *heat flow equation* in the form

$$\chi \nabla^2 T = T \partial_t S \tag{3.40}$$

We now express the increment of entropy density in terms of temperature T and strain $\underline{\varepsilon}$,

$$\Delta S = \left\{ \left(\frac{\partial S}{\partial T}\right)_{\underline{\varepsilon}} \Delta T + \sum_m \left(\frac{\partial S}{\partial \varepsilon_m}\right)_T \Delta \varepsilon_m \right\}, \tag{3.41}$$

with $\underline{\varepsilon}$ in Voigt notation. We next use equations (3.11) and (3.7) in equation (3.41) and isotropy ($c_\varepsilon = c_v$, and $\beta_n = \beta$, equation (3.17)) to obtain

$$\Delta S = \left\{ \frac{c_v}{T} \Delta T + \beta \sum_{m=1}^{3} \Delta \varepsilon_m \right\}. \tag{3.42}$$

Now,

$$\sum_{m=1}^{3} \Delta \varepsilon_m = \Delta \varepsilon_{ii} = \Delta(\vec{\nabla} \cdot \vec{u}); \tag{3.43}$$

see equations (1.44) and (1.5). Thus combining equations (3.40), (3.41), (3.42) and (3.43) we obtain the heat flow equation in final form:

$$\chi \nabla^2 T = \{c_v \partial_t T + T \beta \partial_t (\vec{\nabla} \cdot \vec{u})\}. \tag{3.44}$$

We see that at zero temperature, or if $\vec{\nabla} \cdot \vec{u}$ or its time derivative are negligible, equation (3.44) reduces to the simpler form of the heat flow equation, determining $T(\vec{r}, t)$. It is the same mathematical form as the diffusion equation, Chapter 10. In general, however, we must analyse heat flow and material dynamics, equations (3.44) and (3.32) respectively, as a set of coupled partial differential equations. This we do for a simple, but very instructive, special case in the next section.

3.4   Wave attenuation by thermal conduction

We consider the question whether the wave equation, equation (3.32), coupled to heat flow, equation (3.44), can sustain dilatational plane waves, as discussed in section 2.4.1; see equation (2.45).

     Consider a wave polarized in the x-direction, propagating in the x-direction:

$$u_j(\vec{r}, t) = \delta_{j1} u_0 \exp[\mathrm{i}(kx - \omega t)]. \tag{3.45}$$

Suppose, plausibly, that the temperature deviates from its space-and-time averaged value T_{av} by a similar wave form,

$$[T(\vec{r}, t) - T_{\mathrm{av}}] = T_0 \exp[\mathrm{i}(kx - \omega t)]. \tag{3.46}$$

Then the operator ∂_t applied to $\vec{u}$ gives a factor $(-\mathrm{i}\omega)$, and applied to $T(\vec{r}, t)$ gives $(-\mathrm{i}\omega)(T - T_{\mathrm{av}})$. Similarly $\vec{\nabla}$ reduces to $\partial/\partial x$ which applied to $\vec{u}$ gives a factor $(\mathrm{i}k)$, and applied to $T(\vec{r}, t)$ gives $(\mathrm{i}k)(T - T_{\mathrm{av}})$. Thus equation (3.44) gives

$$(-\chi k^2)(T - T_{\mathrm{av}}) = (-\mathrm{i}\omega c_v)(T - T_{\mathrm{av}}) + (T\beta\omega k)u_1,$$

whence

$$(T - T_{\mathrm{av}}) = \frac{(T\beta k\omega)u_1}{(-k^2\chi + \mathrm{i}\omega c_v)}. \tag{3.47}$$

Thus $(\vec{\nabla}T)$, the right-hand side of equation (3.32), is

$$\vec{\nabla}T = \hat{\varepsilon}_1(\mathrm{i}k)(T - T_{\mathrm{av}}) = \frac{\hat{\varepsilon}_1(\mathrm{i}k^2\omega T\beta)u_1}{(-k^2\chi + \mathrm{i}\omega c_v)}, \tag{3.48}$$

where $\hat{\varepsilon}_1$ is the unit vector in the x direction. We note, parenthetically, that from equation (3.47), $(T - T_{\mathrm{av}})$ is proportional to u_1, but since the proportionality constant is complex, there is a phase difference between the waves u_1 and $(T - T_{\mathrm{av}})$, as we might have anticipated.

     When we substitute from equation (3.48) into the wave equation (3.32), we obtain

$$[-\mu k^2 + \rho_0 \omega^2 - (\lambda^{\mathrm{T}} + \mu)k^2] = \frac{T\beta^2(\mathrm{i}k^2\omega)}{(-k^2\chi + \mathrm{i}\omega c_v)}. \tag{3.49}$$

If we try to cast this in the form of a dispersion relation, $\omega = v'k$, as in equation (2.46), we have

$$\omega^2 = \frac{1}{\rho_0}\left\{ \frac{\mathrm{i}\omega T\beta^2}{(-k^2\chi + \mathrm{i}\omega c_v)} + (\lambda^{\mathrm{T}} + 2\mu) \right\}k^2. \tag{3.50}$$

With $T = 0$, we retrieve equation (2.46) with equation (2.42) for the dilatational-wave dispersion. The additional term with $T \neq 0$ in equation (3.50) cannot be put in that form, both because of its ω- and k-dependence, and because it is a complex number-valued function. We conclude from equation (3.50) that k, or ω, or both must be complex. Consider a wave being generated by periodically applied forces at the surface of a sample of the material. If ω were complex, its imaginary part would need to be negative, so that the wave would go to zero amplitude with time, rather than growing without limit and producing spontaneous breakdown of the material, as would happen if the real part were positive. But the 'dispersion relation', equation (3.50) is independent of position in the medium, and in particular it applies at and near the surface, where the periodic applied forces would prevent attenuation of the wave with time. We therefore conclude that ω must be real, at least for the situation considered above.

It follows that k must be complex,

$$k = (k_1 + ik_2),   (3.51)$$

with both k_1 and k_2 real and positive: then k_1 gives propagation in the positive x-direction, and k_2 gives attenuation rather than buildup to breakdown as the wave propagates through the medium. It will turn out that from equation (3.50), both k_1 and k_2 will be functions of ω. Thus the speed of propagation v is

$$v = \frac{\omega}{k_1(\omega)} = v(\omega),   (3.52)$$

that is, the propagation speed will be frequency dependent, in contrast to the $T = 0$ case, equation (2.42). Correspondingly, the damping coefficient k_2 will be frequency-dependent. Alternatively, we can talk about an attenuation length α,

$$\alpha = \frac{1}{k_2(\omega)} = \alpha(\omega).   (3.53)$$

Note that this is not the same α that comes up in going from equation (3.16) to equation (3.17). Similarly, the wavelength of the wave, λ, is

$$\lambda = \frac{2\pi}{k_1(\omega)} = \lambda(\omega).   (3.54)$$

For the dispersion relation, we have

$$\omega = vk_1.   (3.55)$$

Since we would conventionally plot ω versus k_1 from equation (3.55), we need to express v, equation (3.52), in terms of k_1, rather than in terms of ω. Then we have

$$\omega = v(k_1)k_1 = \omega(k_1).   (3.56)$$

We shall be interested in the frequency dependence of v, equation (3.52), particularly its deviation from v_L, equation (2.42), and the form of $\alpha(\omega)$, equation (3.53), as well as the dispersion $\omega(k_1)$, and particularly its deviation from linearity.

We now return to equation (3.50). First, we can simplify the notation by using some results from classical thermodynamics, section 3.2. In doing so, we must assume that in the irreversible processes represented by thermal conduction in a travelling wave, deviations from equilibrium are small. Consider the factor in { } brackets, equation (3.50). From equation (3.26),

$$T\beta^2 = c_v(B_S - B_T), \tag{3.57}$$

and, from equation (3.22),

$$(\lambda^T + 2\mu) = (B_T + \tfrac{4}{3}\mu). \tag{3.58}$$

Substituting the results of equations (3.57) and (3.58) into equation (3.50) we obtain

$$\omega^2 = \frac{1}{\rho_0} \frac{[i\omega c_v(B_S + \tfrac{4}{3}\mu) - k^2\chi(B_T + \tfrac{4}{3}\mu)]}{(i\omega c_v - k^2\chi)} k^2. \tag{3.59}$$

With simplifying notation,

$$(B_S + \tfrac{4}{3}\mu) = M_S, \qquad (B_T + \tfrac{4}{3}\mu) = M_T, \tag{3.60}$$

we obtain

$$\omega^2 = \frac{1}{\rho_0} \frac{(i\omega c_v M_S - k^2\chi M_T)}{(i\omega c_v - k^2\chi)} k^2. \tag{3.61}$$

From equation (3.51) we now write

$$k^2 \approx (k_1^2 + 2ik_1 k_2), \tag{3.62}$$

where in equation (3.62) we have introduced the linear approximation for weak attenuation: $k_2 \ll k_1$. Using this in equation (3.61), and separating real and imaginary parts we obtain

$$\rho_0\omega^2(\omega c_v - 2\chi k_1 k_2) \approx k_1^2(\omega c_v M_S - 4k_1 k_2 \chi M_T), \tag{3.63}$$

$$\rho_0\omega^2(\chi k_1^2) \approx (k_1^4 \chi M_T + 2k_1 k_2 \omega c_v M_S). \tag{3.64}$$

In equations (3.63) and (3.64), we have only maintained first-order terms in k_2. We replace k_1 in terms of the speed of propagation v,

$$k_1 = \frac{\omega}{v}. \tag{3.65}$$

Then equations (3.63) and (3.64) become

$$\rho_0 v^2 \left(\frac{c_v}{\chi} - \frac{2k_2}{v} \right) = \left(\frac{c_v}{\chi} M_S - 4M_T \frac{k_2}{v} \right), \tag{3.66}$$

$$\rho_0 v^2 = \left(M_T + \frac{2c_v}{\chi} M_S \frac{1}{\omega^2} v^3 k_2 \right). \tag{3.67}$$

We should like to solve equations (3.66) and (3.67) for $v = v(\omega)$ and $k_2 = k_2(\omega)$. Upon eliminating k_2 from equation (3.66) and substituting into equation (3.67) we obtain a cubic polynomial in v^2, equal to zero. We write it as follows:

$$\omega^2 = \left(\frac{c_v}{\chi}\right)^2 M_S v^4 \frac{(\rho_0 v^2 - M_S)}{(\rho_0 v^2 - M_T)(\rho_0 v^2 - 2M_T)}. \tag{3.68}$$

The expression for k_2 in terms of v is

$$k_2 = \left(\frac{c_v}{2\chi}\right) v \frac{(M_S - \rho_0 v^2)}{(2M_T - \rho_0 v^2)}. \tag{3.69}$$

We now discuss equations (3.68) and (3.69) in relation to the experimental fact that for most solids,

$$M_S \approx (1.01) M_T. \tag{3.70}$$

We introduce the notation $(c_v/\chi) = K$, and

$$v_s^2 = \frac{M_S}{\rho_0}, \qquad v_T^2 = \frac{M_T}{\rho_0}. \tag{3.71}$$

Then equations (3.68) and (3.69) take the forms

$$\omega^2 = \frac{K^2 v_S^2 v^4 (v^2 - v_S^2)}{(v^2 - v_T^2)(v^2 - 2v_T^2)}, \tag{3.72}$$

$$k_2 = \frac{K}{2} v \frac{(v_S^2 - v^2)}{(2v_T^2 - v^2)}. \tag{3.73}$$

We note that from equation (3.70)

$$M_T \lesssim M_S < 2M_T. \tag{3.74}$$

Furthermore, since ω must be real, ω^2 must be positive, and thus from equation (3.72) with equation (3.74), we conclude that there are forbidden and allowed regions for v, the allowed region being for $\omega^2 > 0$, namely

$$v_T < v \leq v_S, \qquad \sqrt{2} v_T < v. \tag{3.75}$$

The asymptotes for the plot of v versus ω are easily obtained from equation (3.72): $v \to v_T^+$, $v \to \sqrt{2} v_T^+$ and $v \to [\rho_0/(K^2 M_S)]^{1/2} \omega$, all three as $\omega \to \infty$. Also, $v \to v_S^-$ as $\omega \to 0^+$. It follows that, from equation (3.73), $k_2 \to 0$ as $\omega \to 0^+$ (i.e. as $v \to v_S^-$). Furthermore, as $v \to v_T^+$, since $\omega \to \infty$, we see that $k_2 \to K(v_S^2 - v_T^2)/(2v_T)$ as $\omega \to \infty$. The second branch of k_2, corresponding to $v > \sqrt{2} v_T$, is less easy to analyse qualitatively. All of the above results are illustrated in plots of v versus ω and k_2 versus ω, presented in figures 3.1 and 3.2 respectively, for copper. The asymptotes for v, and the line $v = v_S$, are shown as dashed lines on figure 3.1.

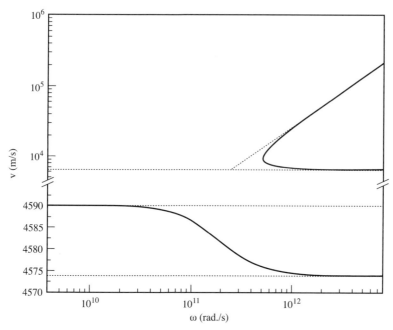

Figure 3.1. Speed of propagation v (m/s) versus angular frequency (rad/s) of a plane dilatational wave in Cu, based on isotropic continuum theory with thermal conduction. See equation (3.72) and discussion following it. Only the lower branch of the solid curve has physical meaning. For $0 < \omega < \infty$, the speed ranges from v_S down to v_T: see equations (3.70) and (3.74). Dashed lines are asymptotes. (Thanks to W A Coish.)

We now discuss figures 3.1 and 3.2. The data used for copper are

$$c_v = 3.44 \times 10^6 \, \text{J/(m}^3 \, {}^\circ\text{C)}$$

$$\chi = 390 \, \text{J/(s m} \, {}^\circ\text{C)}$$

$$B_T = 1.3 \times 10^{11} \, \text{N m}^{-2}$$

$$\rho_0 = 8.89 \times 10^3 \, \text{kg m}^{-3}$$

$$\mu = 4.2 \times 10^{10} \, \text{N m}^{-2}.$$

When these data are used along with the defining equations for M_S, M_T, v_S, v_T, equations (3.60) and (3.71), along with equation (3.70), and the definition $K = (c_v/\chi)$, we obtain the following:

$$(K^2 M_S/\rho_0) = 1.64 \times 10^{15} \, \text{m}^{-2}$$

$$(K/2) = 4.41 \times 10^3 \, \text{s m}^{-2}$$

$$v_S^2 = 2.107 \, \text{m}^2 \, \text{s}^{-2}$$

$$v_T^2 = 2.092 \, \text{m}^2 \, \text{s}^{-2}.$$

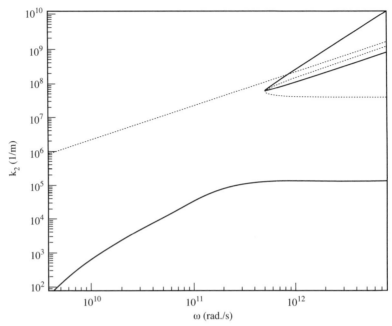

Figure 3.2. Attenuation factor k_2 (m^{-1}) versus angular frequency (rad/s) of a plane dilatational wave in Cu, based on isotropic continuum theory with thermal conduction. See equation (3.73) and discussion following it. Only the lower branch of the solid curve has physical meaning. Dashed lines show k_1 versus ω [see equations (3.41), (3.51) and (3.55)]. (Thanks to W A Coish.)

First, we discuss the approximation $k_2 \ll k_1$, equation (3.62), upon which the curves are based. The plot of $k_1 = (\omega/v)$ is shown by dashed lines on figure 3.2. The straight upper dashed line applies to the lower branch of the solid-line curve for k_2 versus ω. It shows that $k_2 < k_1$ by more than two orders of magnitude for all ω, for this branch. On the other hand, the dashed curve k_1 versus ω lies below the solid curve k_2 versus ω along the whole of the upper branch. This shows that the upper branch is spurious, not conforming to the limitation within which equations (3.72) and (3.73) were derived. Since the upper branch of v versus ω, figure 3.1, corresponds to the upper branch of k_2 versus ω, it is also spurious.

Now to comment briefly on the lower branches of figures 3.1 and 3.2, which constitute the physical results. The speed of propagation varies slightly with frequency, being determined by the adiabatic bulk modulus at low frequency, and by the isothermal bulk modulus at high frequency, for $\omega \gtrsim 10^{12}$ rad/s in the case of copper. Most of the variation, from 4590 m/s to 4574 m/s, occurs for $10^{10} < \omega < 10^{12}$ rad/s. Since v is almost constant, the dispersion relation $\omega = v(k_1)k_1$ is almost linear in k_1: see equation (3.56). The damping factor k_2, figure 3.2, goes to zero as $\omega \to 0$, and approaches an asymptotic high-frequency value of $1.45 \times 10^5 \, \text{m}^{-1}$. The attenuation length which we defined as k_2^{-1} in equation (3.53), is then infinite as $\omega \to 0$, and approaches $6.9 \times 10^{-6} \, \text{m}$ as $\omega \to \infty$. This value, being four orders of magnitude larger than atomic dimensions in the crystal, still lies within the range of the continuum approximation.

In summary, we have considered propagation of a plane dilatational wave in a continuous isotropic medium, subject to thermal conduction. We find that the wave is spatially attenuated, with an attenuation length $\alpha = k_2^{-1}$. We also find that the propagation speed, while frequency dependent, is nearly constant, varying according to equation (3.72) and figure 3.1, for weak attenuation. The attenuation factor k_2 varies from zero at zero frequency to an asymptotic maximum, as determined implicitly from equation (3.73) with equation (3.72), and as illustrated in figure 3.2.

Chapter 4

Surface waves

4.1 Introduction

Surface properties have always been important in materials science. Simply because the topology of a semi-infinite medium, representing a solid with a surface, is different from that of a completely infinite medium, representing the bulk of a solid, there are characteristic features of most phenomena that are specifically related to the surface. In this chapter we show that there is a solution of the wave equation, quite distinct from the travelling waves of Chapter 2, associated with the existence of a surface. This is of significance not only for the real solid, which has surfaces, but for structures on the surface. Such structures include impurities, adsorbed molecules, and artificially produced micron, nanometer, and atomic scale entities, all of which are of technological interest. Detailed study of all such systems, experimental and computational, requires consideration of interaction with the substrate. One aspect of this interaction is the substrate dynamics: its characteristic wave properties. This chapter exemplifies the basis for theoretical understanding of surface elastic wave effects in solids.

Let us consider the surface of an isotropic, homogeneous elastic solid. Choose the surface to be $z = 0$, with the material a semi-infinite continuum $z < 0$. When we apply the free-surface boundary condition with linear elasticity to the equation of motion of the material, we shall find a surface wave, i.e. a wave whose amplitude decreases exponentially with distance from the surface. It will be found to be a specific combination of longitudinal and transverse parts. These Rayleigh waves, so-called after their discoverer (Rayleigh, 1885), are relevant in all surface dynamic processes. Our treatment in this chapter follows that of Landau and Lifshitz (1970), section 24.

In Chapter 2 we showed that the distortion field $\vec{u}(\vec{r}, t)$ in a homogeneous, isotropic elastic solid (or any other vector field) could be written as

$$\vec{u}(\vec{r}, t) = [\vec{u}_{\mathrm{T}}(\vec{r}, t) + \vec{u}_{\mathrm{L}}(\vec{r}, t)],$$

where

$$\vec{\nabla} \cdot \vec{u}_{\mathrm{T}} = 0, \qquad \vec{\nabla} \times \vec{u}_{\mathrm{L}} = 0;$$

48

see equations (2.13) and (2.14). The fields $\vec{u}_T$ and $\vec{u}_L$ were referred to as transverse and longitudinal respectively, which could be taken literally for the case of an infinite medium: see equations (2.40) and (2.47). Furthermore, they have characteristic speeds of propagation v_T and v_L, equations (2.36) and (2.42), with $v_L < v_T$. In this chapter we shall denote these speeds by c_T and c_L.

4.2 Rayleigh waves

We begin with the wave equation,

$$(\partial_t^2 - c_J^2 \nabla^2)\vec{u}_J = 0, \qquad J = T, L, \tag{4.1}$$

where transverse (T) and longitudinal (L) waves satisfy

$$\vec{\nabla} \cdot \vec{u}_T = 0, \qquad \vec{\nabla} \times \vec{u}_L = 0 \tag{4.2a, b}$$

respectively. We consider plane waves, without loss of generality, of the form

$$\vec{u}_J(\vec{r}, t) = \vec{u}_J^{(0)} \, e^{i(k_1 x + k_3 z - \omega t)}, \qquad J = T, L. \tag{4.3}$$

For this to satisfy the wave equation (4.1), we require

$$[-\omega^2 + c_J^2(k_1^2 + k_3^2)] = 0, \qquad J = T, L.$$

The components k_3 and k_1 of the wave vector $\vec{k}$ are therefore related to the angular frequency as follows:

$$k_{J,3}^2 = \left(\frac{\omega^2}{c_J^2} - k_1^2\right), \qquad J = T, L. \tag{4.4}$$

Thus, if $c_J k_1 < \omega$, $k_{J,3}$ is real, and equation (4.3) represents travelling waves in the directions $(k_1, 0, k_{J,3})$; the same longitudinal and transverse waves as are obtained for the bulk material in Chapter 2. However, if $c_J k_1 > \omega$, $k_{J,3}$ is pure imaginary. In that case, denote $k_{J,3} = -iK_J$, and also let $k_1 = k$, with both K_J and k real and positive. The form of equation (4.3) then becomes

$$\vec{u}_J(\vec{r}, t) = \vec{u}_J^{(0)} \, e^{i(kx - \omega t)} \, e^{K_J z}, \qquad J = T, L. \tag{4.5}$$

Since $z < 0$ inside the material, the case $K_J > 0$ corresponds to exponential decay of the wave amplitude with increasing depth, for a wave travelling parallel to the surface with speed (ω/k). The case $K_J < 0$ would be unphysical, with wave amplitude rising exponentially with depth, violating the harmonic approximation for an elastic medium.

4.3 Boundary conditions

We now apply the boundary condition for a free surface, $z = 0$, to the Rayleigh wave solution, equation (4.5); namely from Chapter 1, equation (1.36) with equation (1.33),

$$\sigma_{ij} n_j = 0, \qquad (4.6)$$

where $\hat{n} = (0, 0, 1)$, the unit outward normal vector at the surface, and where the Einstein summation convention is assumed in equation (4.6). Thus,

$$\sigma_{13} = \sigma_{23} = \sigma_{33} = 0. \qquad (4.7)$$

From Hooke's law we have the stress–strain relation in terms of the Lamé elastic constants, equation (1.93):

$$\sigma_{ij} = 2\mu\varepsilon_{ij} + \lambda\delta_{ij}\varepsilon_{kk}. \qquad (4.8)$$

Thus, the boundary condition, equations (4.7), gives us the following constraints on the strain tensor ε_{ij} at the surface $z = 0$:

$$\varepsilon_{13} = \varepsilon_{23} = 0 \qquad (4.9a, b)$$

$$\lambda(\varepsilon_{11} + \varepsilon_{22}) + (\lambda + 2\mu)\varepsilon_{33} = 0. \qquad (4.9c)$$

Let us first consider the transverse component $\vec{u}_T$ of the Rayleigh wave, alone. From the definition of the strain ε_{ij} in terms of the distortion field u_j,

$$\varepsilon_{ij} = \tfrac{1}{2}(\partial_i u_j + \partial_j u_i),$$

equation (1.5), the boundary conditions, equations (4.9) along with equation (4.5) become

$$(iku_{T,3}^{(0)} + K_T u_{T,1}^{(0)}) = 0 \qquad (4.10a)$$

$$K_T u_{T,2}^{(0)} = 0 \qquad (4.10b)$$

$$[ik\lambda u_{T,1}^{(0)} + (\lambda + 2\mu)K_T u_{T,3}^{(0)}] = 0. \qquad (4.10c)$$

The div condition, equation (4.2a), gives

$$(iku_{T,1}^{(0)} + K_T u_{T,3}^{(0)}) = 0. \qquad (4.11)$$

Equations (4.10a) and (4.11) have a non-trivial solution if and only if

$$(K_T^2 + k^2) = 0.$$

This is impossible if both K_T and k are real. Thus the only solution is $u_{T,3}^{(0)} = u_{T,1}^{(0)} = 0$, i.e. $\vec{u}_T = 0$. Similar treatment of $\vec{u}_L$ alone with the curl condition, equation (4.2b), gives $\vec{u}_L = 0$. Thus we conclude that Rayleigh waves cannot be purely longitudinal (curl-less) or transverse (div-less).

We saw in Chapter 2 that the general form of the distortion field $\vec{u}(\vec{r}, t)$ was a linear combination of $\vec{u}_T$ and $\vec{u}_L$ as defined by equations (4.2). With $\vec{u}$

given from equation (4.5) as

$$\vec{u} = (\vec{u}_L + \vec{u}_T),\tag{4.12}$$

the boundary condition, equations (4.9), become

$$ik(u_{T,3}^{(0)} + u_{L,3}^{(0)}) + (K_T u_{T,1}^{(0)} + K_L u_{L,1}^{(0)}) = 0\tag{4.13a}$$

$$(K_T u_{T,2}^{(0)} + K_L u_{L,2}^{(0)}) = 0\tag{4.13b}$$

$$ik\lambda(u_{T,1}^{(0)} + u_{L,1}^{(0)}) + (\lambda + 2\mu)(K_T u_{T,3}^{(0)} + K_L u_{L,3}^{(0)}) = 0.\tag{4.13c}$$

The second of these equations, (4.13b), for the y-component of the wave, is not coupled to the x- and z-components. We may therefore solve for the latter, independent of the former. The simplest case is $u_{T,2}^{(0)} = u_{L,2}^{(0)} = 0$: no y-component of displacement. The other two equations, (4.13a) and (4.13c), will combine with the div and curl conditions, equations (4.2a) and (4.2b), to give four equations in the four amplitudes $u_{J,j}^{(0)}$, $J = T, L, j = 1, 3$.

4.4 Dispersion relation

Let us now apply equations (4.2a) and (4.2b) to the Rayleigh wave, equation (4.5) in the x–z plane. We then have

$$(iku_{T,1}^{(0)} + K_T u_{T,3}^{(0)}) = 0,\tag{4.14a}$$

$$(K_L u_{L,1}^{(0)} - iku_{L,3}^{(0)}) = 0.\tag{4.14b}$$

We eliminate $u_{T,3}^{(0)}$ and $u_{L,3}^{(0)}$ from equations (4.13a) and (4.13c) by using equations (4.14a) and (4.14b). Thus,

$$(k^2 + K_T^2)u_{T,1}^{(0)} + 2K_T K_L u_{L,1}^{(0)} = 0,\tag{4.15a}$$

$$2\mu k^2 u_{T,1}^{(0)} + [(\lambda + 2\mu)K_L^2 - \lambda k^2]u_{L,1}^{(0)} = 0.\tag{4.15b}$$

We now recall the relationship between elastic constants (λ, μ) and wave speeds (c_L, c_T), equations (2.36) and (2.42),

$$(\lambda + 2\mu) = c_L^2 \rho_0,\tag{4.16a}$$

$$\mu = c_T^2 \rho_0,\tag{4.16b}$$

where ρ_0 is the mean mass density of the material. Also recall equations (4.4):

$$K_T^2 = \left(k^2 - \frac{\omega^2}{c_T^2}\right),\tag{4.17a}$$

$$K_L^2 = \left(k^2 - \frac{\omega^2}{c_L^2}\right).\tag{4.17b}$$

When these expressions are substituted into equations (4.15a) and (4.15b), we obtain

$$\left(2k^2 - \frac{\omega^2}{c_T^2}\right)u_{T,1}^{(0)} + 2\left[\left(k^2 - \frac{\omega^2}{c_T^2}\right)\left(k^2 - \frac{\omega^2}{c_L^2}\right)\right]^{1/2}u_{L,1}^{(0)} = 0, \qquad (4.18a)$$

$$2k^2 u_{T,1}^{(0)} + \left(2k^2 - \frac{\omega^2}{c_T^2}\right)u_{L,1}^{(0)} = 0. \qquad (4.18b)$$

The necessary and sufficient condition that equations (4.18a) and (4.18b) have a solution is

$$\left(2k^2 - \frac{\omega^2}{c_T^2}\right)^4 = (4k^2)^2\left(k^2 - \frac{\omega^2}{c_T^2}\right)\left(k^2 - \frac{\omega^2}{c_L^2}\right). \qquad (4.19)$$

This is the dispersion relation, angular frequency ω as a function of wave number k, for Rayleigh waves.

We now solve equation (4.19). Since (ω/k) has the dimensionality of speed, we can write

$$\omega = \xi c_T k, \qquad (4.20)$$

where ξ is a dimensionless parameter. In this notation, equation (4.19) becomes

$$\xi^6 - 8\xi^4 + 8\left(3 - \frac{2c_T^2}{c_L^2}\right)\xi^2 - 16\left(1 - \frac{c_T^2}{c_L^2}\right) = 0. \qquad (4.21)$$

Thus we see that ξ is independent of k, and depends only on the material constant (c_T/c_L). Now consider $(c_T/c_L)^2$,

$$\left(\frac{c_T}{c_L}\right)^2 = \frac{\mu}{(\lambda + 2\mu)} = \frac{1}{\left(\dfrac{\lambda}{\mu} + 2\right)},$$

from equations (4.16a) and (4.16b). From equation (1.67) for Poisson's ratio ν we have

$$\nu = \frac{\lambda}{2(\lambda + \mu)} = \frac{1}{2\left(1 + \dfrac{\mu}{\lambda}\right)}$$

whence

$$\left(\frac{\mu}{\lambda}\right) = \left(\frac{1}{2\nu} - 1\right).$$

It follows that

$$\left(\frac{c_T}{c_L}\right)^2 = \frac{1}{\left[2 + \left(\dfrac{1}{2\nu} - 1\right)^{-1}\right]} = \frac{(1 - 2\nu)}{2(1 - \nu)}.$$

We therefore define a parameter γ as

$$\gamma = \left(\frac{c_T}{c_L}\right)^2 = \frac{(1-2\nu)}{2(1-\nu)}, \tag{4.22}$$

where ν is Poisson's ratio. Since from equations (1.88) and (1.89) we have seen that $-1 \leq \nu \leq \frac{1}{2}$, we see that the corresponding range of γ, equation (4.22), is $\frac{3}{4} \geq \gamma \geq 0$. Furthermore, for Rayleigh waves, following equations (4.17a) and (4.17b), we conclude that $w \leq c_T k$, so from the definition of ξ, equation (4.20), we conclude that $\xi \leq 1$. We are therefore only interested in roots of equation (4.21) with $\xi \leq 1$, where $(c_T/c_L)^2 \leq \frac{3}{4}$. We rewrite equation (4.21) in terms of the following notation:

$$x = \xi^2, \qquad 0 \leq x \leq 1 \tag{4.23a}$$

$$a_1 = 8(3 - 2\gamma), \qquad \gamma = (c_T/c_L)^2 \leq \frac{3}{4} \tag{4.23b}$$

$$a_0 = 16(1 - \gamma). \tag{4.23c}$$

Thus, consider

$$f(x) = (x^3 - 8x^2 + a_1 x - a_0) = 0, \tag{4.24}$$

for $0 \leq x \leq 1$. At $x = 0$, $f(x) = -a_0$, and at $x = 1$, $f(x) = (-7 + a_1 - a_0)$. The possible ranges of these values are determined by the condition $0 \leq \gamma \leq \frac{3}{4}$, from equations (4.23). Thus $(-16) \leq f(0) \leq (-4)$, and $f(1) = 1$. Now the cubic, equation (4.24), has at most three real roots, and since $f(0) < 0$ and $f(1) > 0$, there is at least one root in $0 \leq x \leq 1$. If there is more than one root, it must be that $df/dx = 0$ in $0 \leq x \leq 1$, so that the curve, having crossed the x-axis, can turn back and touch or cross it twice again. For this to happen, $df/dx = 0$ must occur twice in the region $0 \leq x \leq 1$. Thus, consider

$$\frac{df}{dx} = (3x^2 - 16x + a_1). \tag{4.25}$$

At the physical extremities, $df/dx = 0$ has roots $x = (4.431, 0.903)$ for $\gamma = \frac{3}{4}$, and has imaginary roots for $\gamma = 0$. Thus there is only one extremum of $f(x)$ in $0 \leq x \leq 1$, and therefore there is only one real root of $f(x)$ within the range of physical possibility, $-1 \leq \nu \leq \frac{1}{2}$. This root lies in the range

$$0.475 \leq \xi \leq 0.913 \qquad \text{for } -1 \leq \nu \leq \frac{1}{2}. \tag{4.26}$$

In summary to this point, we have found that Rayleigh surface waves are of the form

$$\vec{u}(\vec{r}, t) = [\vec{u}_T(\vec{r}, t) + \vec{u}_L(\vec{r}, t)], \tag{4.12}$$

where

$$\vec{u}_J(\vec{r}, t) = \vec{u}_J^{(0)} e^{i(kx - \omega t)} e^{K_J z}, \qquad J = T, L, \quad z \leq 0, \tag{4.5}$$

with

$$\omega = \xi c_T k, \tag{4.20}$$

and

$$K_J = \left(k^2 - \frac{\omega^2}{c_J^2}\right)^{1/2} = k\left(1 - \xi^2 \frac{c_T^2}{c_J^2}\right)^{1/2}. \tag{4.17}$$

In equation (4.20), ξ is material-dependent: specifically, it is a function of Poisson's ratio: see equation (4.22). Thus, in equation (4.5), the damping factors K_J for transverse and longitudinal parts of the wave depend exclusively on ν, while the frequency for a given wave number is a function of ν and c_T: see equation (4.20).

4.5   Character of the wave motion

Now let us examine the amplitudes $\vec{u}_J^{(0)}$ in equation (4.12), with equation (4.5). From equations (4.14a), (4.14b) and (4.15a),

$$u_{L,1}^{(0)} = \frac{-(k^2 + K_T^2)}{2K_T K_L} u_{T,1}^{(0)}, \tag{4.27a}$$

$$u_{L,3}^{(0)} = \frac{i(k^2 + K_T^2)}{2K_T k} u_{T,1}^{(0)}, \tag{4.27b}$$

$$u_{T,3}^{(0)} = \frac{-ik}{K_T} u_{T,1}^{(0)}. \tag{4.27c}$$

Thus, for arbitrary $u_{T,1}^{(0)}$, the other amplitudes are determined by equations (4.27). Denote

$$u_{T,1}^{(0)} = A. \tag{4.28}$$

Then from equations (4.27a) and (4.27b),

$$\frac{\vec{u}_L}{A} = \frac{-(k^2 + K_T^2)}{2K_T K_L} \left\{\hat{i} - \frac{iK_L \hat{k}}{k}\right\} e^{i(kx - \omega t)} e^{K_L z}, \tag{4.29a}$$

and, from equation (4.27c),

$$\frac{\vec{u}_T}{A} = \left\{\hat{i} - \frac{ik}{K_T}\hat{k}\right\} e^{i(kx - \omega t)} e^{K_T z}. \tag{4.29b}$$

In equations (4.29), $\hat{i}$ and $\hat{k}$ are unit vectors in the x and z directions respectively.

Since the distortion field $\vec{u}(\vec{r}, t)$ is real, we must take real or imaginary parts of $\vec{u}_J$ in equation (4.12), $J = T, L$. We consider the real parts. Noting in equations (4.29) that

$$\pm i = \exp(\pm i\pi/2), \qquad (4.30)$$

and

$$\cos\left(\theta \pm \frac{\pi}{2}\right) = \mp \sin\theta, \qquad (4.31)$$

we obtain

$$\frac{\vec{u}_L}{A} = \frac{-(k^2 + K_T^2)}{2K_T K_L} e^{K_L z}\left\{\hat{i}\cos(kx - \omega t) + \frac{\hat{k}K_L}{k}\sin(kx - \omega t)\right\}, \qquad (4.32a)$$

$$\frac{\vec{u}_T}{A} = e^{K_T z}\left\{\hat{i}\cos(kx - \omega t) + \frac{\hat{k}\cdot k}{K_T}\sin(kx - \omega t)\right\}, \qquad (4.32b)$$

or

$$\frac{\vec{u}(\vec{r}, t)}{A} = \hat{i}\left\{\frac{-(k^2 + K_T^2)}{2K_T K_L} e^{K_L z} + e^{K_T z}\right\}\cos(kx - \omega t)$$

$$+ \hat{k}\left\{\frac{-(k^2 + K_T^2)}{2K_T k} e^{K_L z} + \frac{k}{K_T} e^{K_T z}\right\}\sin(kx - \omega t). \qquad (4.33)$$

We note from equations (4.32a) and (4.32b) that the longitudinal component contains a contribution from the z-direction, while the transverse component contains one from the x-direction. This illustrates that the terms 'longitudinal' and 'transverse' are not to be taken literally. From equation (4.33) we see that x and z components of the wave are out of phase by $\pi/2$ radians.

Finally, consider the relative amplitudes of x and z components. From equation (4.33), their ratio R is

$$R = \frac{k}{K_L} \cdot \frac{[-(k^2 + K_T^2)e^{K_L z} + 2K_T K_L e^{K_T z}]}{[-(k^2 + K_T^2)e^{K_L z} + 2k^2 e^{K_T z}]}. \qquad (4.34)$$

If we rewrite equation (4.34) in terms of the notation of equations (4.17), (4.20) and (4.22), we have

$$R = (1 - \xi^2)^{1/2} \frac{\left\{1 - \dfrac{(1 - \xi^2/2)e^{(K_L - K_T)z}}{(1 - \xi^2)^{1/2}(1 - \xi^2\gamma)^{1/2}}\right\}}{\{1 - (1 - \xi^2/2)e^{(K_L - K_T)z}\}}. \qquad (4.35)$$

Recall that ξ is a function only of γ. At the surface, $z = 0$, we can evaluate R at the limits of the range $0 \leq \gamma \leq \frac{3}{4}$, corresponding to $0.913 \geq \xi \geq 0.475$. The results are

$$R(\gamma = 0, z = 0) = -0.42,$$

$$R(\gamma = \tfrac{3}{4}, z = 0) = -0.83.$$

In fact, the ratio is negative, and $|R| < 1$, for all values of γ, i.e. for all values of Poisson's ratio ν. Referring to equation (4.33), we can without loss of generality write $\vec{u}(\vec{r}, t)$ for an arbitrary value of x, say $x = 0$. Then near $z = 0$ ($z < 0$), with a particular choice of A we have

$$\vec{u}(0, t) = \hat{i} u_x^{(0)} \cos(-\omega t) + \hat{k} u_z^{(0)} \sin(-\omega t)$$

$$= \hat{i} u_x^{(0)} \cos(\omega t) - \hat{k} u_z^{(0)} \sin(\omega t)$$

$$= \hat{i} u_x^{(0)} \cos(\omega t) + \hat{k} |u_z^{(0)}| \sin(\omega t), \qquad (4.36)$$

where $u_x^{(0)}$ is positive due to the choice of A. From equation (4.36) we see that the motion of material in the surface due to a Rayleigh wave is on a counterclockwise closed path. This path is described by

$$x = u_x^{(0)} \cos(\omega t), \qquad z = |u_z^{(0)}| \sin(\omega t)$$

which can be written as

$$\left\{ \frac{x^2}{u_x^{(0)^2}} + \frac{z^2}{u_z^{(0)^2}} \right\} = 1.$$

This is the formula of an ellipse, with its major axis in the z direction, since the amplitudes perpendicular to the surface, $u_z^{(0)}$, are greater than those parallel to it, $u_x^{(0)}$, from $|R| < 1$. This conforms to the fact that the material is less constrained in the z than in the x direction. However, referring to equation (4.35), we see that there is a depth $z < 0$, at which $u_x^{(0)} = 0$, beyond which $R > 0$. When $u_x^{(0)} = 0$, the material oscillates vertically, in the z direction only, and when $R > 0$, it rotates *clockwise* along an elliptic path.

Chapter 5

Dislocations

5.1   Introduction

We now enter the realm of plastic deformation of a solid material. Plastic deformation is not elastic: when the stresses that cause the deformation are removed, the deformation does not disappear. For crystalline materials, plastic deformation is associated with the motion of a particular kind of crystal lattice defect called a dislocation. Although this type of defect can be described up to a point by the continuum theory of solids, and will be so described in this chapter, more detailed understanding requires atomistic considerations.

A dislocation is essentially a linear or filamentary defect in a crystal. The variety of crystal structures that can exist, coupled with the topological variety associated with filamentary systems, leads to a huge variety of configurations, motions, and physical effects. Beyond that, when one considers dynamical aggregates of dislocations as they occur in materials, and the interactions of dislocations with other crystal lattice defects such as point defects, interfaces and surfaces, the variety of concepts and phenomena burgeons much further. In this chapter we limit the discussion to introducing the basic terms and concepts by which dislocations are described (section 5.2), and to the application of continuum elasticity theory (Chapter 1) to determine the equilibrium deformation field at some distance from a dislocation (section 5.3). In section 5.4 we analyse an aspect of dislocation motion, and in section 5.5 we give some view of the range of topics not to be discussed here, as well as some comments on the textbook/monograph literature.

5.2   Description of dislocations

The textbook by the Weertmans (1964) gives an extraordinarily lucid introduction to the nature of dislocations, and we follow their approach closely.

We can think of creating a dislocation in a crystal through a process of plastic deformation that is illustrated in an extremely schematic

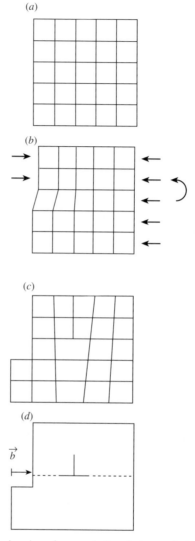

Figure 5.1. (*a*) Schematic drawing of segment of perfect crystal, showing atomic planes. (*b*) Segment of crystal subjected to a combination of shear, compressional and torque stresses: elastic deformation. (*c*) Edge dislocation that is left after forces in part (*b*) have produced plastic deformation. (*d*) Continuum picture of part (*c*), indicating edge dislocation symbol (inverted T) and Burgers vector $\vec{b}$.

(oversimplified) way in figure 5.1. Figure 5.1*a* represents a perfect crystal, with the lines representing planes of atoms seen edge-on, the intersections representing rows of atoms perpendicular to the page. Forces represented by straight arrows are applied in figure 5.1*b* to produce a shearing effect (along with some compression). A torque represented by the curved arrow prevents rotation of the sample. If the forces are built up to a sufficient strength, plastic deformation shown in figure 5.1c results, and remains after removal of forces and torques. At the atomic scale, what has happened is that a vertical plane of atoms has broken, one edge remaining discon-nected, while the other edge has connected up with the partial plane above it that was originally to its left. In fact, part of a horizontal plane of atoms has slipped to the right, its atoms coming into register with those that were originally below them to the right. The disconnected edge of the vertical plane is called an *edge dislocation*, and the plane where the slipping has occurred is called the *slip plane*. If we wipe out the lattice planes from figure 5.1c, we can identify the edge dislocation by the conventional T-shaped symbol (inverted here), and the slip plane by a dashed line, in figure 5.1d. Note that in this illustration the slip plane and the dislocation symbol for the edge are interstitial. The displacement along the slip plane equal to an interplanar distance in the crystal, denoted $\vec{b}$ in figure 5.1d, is the *Burgers vector* of this edge dislocation. We note that, from figure 5.1c, an edge dislocation looks like the edge of a partial plane of atoms that is inserted into an otherwise perfect segment of crystal. Such an atomistic configuration can occur with or without the stepped configuration of the left-hand face of crystal shown in figure 5.1c.

For a continuum picture of an edge dislocation, we can think of figure 5.1a as having been created by cutting horizontally into a block of material from the left, displacing the edge of the left face above the cut by $\vec{b}$ relative to the edge of the face below the cut, and then rejoining the material inside the sample. The edge dislocation then coincides with the edge of the cut inside the sample, the cut defining the slip plane. The only atomistic information in this picture is that $|\vec{b}|$ is an interatomic distance. In a similar way we can create a *screw dislocation*, by displacing the two parts of the left-hand face by a Burgers vector $\vec{b}'$ *parallel* to the edge of the cut. These two cases are illustrated in figures 5.2a and 5.2b respectively. The screw dislocation is defined to be the edge of the cut inside the sample, just as for the edge dislocation. Using the word screw to describe this case can best be illustrated with a cylindrical sample of material, as in figure 5.2c. The shaded end of the cylinder that was originally an atomic plane, is deformed into a helical ramp, or screw-type configuration. All atomic planes that were originally parallel to this one inside the sample join up to make the helical ramp continuous. A *mixed dislocation* can also be created by making the edge of the cut within the sample curved, as shown in figure 5.2d. The dashed dislocation line, where it emerges from the right-side face of the sample is screw-like, the relative displacement of material above and below the cut being *parallel* to the dislocation. Where it emerges from the front face (partly shaded), however, the displacement is *perpendicular* to the dislocation line, rendering the latter edge-like.

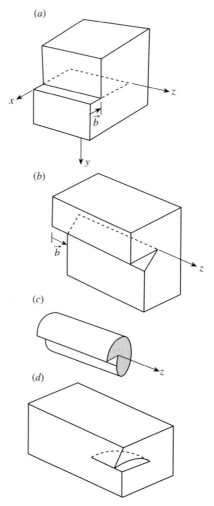

Figure 5.2. (*a*) Creation of an edge dislocation by cutting into the sample and displacing an outer edge of the cut perpendicular to the inner edge of the cut. (*b*) Creation of a screw dislocation by cutting into the sample and displacing an outer edge of the cut parallel to the inner edge. (*c*) Illustrating the process of creating a screw dislocation as in part (*b*), but in a cylindrical sample of material. (*d*) Creation of a mixed dislocation by cutting into the sample and displacing one edge of the cut perpendicular to one part of the inner edge and parallel to another part.

We note that a dislocation line, edge or screw, cannot end inside the material because it is the edge of a cut coming from the outside surfaces. We can image a cut, or separation, between two atomic planes, such that the cut is completely inside the material. We then create a closed *dislocation loop* entirely inside the material by moving material above the cut by an interatomic displacement relative to the material below the cut. Dislocation loops are extremely common and important in materials science. They are very well illustrated and discussed by the Weertmans (1964) in their figure 1.6.

If we return to figure 5.1, we can perhaps see that reapplication of shear could cause the edge dislocation to move to the right in steps of $\vec{b}$ (see figure 5.1a) until it disappears at the right-hand face, which at that point would have an overhanging step above the slip plane. Thus if a material contains a reasonable density of dislocations (edge and/or screw) then plastic shear deformation takes place by the migration of the dislocations across their slip planes, rather than by displacement of one whole atomic plane relative to an adjacent one all in one movement. The energy required for the latter process is much larger than that for the former. The relative ease of plastic deformation compared with what would be expected for a perfect crystal was evidently a key fact that led to the discovery of the dislocation.

In figure 5.1a we introduced the Burgers vector $\vec{b}$ for that edge dislocation. Although the idea of the Burgers vector is fairly simple, its precise definition is tricky as the reader will find on consulting some of the standard references cited later in this chapter. In terms of an atomistic model, one compares a closed circuit in a perfect, dislocation-free crystal with the corresponding circuit around a dislocation line, stepping from one atomic site to the next one. By 'the corresponding circuit' we mean the same sequence of site-to-site steps, each in the same crystal direction as before. If the circuit is not too close to the dislocation line, the crystal directions of the perfect crystal are only slightly perturbed, locally, in the presence of the dislocation. Consider first a closed circuit in the perfect crystal. Let the atomic sites involved be denoted by position vectors $\vec{R}_{J0}$, $J = 1, 2, \ldots, n$. Then the total displacement around the circuit is, trivially,

$$(\vec{R}_{20} - \vec{R}_{10}) + (\vec{R}_{30} - \vec{R}_{20}) + \cdots + (\vec{R}_{n0} - R_{n-1,0}) + (\vec{R}_{10} - \vec{R}_{n0}) = 0. \quad (5.1)$$

In the presence of the dislocation, atomic positions that are not close to the dislocation line will be slightly displaced from perfect crystal sites $\vec{R}_{J0}$ to $(\vec{R}_{J0} + \vec{u}_J)$. The sequence of site-to-site steps defined by the perfect-crystal circuit will not now close, because one of the atoms along the original circuit will have been replaced by another atom in the 'cutting and displacement' process upon which, in principle, the dislocation is based. Neglecting small site displacements $\vec{u}_J$, the last step of the sequence will end up at $\vec{R}_{n+1,0}$ where $(n + 1) \neq 1$, and where in fact

$$(\vec{R}_{n+1,0} - \vec{R}_{10}) \approx (-\vec{b}). \quad (5.2)$$

For the closed circuit in the presence of the dislocation, the closure being provided by the Burgers vector $\vec{b}$, if we include the small displacements $\vec{u}_J$, we have

$$(\vec{u}_2 - \vec{u}_1) + (\vec{u}_3 - \vec{u}_2) + \cdots + (\vec{u}_{n+1} - \vec{u}_n) \approx -\vec{b}. \qquad (5.3)$$

or

$$\sum_{J=1}^{n} (\Delta\vec{u}_J) \approx (-\vec{b}). \qquad (5.4)$$

The discussion above, for a crystal of discrete atoms, requires that the circuit in the presence of the dislocation should not pass too close to the dislocation line. In fact the discussion is valid, in the sense of the approximate equalities of equations (5.2)–(5.4), to fair accuracy, provided the circuit does not intersect the dislocation line. If the circuit stays well away from the dislocation line, then all of the displacements $\vec{u}_J$ in equation (5.4) are small in the sense of linear elasticity theory (Chapters 1 and 2), and the number of atomic sites on the circuit is large, i.e. the size of the circuit is large compared with atomic dimensions. In that case, equation (5.4) reduces to

$$\int_c d\vec{u} = (-\vec{b}), \qquad (5.5)$$

the equality being highly accurate.

For a given circuit c in equation (5.5), and correspondingly in the discrete-atom picture of equations (5.3) and (5.4) with equation (5.1), the direction around the circuit is arbitrary, and so therefore is the direction of $\vec{b}$. The ambiguity can be resolved as follows. Arbitrarily define a positive direction, represented by the unit vector $\hat{\tau}$, along a dislocation line. Define the direction around the circuit to establish, by the right-hand thumb rule, the orientation of the circuit in the perfect crystal to be perpendicular to $\hat{\tau}$. Then, as above, the Burgers vector $\vec{b}$ is the closure of the circuit in the presence of the dislocation. In this case, the vector $(\vec{b} \times \hat{\tau})$ lies in the extra plane of atoms that defines an edge dislocation, and $\vec{b}$ is parallel to the positive direction of a screw dislocation for which the screw is *left-handed*, as is the screw dislocation illustrated in figure 5.2c.

5.3   Deformation fields of dislocations

We now consider dislocations at equilibrium in a solid on the basis of continuum mechanics. From the last section we realize that such a treatment will need to be limited to the region not too close to the dislocation line. The analysis will be based on the equilibrium condition for an isotropic continuous solid, equation (1.94), Chapter 1,

$$(\lambda + \mu)\partial_i\partial_k u_k + \mu\partial_k\partial_k u_i = 0, \qquad (5.6)$$

along with the requirement for a dislocation, equation (5.5).

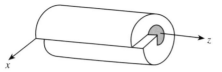

Figure 5.3. Creating a screw dislocation by *rigidly* displacing that part of the cut in the elastically deformed non-core region, the core region being indicated by shading.

5.3.1 Screw dislocation

For a screw dislocation, we introduce the Burgers vector explicitly by considering a configuration of material similar to that in figure 5.2c. We cannot apply continuum theory right down to the dislocation line, as illustrated in figure 5.2c. Instead, we consider a *uniform* displacement over that part of the slip plane that lies at a sufficient distance from the dislocation line; that is, outside the dislocation core. The proposed configuration is illustrated in figure 5.3, where the intersection of the core with the end of the cylindrical sample of material is shown shaded. In figure 5.3 we have not shown the shear deformation within the core, since it cannot be accurately represented by continuum theory. The system shown in figure 5.3, having been subjected to a shear, will not change volume in first order: see the end of section 1.2.1. Thus if we identify the direction of the dislocation line as the z direction (indices i or k equal to 3 in equation (5.6)), then there is no deformation in the x–y plane: $u_1 = u_2 = 0$. From figure 5.3, outside the core u_3 will be independent of the radial distance r, because the displacement in this region is uniform with respect to r in the slip plane. From the isotropy of the material, we conclude that

$$u_3 = -\frac{b}{2\pi} \cdot \theta, \qquad u_1 = u_2 = 0, \tag{5.7}$$

where θ is the angular position relative to the slip plane in the right-handed sense relative to the z axis. The negative sign in equation (5.7) comes from this choice. Equation (5.7) represents a multi-valued distortion field $\vec{u}(\vec{r})$, in the sense that θ and $(\theta + 2\pi)$ represent the same point in the material, for given (r, z). This conforms with the earlier representation of the Burgers vector, equation (5.5), with $\vec{u}$ increasing by $(-\vec{b})$ with each passage around the closed path c. Because the slip plane defines a transverse reference direction the system does not have cylindrical symmetry, so a cartesian coordinate system is as convenient as a cylindrical coordinate system, for purposes of visualization. If $\theta = 0$ defines the x–z plane, then from equation (5.7),

$$u_3 = -\frac{b}{2\pi} \tan^{-1}\left(\frac{y}{x}\right). \tag{5.8}$$

Let us test equations (5.7) and (5.8) as a solution for the equilibrium condition, equation (5.6). Since u_3 is independent of z, equation (5.8), and $u_1 = u_2 = 0$, equation (5.7), we have $(\vec{\nabla} \cdot \vec{u}) = 0$, so the first term in equation (5.6) is zero ($\partial_k u_k = \vec{\nabla} \cdot \vec{u}$). We are left, in equation (5.6), with

$$\mu(\partial_1^2 + \partial_2^2)u_3 = 0. \tag{5.9}$$

From equation (5.7), we must therefore evaluate

$$(\partial_1^2 + \partial_2^2)u_3 = -\frac{b}{2\pi}(\partial_1^2 + \partial_2^2)\theta. \tag{5.10}$$

From $\tan \theta = (y/x)$ we easily conclude that

$$\partial_1^2 \theta = \frac{2xy}{(x^2 + y^2)} = -\partial_2^2 \theta, \tag{5.11}$$

so that the equilibrium condition, equation (5.9), is satisfied.

We can now evaluate the strain tensor ε_{ij}, equation (1.5). From equations (5.7) and (5.8) we easily see that

$$\varepsilon_{11} = \varepsilon_{22} = \varepsilon_{33} = 0 = \varepsilon_{12}. \tag{5.12}$$

Furthermore:

$$\varepsilon_{31} = \tfrac{1}{2}(\partial_3 u_1 + \partial_1 u_3) = \tfrac{1}{2}\partial_1 u_3 \tag{5.13}$$

$$\varepsilon_{23} = \tfrac{1}{2}(\partial_2 u_3 + \partial_3 u_2) = \tfrac{1}{2}\partial_2 u_3 \tag{5.14}$$

$$\partial_1 u_3 = -\frac{b}{2\pi}\partial_1 \theta = -\frac{b}{2\pi}(\cos^2 \theta)\left(-\frac{y}{x^2}\right)$$
$$= \frac{b}{2\pi}\frac{y}{(x^2 + y^2)} = \frac{b}{2\pi}\frac{\sin \theta}{r}, \tag{5.15}$$

$$\partial_2 u_3 = -\frac{b}{2\pi}(\cos^2 \theta)\frac{1}{x} = -\frac{b}{2\pi}\frac{x}{(x^2 + y^2)} = -\frac{b}{2\pi}\frac{\cos \theta}{r}. \tag{5.16}$$

We now combine equations (5.13)–(5.16), writing the result in cylindrical (r, θ, z) coordinates, based on $x = r\cos\theta$, $y = r\sin\theta$:

$$\varepsilon_{31} = \frac{b}{4\pi}\frac{\sin \theta}{r}, \tag{5.17}$$

$$\varepsilon_{23} = -\frac{b}{4\pi}\frac{\cos \theta}{r}. \tag{5.18}$$

The characteristic r^{-1} behavior of the strain at large distance from a dislocation is evident here, along with asymmetry associated with the slip plane. Equations (5.17) and (5.18) also illustrate the problem with the continuum approximation in relation to the core region, where extrapolation $r \to 0$ would lead to infinite strain, contrary to the assumptions of linear elasticity theory. Although we have obtained a multi-valued distortion field $\vec{u}(\vec{r})$, equation (5.7), we see that the strain field ε_{ij}, equations (5.12), (5.17) and (5.18), is single valued. Finally, we note that for an isotropic material, from Chapter 1 following equation (1.62) and translating from Voigt to cartesian tensor notation, the stress tensor field is

$$\sigma_{31} = \frac{\mu b}{2\pi} \frac{\sin \theta}{r}, \tag{5.19}$$

$$\sigma_{23} = -\frac{\mu b}{2\pi} \frac{\cos \theta}{r}. \tag{5.20}$$

Let us now consider the resultant force and torque on a cylinder of radius R containing the screw dislocation whose stress tensor is given by equations (5.19) and (5.20). We can do this from the relationship between external surface forces per unit area P_i and stress σ_{ij} at the surface of the material, from Chapter 1,

$$P_i = n_j \sigma_{ij} \tag{1.36}$$

where n_j is the unit outward normal at the surface. On the cylinder,

$$n_1 = \cos \theta, \qquad n_2 = \sin \theta, \qquad n_3 = 0. \tag{5.21}$$

From equations (5.19)–(5.21) we see that $P_i = 0$, $i = 1, 2, 3$. Since the external force per unit area vanishes everywhere on the cylinder's surface, there can be no net force or net torque on it.

5.3.2 Edge dislocation

To study an edge dislocation in the continuum approximation we use the same concepts as in the previous section: that is, we produce uniform displacement over that part of the slip plane that lies outside the core: see figure 5.4. The non-uniform displacement occurs entirely in the core region. For a long dislocation line, the properties of the system are independent of z. We now conjecture that, outside the core, u_1, varying from zero to $(-b)$, does so linearly with angular position θ, independent of r. Presumably, if this conjecture is valid, it will be so as a result of very special boundary

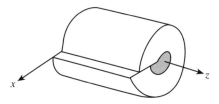

Figure 5.4. Creating an edge dislocation by *rigidly* displacing that part of the cut in the elastically deformed non-core region, the core region being indicated by shading.

conditions. We then have

$$u_1 = -\frac{b}{2\pi}\theta = -\frac{b}{2\pi}\tan^{-1}\left(\frac{y}{x}\right). \tag{5.22}$$

The thought that u_2, as well as u_3, might be zero, is quickly dispelled: the equilibrium equation (5.6) would not be satisfied. Specifically, its y component, $i = 2$, becomes

$$(\lambda + \mu)\partial_2\partial_1 u_1 = 0.$$

But from equation (5.22),

$$\partial_2\partial_1 u_1 = \left(\frac{-b}{2\pi}\right)\frac{(y^2 - x^2)}{(x^2 + y^2)^2},$$

which is not zero in general. Nonetheless, let us persist with $u_3 = 0$, u_1 given by equation (5.22), and u_2 to be determined. Then the equilibrium equation (5.6) gives

$$\{(\lambda + 2\mu)\partial_1^2 + \mu\partial_2^2\}u_1 + (\lambda + \mu)\partial_1\partial_2 u_2 = 0, \qquad i = 1 \tag{5.23}$$

$$\{(\lambda + 2\mu)\partial_2^2 + \mu\partial_1^2\}u_2 + (\lambda + \mu)\partial_2\partial_1 u_1 = 0, \qquad i = 2. \tag{5.24}$$

Again from equation (5.22), these reduce to

$$\partial_1\partial_2 u_2 = \frac{xy}{(x^2 + y^2)^2}\left(\frac{b}{\pi}\right) \tag{5.25}$$

and

$$\{(\lambda + 2\mu)\partial_2^2 + \mu\partial_1^2\}u_2 = (\lambda + \mu)\frac{(x^2 - y^2)}{(x^2 + y^2)^2}\left(-\frac{b}{2\pi}\right). \tag{5.26}$$

Now we have chosen $u_1 = u_1(\theta)$. Let us investigate whether a solution exists with $u_2 = u_2(r)$, where

$$r = (x^2 + y^2)^{1/2}. \tag{5.27}$$

We then have

$$\partial_1 u_2 = \frac{du_2}{dr}\,\partial_1 r = \frac{x}{(x^2 + y^2)^{1/2}}\frac{du_2}{dr} \tag{5.28}$$

$$\partial_2\partial_1 u_2 = \left\{-\frac{xy}{(x^2 + y^2)^{3/2}}\frac{du_2}{dr} + \frac{x}{(x^2 + y^2)^{1/2}}\frac{d^2 u_2}{dr^2}\,\partial_2 r\right\}$$

$$= \left\{-\frac{xy}{r^3}\frac{du_2}{dr} + \frac{xy}{r^2}\frac{d^2 u_2}{dr^2}\right\}. \tag{5.29}$$

Substituting equation (5.29) into equation (5.25), we have

$$\frac{xy}{r^2}\left(\frac{d^2u_2}{dr^2} - \frac{1}{r}\frac{du_2}{dr}\right) = \left(\frac{b}{\pi}\right)\frac{xy}{r^4} \tag{5.30}$$

or

$$\left(\frac{d}{dr} - \frac{1}{r}\right)\left(\frac{du_2}{dr}\right) = \left(\frac{b}{\pi}\right)\frac{1}{r^2}. \tag{5.31}$$

If we denote $du_2/dr = f$, then this is

$$\left(\frac{d}{dr} - \frac{1}{r}\right)f = \left(\frac{b}{\pi}\right)\frac{1}{r^2}. \tag{5.32}$$

Seeking a solution $f = cr^n$, we find

$$f \equiv \frac{du_2}{dr} = \left(-\frac{b}{2\pi}\right)\frac{1}{r} \tag{5.33}$$

whose solution is

$$u_2 = \left(-\frac{b}{2\pi}\right)\ln\left(\frac{r}{R}\right) = \left(-\frac{b}{4\pi}\right)\ln\left(\frac{x^2+y^2}{R^2}\right). \tag{5.34}$$

The additive constant of integration expressed by R represents the boundary condition that when $r = R$, $u_2(r = R) = 0$. This condition, $u_2 = 0$, will only be rigorously satisfied, independent of θ, as $R \to \infty$ for an infinitely thick cylinder of material. It will be approximately valid for a large enough sample, at and beyond large but finite R. The solution that we have found is approximate in this sense.

We must now see whether this solution, equation (5.34), satisfies the other equilibrium equation (5.26). Now we have

$$\partial_2^2 u_2 = \left(-\frac{b}{4\pi}\right)\partial_2\left(\frac{2y}{x^2+y^2}\right) = \left(-\frac{b}{2\pi}\right)\frac{(x^2-y^2)}{(x^2+y^2)^2} \tag{5.35}$$

$$\partial_1^2 u_2 = \left(-\frac{b}{4\pi}\right)\partial_1\left(\frac{2x}{x^2+y^2}\right) = \left(-\frac{b}{2\pi}\right)\frac{(y^2-x^2)}{(x^2+y^2)^2} \tag{5.36}$$

whence the left-hand side of equation (5.26) is

$$[(\lambda + 2\mu) - \mu]\frac{(x^2-y^2)}{(x^2+y^2)^2}\left(-\frac{b}{2\pi}\right), \tag{5.37}$$

in agreement with the right-hand side. We have therefore found a solution

$$u_1 = -\frac{b}{2\pi}\theta, \qquad u_2 = -\frac{b}{2\pi}\ln\left(\frac{r}{R}\right), \qquad u_3 = 0. \tag{5.38}$$

We may use this solution to obtain the strain tensor and the stress tensor. This will lead us to discover the boundary conditions and externally applied stresses to which the solution is subject. Rewrite equations (5.38) in cartesian coordinates:

$$u_1 = -\frac{b}{2\pi}\tan^{-1}\left(\frac{y}{x}\right), \qquad u_2 = -\frac{b}{4\pi}\ln\left(\frac{x^2+y^2}{R^2}\right), \qquad u_3 = 0. \quad (5.39)$$

We readily evaluate the strain tensor, defined in equation (1.5), using equations (5.39):

$$\varepsilon_{11} = -\varepsilon_{22} = \frac{b}{2\pi}\frac{y}{(x^2+y^2)} = \frac{b}{2\pi}\frac{\sin\theta}{r}, \qquad (5.40)$$

$$\varepsilon_{12} = \varepsilon_{21} = -\frac{b}{2\pi}\frac{x}{(x^2+y^2)} = -\frac{b}{2\pi}\frac{\cos\theta}{r}, \qquad (5.41)$$

with all other elements zero. We note from equation (5.40) that $\mathrm{Tr}\,\varepsilon = 0$, so there is no dilatation or compression in the given strain, only shear: see equations (1.14) and (1.19), and also equation (1.25): no rotation. We also note that the radius R defined for equation (5.39) does not enter the strain field equations (5.40) and (5.41). All that is required is that such a radius exist.

For the stress in this isotropic medium we use equations (1.60)–(1.62) and the following equation in Chapter 1, along with equation (1.44), to obtain

$$\sigma_{11} = -\sigma_{22} = \frac{\mu b}{\pi}\frac{\sin\theta}{r}, \qquad \sigma_{12} = -\frac{\mu b}{\pi}\frac{\cos\theta}{r}. \quad (5.42)$$

In equation (5.42) the Lamé elastic constant μ is known to be the shear modulus: see equation (1.79). This reflects our earlier statement that only shear is involved, and not dilatation or compression.

We are now in a position to determine the external forces that are necessary to maintain the deformation field that we have derived, equations (5.39). As for the screw dislocation of the previous section, we evaluate the external force per unit area $P_j = n_i\sigma_{ij}$ on a cylindrical sample of radius R, now containing the edge dislocation, for which the stress tensor σ_{ij} is given by equations (5.42). Again with $n_1 = \cos\theta$, $n_2 = \sin\theta$, we obtain

$$P_1 = P_3 = 0, \qquad P_2 = -\frac{\mu b}{\pi R}. \quad (5.43)$$

The net force per unit length of cylinder is therefore given by

$$F_1 = F_3 = 0, \qquad F_2 = \int_0^{2\pi} d\theta R P_2(R,\theta) = -2\mu b. \quad (5.44)$$

The torque per unit area $\vec{t}$ is given by

$$\vec{t} = \hat{n}R \times \vec{P} = \det \begin{pmatrix} \hat{i} & \hat{j} & \hat{k} \\ R\cos\theta & R\sin\theta & 0 \\ 0 & -(\mu b/\pi R) & 0 \end{pmatrix} = -\hat{k}\frac{\mu b}{\pi}\cos\theta. \qquad (5.45)$$

The net torque $\vec{T}$ per unit length of cylinder is therefore

$$\vec{T} = \int_0^{2\pi} d\theta\, R\vec{t}(\theta) = 0. \qquad (5.46)$$

This result may be obvious from the symmetry of P_2 with respect to the y–z plane in equation (5.43). If r_0 is the radius of the core region, then there is a further force per unit area $(\mu b)/(\pi r_0)$ on the *inner* cylindrical boundary of the non-core region where $n_1 = -\cos\theta$, $n_2 = -\sin\theta$, resulting in a further net force $F_2' = (2\mu b)$ on that boundary. The force F_2 on the outer surface of the cylinder must come from an outside object. The equal and opposite force F_2' must come from the core of the dislocation. Some external agent must be pushing on the core with this force, though experimentally such a force is not realizable. The fully correct deformation field would not have any such forces: the external forces on core and non-core regions would be zero. Weertman and Weertman (1964) present a solution (their equation (2.13)) that eliminates the need for external forces applied to the core or to the outer surface of the cylinder. Nabarro (1967) discusses this solution in detail. (See also Exercise 5.1). We note that it involves Poisson's ratio, and hence dilatation as well as shear. The solution that we have worked with, equations (5.37), omits a term ($\sim \sin\theta\cos\theta$) for u_1 and a term ($\sim \sin^2\theta$) for u_2, and it overestimates the strength of the logarithmic term.

5.4   Uniform dislocation motion

We now want to examine the condition for the screw dislocation of figure 5.3 to propagate with uniform speed, say V, in the x direction. The static solution for the deformation field $\vec{u}(\vec{r})$ is given by equations (5.7) and (5.8):

$$u_1(\vec{r}) = u_2(\vec{r}) = 0, \qquad u_3(\vec{r}) = -\frac{b}{2\pi}\theta = -\frac{b}{2\pi}\tan^{-1}\left(\frac{y}{x}\right). \qquad (5.47)$$

We recall that, in figure 5.3 the deformation in the non-core region is pure shear, and that only the shear modulus comes into the stress–strain relations in this case. The equation of motion for deformation is given by equation (2.33). If we look for a shear wave solution, noting that the deformation, equation (5.47), satisfies $\vec{\nabla}\cdot\vec{u} = 0$, then from equation (2.35) with equation (2.36),

$$\left(\nabla^2 - \frac{1}{v_T^2}\partial_t^2\right)\vec{u} = 0.$$

From equation (5.47), this reduces to

$$\left\{(\partial_1^2 + \partial_2^2) - \frac{1}{v_T^2}\,\partial_t^2\right\}u_3 = 0. \tag{5.48}$$

Let us seek a solution $u_3(\vec{r}, t) = u_3(x, y, t)$ which propagates in the x direction, perpendicular to the screw dislocation line, without change of shape:

$$u_1 = u_2 = 0, \qquad u_3(x, y, t) = f(x - Vt, y). \tag{5.49}$$

Then

$$\partial_t^2 f = V^2 \partial_x^2 f. \tag{5.50}$$

Combining equations (5.48)–(5.50),

$$\left\{\left(1 - \frac{V^2}{v_T^2}\right)\partial_{x'}^2 + \partial_y^2\right\}f = 0, \tag{5.51}$$

where $x' = (x - Vt)$. From equation (5.51) we can see that the requirement is

$$(\partial_{x''}^2 + \partial_y^2)f = 0, \tag{5.52}$$

where

$$x'' = \frac{x'}{\beta} = \frac{(x - Vt)}{\beta}, \qquad \beta = \left(1 - \frac{V^2}{v_T^2}\right)^{1/2}. \tag{5.53}$$

Now we have seen that our solution, equation (5.47), for the static deformation field of a screw dislocation as in figure 5.3, satisfies

$$(\partial_x^2 + \partial_y^2)u_3 = 0. \tag{5.54}$$

Thus for the time-dependent deformation, $u_3 = f$, equation (5.49), propagating with speed V in the positive x-direction without change of shape, we may take, from equation (5.52),

$$u_3(x, y, t) = f(x'', y), \tag{5.55}$$

where

$$f(x'', y) = -\frac{b}{2\pi}\tan^{-1}\left(\frac{y}{x''}\right), \tag{5.56}$$

with x'' given in equation (5.53). This form will propagate without change of shape. From equations (5.55), (5.56) and (5.53), we conclude that

$$u_1 = u_2 = 0, \qquad u_3 = -\frac{b}{2\pi}\tan^{-1}\left(\frac{\beta y}{x - Vt}\right). \tag{5.57}$$

The relation given in equation (5.53), that at $t = 0$,

$$x = \beta x'', \qquad \beta = \left(1 - \frac{V^2}{v_T^2}\right)^{1/2},$$

is analogous to the Lorentz–Fitzgerald contraction in special relativity, with the transverse speed of propagation v_T now playing the role of a limiting speed for the relative motion. We do not take the analogy to be particularly meaningful, however, except that there is a limiting speed for the motion described here. It is straightforward to pass from the deformation field $\vec{u}$, equation (5.57) to strain ε_{ij}, to stress σ_{ij}, as in equations (5.13), (5.14), and (5.17)–(5.20). The resultant stress, in cartesian coordinates, is

$$\sigma_{31} = \frac{\mu b}{2\pi} \frac{\beta y}{[(x - Vt)^2 + \beta^2 y^2]}, \tag{5.58}$$

$$\sigma_{23} = -\frac{\mu b}{2\pi} \frac{(x - Vt)}{[(x - Vt)^2 + \beta^2 y^2]}. \tag{5.59}$$

Consider the plane $x = Vt$, perpendicular to the x axis, containing the dislocation line, moving in the x direction with speed V. In the non-core region, on this plane

$$\sigma_{31}(x = Vt, y) = \frac{\mu b}{2\pi} \frac{1}{\left(1 - \dfrac{V^2}{v_T^2}\right)^{1/2} y}. \tag{5.60}$$

Thus as $V \to v_T^-$, $\sigma_{31} \to \infty$ for finite y; the shear stress becomes infinite. Thus $V = v_T$, the speed of shear waves in the medium, is an upper bound for the speed of propagation of a straight screw dislocation perpendicular to the dislocation line in the slip plane. It turns out that the same limit applies to edge dislocations, from this level of theory (see for example Nabarro (1967), section 7.1.1.1).

5.5 Further study of dislocations

In this chapter we have discussed only the simplest aspects of dislocations. Even an elementary course devoted to dislocations will cover a much wider range of topics. In this context we have already mentioned the book by Weertman and Weertman (1964), which has a theoretical orientation. We also mention Hull and Bacon (1995) at a similar level. At the advanced level, we mention three works, each with somewhat different orientation, and therefore complementary: Friedel (1964), Nabarro (1967) and Nadgorny (1988). No doubt there are many other worthy books at both elementary and advanced levels. Perhaps the best pedagogical work on fundamental properties known to the author is the five-volume series entitled *Dislocations in*

Solids edited by Nabarro (1979 and later). This series in fact extends to at least ten volumes with more specialized topics.

The reader can readily imagine the range of topics through which this subject extends. For example: point defect interactions with dislocations, including aggregation and stability; interactions of dislocations with surfaces and interfaces; interactions amongst dislocations; configurations, dynamics, topology and evolution of dislocation networks; optical, magnetic, electric and chemical, as well as mechanical properties of dislocations; rate processes involving dislocations and other crystal defects in interaction; the role of dislocations in plastic deformation, cracking and fracture; and combinations from the preceding list.

While we have emphasized the continuum description of dislocations in this chapter, there is a substantial and growing literature in which atomistic (i.e. discrete atom) models are used, either exclusively or in a hybrid combination of atomistic core region with continuum non-core region. In the former context, we mention two approaches. An atomistic modelling scheme based on the shell model for insulating crystals has been developed (Puls *et al.*, 1977) and extensively applied (see for example Rabier and Puls, 1987). The shell model will be introduced in Chapter 9. The other approach is based on the embedded atom method (EAM) of Daw and Baskes (1983, 1984); see also Daw *et al.* (1993). EAM incorporates the inescapable many-body quantum-mechanical features of interatomic interaction in metallic systems. Its applications to dislocations are exemplified by the works of Simmons *et al.* (1997), and of Rao *et al.* (1999).

Chapter 6

Classical theory of the polaron

6.1 Introduction

The concept of a polaron arises from considering the motion of an electron in a dielectric medium. The electron carries a characteristic field of dielectric polarization along with it, leaving behind a decaying wake of polarization. One can then think of the electron and its dynamic polarization field combined as a quasiparticle. It is this quasiparticle that is called a polaron. The properties of polarons are important not only for electron motion in insulators and semiconductors [see for example, Appel (1968), especially sections 10 and 20 entitled 'Experimental Situation'], but for highly diffuse bound states, such as the excited state of an electron in the presence of an anion vacancy (called an F center) in an ionic crystal [see for example Fowler (1964)]. In fact, the concept generalizes widely, to the description of various quasiparticles in condensed matter interacting with a self-created field in the medium.

Apart from the importance of the polaron and of polaronic effects in condensed matter, the effect of a particle's self-created field on its dynamical state is of major importance in the history of elementary particle field theory. Specifically, the non-zero interaction energy of a hydrogen atom electron in its $2s\frac{1}{2}$ state with its self-field (electromagnetic), compared with that in its $2p\frac{1}{2}$ state, which is zero, accounts for the Lamb–Retherford effect (1947). This $2s\frac{1}{2}$–$2p\frac{1}{2}$ energy level difference is zero at the level of Dirac electron theory when self-interaction is omitted. In addition to an energy level shift (the Lamb shift), the self-interaction contributes to the particle's kinetic energy, or to its observable mass. In quantum field theory, the numerical values of both the mass contribution and the level shift are infinite. Consistent and meaningful ways of dealing with these infinities to give finite results in agreement with experiment have been developed: so-called renormalization theory. Excellent discussions of the Lamb–Retherford effect are given by Sakurai (1967, section 2.8), and by Baym (1969, p. 574). To the uninitiated, it is very difficult to have an intuitive picture of mass renormalization and energy-level shift due to self-interaction for an electron in vacuum, particularly when attempts at an analytical description yield divergent integrals. On the other hand, the classical picture of the polaron to be presented here seems to the author to be much simpler to follow and to visualize, and it leads to explicit, finite expressions for mass renormalization, or effective mass, and for the self-energy.

We shall consider a classical electron moving at constant speed (corresponding to a quantum-mechanical momentum eigenstate) in a classical homogeneous isotropic dielectric continuum. In section 6.2 we derive the classical equations of motion of a polaron, and in section 6.3 we solve for the properties of a polaron with constant velocity. In section 6.4 the momentum dependence of the effective potential for the polaron quasiparticle will be used to discuss briefly an interesting aspect of the quantization process for such a system in a magnetic field. We remark that, in a sense, this chapter contains an extension of the continuum theory of elasticity upon which Chapters 1 to 5 are based.

6.2   Equations of motion

We model this section on an introduction to polaron theory by Fröhlich (1963). It uses the hamiltonian formulation of mechanics: see Goldstein (1980), especially sections 1.4 and 8.1. For polarization that is not too strong, we can take the energy density of the field to be harmonic. The polarization $\vec{P}(\vec{r}, t)$ is the electric dipole moment per unit volume of dielectric. In Chapter 9 we shall show, in the context of a classical atomistic model of an insulator, that $\vec{P}$ is proportional to the deformation field $\vec{u}(\vec{r}, t)$, introduced in section 1.2. In harmonic approximation (i.e. within the limits of linear elasticity theory), therefore, the lagrangian density of the polarization field is of the form

$$l(\vec{r}, t) = \frac{\mu}{2} [\dot{P}^2(\vec{r}, t) - \omega^2 P^2(\vec{r}, t)]. \tag{6.1}$$

In equation (6.1), the dot indicates time derivative. Fröhlich (1963) shows that the effective 'force constant' $(\mu\omega^2)$ is simply related to the low and high frequency dielectric constants of the material. We put 'force constant' in quotation marks because it does not have the dimensionality of a force constant. The parameter μ is proportional to the mass density, through the linear relationship between $\vec{P}$ and $\vec{u}$.

We now Fourier analyse the polarization field subject to periodic boundary conditions on a large cubical volume V,

$$P_\alpha(\vec{r}, t) = V^{-1/2} \sum_{\vec{k}} q_{\vec{k},\alpha}(t) \, e^{i\vec{k}\cdot\vec{r}} \tag{6.2}$$

where

$$k_\alpha = 2\pi V^{-1/3} n_\alpha, \qquad n_\alpha = 0, \pm 1, \pm 2, \ldots \tag{6.3}$$

In equations (6.2) and (6.3), α refers to cartesian components. Substituting equation (6.2) into equation (6.1) we have for the lagrangian density

$$l(\vec{r}, t) = \frac{\mu}{2} \sum_{\vec{k}, \vec{k}', \alpha} \{\dot{q}_{\vec{k}, \alpha}(t) \dot{q}_{\vec{k}', \alpha}(t) - \omega^2 q_{\vec{k}, \alpha}(t) q_{\vec{k}', \alpha}(t)\} V^{-1} \exp[i(\vec{k} + \vec{k}') \cdot \vec{r}].$$
(6.4)

The total lagrangian for the field is therefore

$$L_P(t) = \int_V d^3r\, l(\vec{r}, t).$$
(6.5)

We note that in view of equation (6.3)

$$V^{-1} \int_V d^3r \exp[i(\vec{k} + \vec{k}') \cdot \vec{r}] = \delta_{\vec{k}, -\vec{k}'}.$$
(6.6)

Substituting equation (6.4) into equation (6.5) and using equation (6.6) we obtain

$$L_P = \frac{\mu}{2} \sum_{\vec{k}', \alpha'} \{\dot{q}_{\vec{k}', \alpha'} \dot{q}_{-\vec{k}', \alpha'} - \omega^2 q_{\vec{k}', \alpha'} q_{-\vec{k}', \alpha'}\}.$$
(6.7)

We note that L_P in equation (6.7) is real because $\vec{q}_{-\vec{k}} = \vec{q}_{\vec{k}}^*$ is required so that the polarization $\vec{P}$ in equation (6.2) will be real.

　　We must identify the Fourier coefficients $q_{\vec{k}, \alpha}$ as the generalized coordinates for the mechanical system consisting of the polarization field. The canonical momenta are

$$\Pi_{\vec{k}, \alpha}(t) = \frac{\partial}{\partial \dot{q}_{\vec{k}, \alpha}} L_P = \mu \dot{q}_{-\vec{k}, \alpha}.$$
(6.8)

We note that in differentiating L_P, equation (6.7), with respect to $\dot{q}_{\vec{k}, \alpha}$, the particular vector $\vec{k}$ occurs in two terms in the sum over $\vec{k}'$. From equations (6.7) and (6.8) we obtain the hamiltonian for the field,

$$H_P = \left\{ \sum_{\vec{k}, \alpha} \dot{q}_{\vec{k}, \alpha} \Pi_{\vec{k}, \alpha} - L_P \right\}.$$
(6.9)

Using equation (6.8) to eliminate $\dot{q}_{\vec{k}, \alpha}$ in favor of $\Pi_{\vec{k}, \alpha}$, we obtain from equation (6.9) with equation (6.7)

$$H_P = \sum_{\vec{k}', \alpha'} \left\{ \frac{1}{2\mu} \Pi_{\vec{k}', \alpha'} \Pi_{-\vec{k}', \alpha'} + \frac{\mu \omega^2}{2} q_{\vec{k}', \alpha'} q_{-\vec{k}', \alpha'} \right\}.$$
(6.10)

The electron part of the hamiltonian is trivial,

$$H_{el} = \frac{1}{2m} p^2,$$
(6.11)

where $\vec{p}$ is the canonical momentum of the electron and m is its mass. In fact, if $\vec{r}_{el}$, the position of the electron, has its components identified as the electron's canonical coordinates, then

$$p_\alpha = m\dot{r}_{el,\alpha}, \qquad \alpha = 1, 2, 3, \tag{6.12}$$

or

$$\vec{p} = m\vec{v}, \qquad \vec{v} = \dot{\vec{r}}_{el}, \tag{6.13}$$

where $\vec{v}$ is the electron's velocity.

We now consider the hamiltonian term representing electron–field interaction. If the electron were at rest at $\vec{r}_{el}$, its displacement field $\vec{D}(\vec{r}; \vec{r}_{el})$ in the medium would interact with the polarization that it produces with an energy which, in linear approximation, is

$$H_{P,el} = - \int d^3 r' \, \vec{P}(\vec{r}') \cdot \vec{D}(\vec{r}'; \vec{r}_{el})/\varepsilon_0. \tag{6.14}$$

When the electron is in uniform motion, a dissipative effect arising from dynamic polarization, like a viscous drag, acts on the electron. We assume that the electron is moving so slowly that this dissipative effect is negligible. This validates equation (6.12) above. We Fourier analyse $\vec{P}$ as in equation (6.2) and from equation (6.14) obtain

$$H_{P,el} = - \sum_{\vec{k},\alpha} d_{-\vec{k},\alpha} q_{\vec{k},\alpha} \tag{6.15}$$

where

$$d_{\vec{k},\alpha} = V^{-1/2} \int d^3 r \, e^{-i\vec{k}\cdot\vec{r}} D_\alpha(\vec{r}; \vec{r}_{el}(t))/\varepsilon_0, \tag{6.16}$$

is the Fourier transform of $\vec{D}/\varepsilon_0$ in V with periodic boundary conditions. We now combine equations (6.10), (6.11) and (6.15) to get the hamiltonian for the whole system:

$$H = (H_P + H_{el} + H_{P,el})$$

$$= \left[\sum_{\vec{k}',\alpha'} \left\{ \frac{1}{2\mu} \Pi_{\vec{k}',\alpha'} \Pi_{-\vec{k}',\alpha'} + \frac{\mu\omega^2}{2} q_{\vec{k}',\alpha'} q_{-\vec{k}',\alpha'} \right\} \right.$$

$$\left. + \frac{p^2}{2m} - \sum_{\vec{k}',\alpha'} d_{-\vec{k}',\alpha'}(\vec{r}_{el}) q_{\vec{k}',\alpha'} \right]. \tag{6.17}$$

We note that, from equation (6.12), $d_{\vec{k},\alpha}$ is a function of $\vec{r}_{el}$. Here, ε_0 is the permittivity of free space.

Perhaps we should comment on the use of periodic boundary conditions for a system of net charge $(-e)$ due to the excess electron $(e > 0)$. We assume that the periodically repeated volume V is so large that the effect of one such volume on an adjacent one is negligible: the polarons in different regions of volume V are essentially non-interacting. In the infinite medium represented by the periodic boundary conditions there is an infinite number of electrons whose repulsive interaction energy is infinite, but unchanging as the system evolves in time. Since it contributes nothing to the dynamics, we ignore it as a harmless artifact of the formulation.

We now evaluate the hamiltonian equations of motion, from equation (6.17):

$$\dot{\Pi}_{\vec{k},\alpha} = -\frac{\partial H}{\partial q_{\vec{k},\alpha}} = \{-\mu\omega^2 q_{-\vec{k},\alpha} + d_{-\vec{k},\alpha}(\vec{r}_{\text{el}})\}, \tag{6.18}$$

$$\dot{p}_\alpha = -\frac{\partial H}{\partial r_{\text{el},\alpha}} = \sum_{\vec{k}',\alpha'} \left(\frac{\partial d_{-\vec{k}',\alpha'}}{\partial r_{\text{el},\alpha}}\right) q_{\vec{k}',\alpha'}, \tag{6.19}$$

$$\dot{q}_{\vec{k},\alpha} = \frac{\partial H}{\partial \Pi_{\vec{k},\alpha}} = \frac{1}{\mu}\Pi_{-\vec{k},\alpha}, \tag{6.20}$$

$$\dot{r}_{\text{el},\alpha} = \frac{\partial H}{\partial p_\alpha} = \frac{1}{m}p_\alpha. \tag{6.21}$$

We combine equations (6.18) and (6.20),

$$\ddot{q}_{\vec{k},\alpha} = \frac{1}{\mu}\dot{\Pi}_{-\vec{k},\alpha} = \left\{-\omega^2 q_{\vec{k},\alpha} + \frac{1}{\mu}d_{\vec{k},\alpha}(\vec{r}_{\text{el}})\right\},$$

or

$$(\ddot{q}_{\vec{k},\alpha} + \omega^2 q_{\vec{k},\alpha}) = \frac{1}{\mu}d_{\vec{k},\alpha}(\vec{r}_{\text{el}}). \tag{6.22}$$

Let us evaluate $d_{\vec{k},\alpha}(\vec{r}_{\text{el}})$ in equation (6.22) for the point charge electron at $\vec{r}_{\text{el}}$, using the definition, equation (6.16). From Maxwell's equation,

$$\vec{\nabla} \cdot \vec{D} = \rho \tag{6.23}$$

with

$$\rho(\vec{r}) = -e\delta(\vec{r} - \vec{r}_{\text{el}}), \tag{6.24}$$

for a classical point-charge electron, and applying Gauss's theorem to equation (6.23), we obtain the Coulomb result,

$$\vec{D}(\vec{r}, \vec{r}_{\text{el}}) = -\frac{e}{4\pi}\frac{(\vec{r} - \vec{r}_{\text{el}})}{|\vec{r} - \vec{r}_{\text{el}}|^3}. \tag{6.25}$$

We substitute from equation (6.25) into equation (6.16),

$$d_{\vec{k},\alpha} = -\frac{V^{-1/2}e}{4\pi\varepsilon_0}\int d^3r\,\frac{(\vec{r} - \vec{r}_{\text{el}})}{|\vec{r} - \vec{r}_{\text{el}}|^3}\,e^{-i\vec{k}\cdot\vec{r}}. \tag{6.26}$$

We make a change of variable,

$$\vec{r}' = (\vec{r} - \vec{r}_{el}),$$

and perform the integration in spherical polar coordinates with $\vec{k}$ defining the polar axis. Then $d_{\vec{k},1} = d_{\vec{k},2} = 0$, and

$$d_{\vec{k},3} = \frac{ieV^{-1/2}}{\varepsilon_0} e^{-i\vec{k}\cdot\vec{r}_{el}} \int_0^\infty dr' \left\{ -\frac{\cos(kr')}{(kr')} + \frac{\sin(k'r)}{(kr')^2} \right\}. \tag{6.27}$$

Now integrating the first term by parts, with the change of variable $x = kr'$,

$$\frac{1}{k}\left\{ -\int_0^\infty dx \frac{\cos x}{x} \right\} = -\frac{1}{k}\left\{ \left[\frac{\sin x}{x}\right]_0^\infty + \int_0^\infty dx \frac{\sin x}{x^2} \right\}.$$

Thus the integral in equation (6.27) has the value

$$-\frac{1}{k}\left[\frac{\sin x}{x}\right]_0^\infty = \frac{1}{k}, \tag{6.28}$$

and equation (6.27) reduces to

$$d_{\vec{k},\alpha} = \frac{ieV^{-1/2}}{\varepsilon_0 k} e^{-i\vec{k}\cdot\vec{r}_{el}}\delta_{\alpha,3}, \tag{6.29}$$

where we recall that the cartesian coordinate axis $\alpha = 3$ is in the direction $\vec{k}$. We can therefore write

$$\vec{d}_{\vec{k}} = \mu\vec{\varphi}_{\vec{k}}\, e^{-i\vec{k}\cdot\vec{r}_{el}}. \tag{6.30}$$

where

$$\vec{\varphi}_{\vec{k}} = \frac{ieV^{-1/2}}{\varepsilon_0\mu k}\frac{\vec{k}}{k} = \vec{\varphi}^*_{-\vec{k}}. \tag{6.31}$$

With equation (6.29) we obtain the equation of motion for the polarization field from equation (6.22),

$$(\ddot{q}_{\vec{k},\alpha} + \omega^2 q_{\vec{k},\alpha}) = \varphi_{\vec{k},\alpha}\, e^{-i\vec{k}\cdot\vec{r}_{el}}. \tag{6.32}$$

Schultz (1963) gives an excellent discussion of the Green's function solution of equation (6.32).

6.3 The constant-velocity polaron

Suppose the polaron is moving with constant velocity $\vec{v}$,

$$\vec{r}_{el} = \vec{v}\cdot t. \tag{6.33}$$

Equation (6.32) becomes

$$(\ddot{q}_{\vec{k},\alpha} + \omega^2 q_{\vec{k},\alpha}) = \varphi_{\vec{k},\alpha}\, e^{-i\Omega_{\vec{k}} t} \tag{6.34}$$

where

$$\Omega_{\vec{k}} = (\vec{k}\cdot\vec{v}) = -\Omega_{-\vec{k}}. \tag{6.35}$$

Equation (6.34) has a solution $q_{\vec{k},\alpha} \sim e^{-i\Omega_{\vec{k}} t}$, namely,

$$q_{\vec{k},\alpha}(t) = \frac{\varphi_{\vec{k},\alpha}\, e^{-i\Omega_{\vec{k}} t}}{(\omega^2 - \Omega_{\vec{k}}^2)}. \tag{6.36}$$

Next consider the equations of motion of the electron, equations (6.19) and (6.21),

$$\ddot{r}_{\mathrm{el},\alpha} = \frac{1}{m}\dot{p}_{\alpha} = \frac{1}{m}\sum_{\vec{k}',\alpha'}\frac{\partial}{\partial r_{\mathrm{el},\alpha}}(d_{-\vec{k}',\alpha'})q_{\vec{k}',\alpha'}. \tag{6.37}$$

We ask whether this is compatible with our assumed solution $\vec{r}_{\mathrm{el}} = \vec{v}\cdot t$, i.e.

$$\ddot{r}_{\mathrm{el},\alpha} = 0. \tag{6.38}$$

From equations (6.30) and (6.36) we can evaluate the right hand side of equation (6.37), to see if it is zero. We have

$$\frac{1}{m}\sum_{\vec{k}',\alpha'}\frac{\partial}{\partial r_{\mathrm{el},\alpha}}(\mu\varphi_{-\vec{k}',\alpha'}\, e^{i\vec{k}'\cdot\vec{r}_{\mathrm{el}}})\frac{\varphi_{\vec{k}',\alpha'}\, e^{-i\vec{k}'\cdot\vec{r}_{\mathrm{el}}}}{(\omega^2 - \Omega_{\vec{k}'}^2)} = \frac{\mu}{m}\sum_{\vec{k}',\alpha}(ik_{\alpha'})\frac{|\varphi_{\vec{k}'}|^2}{(\omega^2 - \Omega_{\vec{k}'}^2)}. \tag{6.39}$$

We introduce spherical polar coordinates (k,θ,φ) in $\vec{k}$-space, with the polar direction given by $\vec{v}$. We replace the discrete variable $\vec{k}$, equation (6.3), by a continuous variable, and use the density of points ρ_k in $\vec{k}$-space that follows from equation (6.3):

$$\rho_k = V/(2\pi)^3. \tag{6.40}$$

Equation (6.39) then becomes

$$\frac{\mu}{m}\frac{V}{(2\pi)^3}i\sum_{\alpha=1}^{3}\int d^3k'\, k'_{\alpha}\frac{|\varphi_{\vec{k}'}|^2}{(\omega^2 - \Omega_{\vec{k}'}^2)}, \tag{6.41}$$

which with the definitions of $\varphi_{\vec{k}}$ and $\Omega_{\vec{k}}$, equations (6.31) and (6.35), becomes

$$\frac{\mu}{m}\frac{V}{(2\pi)^3}i\int_0^{\infty}dk'\, k'^2\int_0^{2\pi}d\varphi\int_{-1}^{1}d\lambda\frac{\left(\frac{ie}{\varepsilon_0\mu k'}\right)^2\frac{1}{V}k'_{\alpha}}{(\omega^2 - k'^2 v^2\lambda^2)}, \tag{6.42}$$

where we have introduced $\lambda = \cos\theta$. The expression in equation (6.42) is proportional to

$$\int_0^\infty dk' \int_0^{2\pi} d\varphi \int_{-1}^1 d\lambda \, \frac{k'_\alpha}{(\omega^2 - k'^2 v^2 \lambda^2)}. \tag{6.43}$$

Now the x, y, z components of $\vec{k}'$ in polar coordinates are

$$k'_1 = k' \sin\theta \cos\varphi \tag{6.44}$$

$$k'_2 = k' \sin\theta \sin\varphi \tag{6.45}$$

$$k'_3 = k' \cos\theta = k'\lambda. \tag{6.46}$$

Thus the expressions in equation (6.43) for $\alpha = 1$ and $\alpha = 2$ are zero from the integral over φ, and that for $\alpha = 3$ is zero because the integrand is odd in λ, and the integration is over a symmetric region about $\lambda = 0$. We therefore conclude that the equation of motion, equation (6.37) and the assumed solution $\vec{r}_{el} = \vec{v} \cdot t$ are compatible.

We can now evaluate the total energy from the system's hamiltonian, equation (6.17), using equations (6.20), (6.36) and (6.30). We have

$$\Pi_{\vec{k}',\alpha'} \Pi_{-\vec{k}',\alpha'} = \mu^2 \dot{q}_{-\vec{k}',\alpha'} \dot{q}_{\vec{k}',\alpha'}, \tag{6.47}$$

$$q_{\vec{k}',\alpha'} q_{-\vec{k}',\alpha'} = \frac{|\varphi_{\vec{k}',\alpha'}|^2}{(\omega^2 - \Omega^2_{\vec{k}'})^2}, \tag{6.48}$$

$$d_{-\vec{k}',\alpha'} q_{\vec{k}',\alpha'} = \frac{\mu|\varphi_{\vec{k}',\alpha'}|^2}{(\omega^2 - \Omega^2_{\vec{k}'})}. \tag{6.49}$$

In arriving at equation (6.49) we have used the result from equations (6.30) and (6.31) that

$$\vec{d}_{\vec{k}} = \mu \vec{\varphi}_{\vec{k}} \, e^{-i\vec{k} \cdot \vec{r}_{el}} = \vec{d}^*_{\vec{k}}. \tag{6.50}$$

Now substituting from equations (6.47)–(6.49) into equation (6.17) we obtain

$$H = \left\{ \frac{p^2}{2m} + \sum_{\vec{k}',\alpha'} \left[\frac{\mu}{2}(-i\Omega_{-\vec{k}'})(-i\Omega_{\vec{k}'}) + \frac{\mu\omega^2}{2} \right] \frac{|\varphi_{\vec{k}',\alpha'}|^2}{(\omega^2 - \Omega^2_{\vec{k}'})^2} \right.$$
$$\left. - \mu \sum_{\vec{k}',\alpha'} \frac{|\varphi_{\vec{k}',\alpha'}|^2}{(\omega^2 - \Omega^2_{\vec{k}'})} \right\}. \tag{6.51}$$

Simplifying equation (6.51) we obtain

$$H = \left\{ \frac{p^2}{2m} - \frac{\mu}{2} \sum_{\vec{k}} \frac{(\omega^2 - 3\Omega^2_{\vec{k}})}{(\omega^2 - \Omega^2_{\vec{k}})^2} |\vec{\varphi}_{\vec{k}}|^2 \right\}, \tag{6.52}$$

where we have used $\Omega_{-\vec{k}} = -\Omega_{\vec{k}}$. To see this in terms of the canonical variables, we must express $\Omega_{\vec{k}}$ in terms of $\vec{p}$ through equations (6.35), (6.33) and (6.21):

$$\Omega_{\vec{k}} = \left(\frac{\vec{k} \cdot \vec{p}}{m} \right). \tag{6.53}$$

From equations (6.52) and (6.53) we then have

$$H = \left\{ \frac{p^2}{2m} - \frac{\mu}{2} \sum_{\vec{k}} |\vec{\varphi}|^2 \frac{\left[\omega^2 - 3\left(\frac{\vec{k} \cdot \vec{p}}{m} \right)^2 \right]}{\left[\omega^2 - \left(\frac{\vec{k} \cdot \vec{p}}{m} \right)^2 \right]^2} \right\}. \tag{6.54}$$

For low velocity $\vec{v}$, or small $\vec{p}$, we can expand the second term in equation (6.54) to first order in $[(\vec{k} \cdot \vec{p})/(m\omega)]^2$:

$$H \approx \left\{ \frac{p^2}{2m} + \frac{\mu}{2\omega^2} \sum_{\vec{k}} |\vec{\varphi}_{\vec{k}}|^2 \left[\left(\frac{\vec{k} \cdot \vec{p}}{m\omega} \right)^2 - 1 \right] \right\}. \tag{6.55}$$

Since $\vec{v}$ is in the z direction, so is $\vec{p}$, i.e. $p_\alpha = p\delta_{\alpha,3}$. Then

$$H \approx \left\{ \frac{p^2}{2m^*} - \frac{\mu}{2\omega^2} \sum_{\vec{k}} |\vec{\varphi}_{\vec{k}}|^2 \right\}, \tag{6.56}$$

where the *effective mass* of the electron, m^*, is given by

$$\frac{1}{m^*} = \left\{ \frac{1}{m} + \frac{\mu}{m^2\omega^4} \sum_{\vec{k}} |\vec{\varphi}_{\vec{k}}|^2 k_3^2 \right\}, \tag{6.57}$$

and the *effective potential* E_s due to the electron's interaction with its self-induced polarization field is

$$E_s \approx -\frac{\mu}{2\omega^2} \sum_{\vec{k}} |\vec{\varphi}_{\vec{k}}|^2. \tag{6.58}$$

With this lowest-order approximation, there is no momentum dependence in E_s. Without the approximation, we can see from equations (6.54) and (6.57) that

$$H = \frac{p^2}{2m^*} + \left\{ -\frac{\mu}{2\omega^2} \sum_{\vec{k}} |\vec{\varphi}_{\vec{k}}|^2 \frac{\left[1 - 3\left(\frac{k_3 p}{m\omega} \right)^2 \right]}{\left[1 - \left(\frac{k_3 p}{m\omega} \right)^2 \right]^2} - \frac{p^2}{2} \left(\frac{1}{m^*} - \frac{1}{m} \right) \right\}. \tag{6.59}$$

The momentum dependence of the effective potential that comes from the denominator will be further discussed in the next section.

In equation (6.57) the term added to $(1/m)$ increases the kinetic energy, for given momentum $\vec{p}$ *of the electron*. The definition of m^* given there is an effective mass *less than* the free electron mass m. A more satisfactory expression for effective mass comes from identifying the electron's momentum as $\vec{p} = m\vec{v}$, where $\vec{v}$ is the velocity both of the electron and of the polaron quasiparticle. When this substitution is made in equation (6.55) we obtain

$$H \approx \left\{ \tfrac{1}{2} m^{*\prime} v^2 - \frac{\mu}{2\omega^2} \sum_{k} |\vec{\varphi}_{\vec{k}}|^2 \right\} \tag{6.60}$$

where

$$m^{*\prime} = \left\{ m + \frac{\mu}{\omega^4} \sum_{k} |\vec{\varphi}_{\vec{k}}|^2 k_3^2 \right\} > m. \tag{6.61}$$

Now, in equation (6.60), the term $(m^{*\prime} v^2 / 2)$, the kinetic energy, is clearly that of a *quasiparticle* with velocity $\vec{v}$ and effective mass $m^{*\prime} > m$, the latter inequality expressing the effect of the polarization field's inertia. The corresponding new expression for the velocity-dependent effective potential E_s' is

$$E_s' = \left\{ -\frac{\mu}{2\omega^2} \sum_{k} |\vec{\varphi}_{\vec{k}}|^2 \frac{\left[1 - 3\left(\frac{k_2 v}{\omega} \right)^2 \right]}{\left[1 - \left(\frac{k_3 v}{\omega} \right)^2 \right]^2} - \frac{v^2}{2} (m^{*\prime} - m) \right\} \tag{6.62}$$

whence the total energy is

$$H = (\tfrac{1}{2} m^{*\prime} v^2 + E_s'). \tag{6.63}$$

We note, from equations (6.62) and (6.61), that the effective potential E_s' is independent of electron mass, unlike the form E_s given in equation (6.59) with equation (6.57).

The lowest-order approximation to E_s', to order v^2, is the same as E_s, equation (6.58). To evaluate this self energy we must consider

$$\sum_{k} |\varphi_{\vec{k}}|^2 = \frac{1}{V} \left(\frac{e}{\varepsilon_0 \mu} \right)^2 \sum_{k} \frac{1}{k^2}. \tag{6.64}$$

We can convert the summation over $\vec{k}$ in equation (6.64) to an integral, as we did in going from equation (6.39) to equation (6.41):

$$\sum_{\vec{k}} \frac{1}{k^2} = \int d^3k \, \rho_k \frac{1}{k^2} = \frac{V}{(2\pi)^3} 4\pi \int_0^\infty dk \, k^2 \frac{1}{k^2}. \tag{6.65}$$

The result diverges. When this integral is encountered in quantum electrodynamics (Sakurai, 1967, section 2.8), a finite upper limit is introduced on the grounds that the theory becomes meaningless when virtual photons that are capable of producing more electrons (along with positrons) are included. While this is easy to grasp, it is by no means clear that introducing an upper limit in this way should lead to physically correct results. In the present case, however, there is a good reason why the integral should be cut off at a finite value. It is because, in a real material, the normal modes of vibration represented by $\vec{q}_{\vec{k}}$ in equation (6.22) are limited to a shortest wavelength, or largest $\vec{k}$-value, by the finite interatomic spacing. For the free field $\vec{q}_{\vec{k}}$, i.e. with $\vec{d}_{\vec{k}} = 0$ in equation (6.22) the dynamical equations determine a dispersion relation $\omega = \omega(\vec{k})$ [see equations (2.39) and (2.46), for the continuum case]. Thus we conclude that the model adopted in this chapter, represented by equation (6.1), replaces $\omega(\vec{k})$ by a single value, some sort of average value, for example. In Chapter 8, section 8.2, we see how the dispersion relation for a classical linear chain acquires an upper limit in $\vec{k}$, as well as a non-linear ω versus k dependence. Returning to equation (6.22) with $\vec{D} \neq 0$, where the vibrations of the material are determined by the driving source $\vec{d}_{\vec{k}}$ due to the electron's charge, the same atomistic consideration applies.

The simplest way to take account of the atomistic nature of a solid is to let n_0 be the number of atoms per unit volume. Then there are $(3n_0 V)$ degrees of freedom for the $(n_0 V)$ atoms in volume V of the material, and correspondingly $(3n_0 V)$ normal modes, and $(n_0 V)$ normal-mode wave vectors $\vec{k}$. Let k_M be the largest k-value. Then

$$n_0 V = \int_{k \leq k_M} d^3k\, \rho_k = \frac{V}{(2\pi)^3} \frac{4\pi}{3} k_M^3 \tag{6.66}$$

whence we estimate the cutoff at

$$k_M = (6\pi^2 n_0)^{1/3}. \tag{6.67}$$

Combining equations (6.64), (6.65) cut off at $k = k_M$, and (6.67), we find

$$\sum_{\vec{k}} |\vec{\varphi}_{\vec{k}}|^2 = \frac{1}{V}\left(\frac{e}{\varepsilon_0 \mu}\right)^2 \frac{V}{2\pi^2} \int_0^{k_M} dk = \frac{1}{2}\left(\frac{e}{\varepsilon_0 \mu \pi}\right)^2 (6\pi^2 n_0)^{1/3}. \tag{6.68}$$

Referring to equation (6.58), we find

$$E_s' \approx -\frac{1}{4\mu}\left(\frac{e}{\varepsilon_0 \pi \omega}\right)^2 (6\pi^2 n_0)^{1/3}. \tag{6.69}$$

We can similarly evaluate the mass renormalization given by the second term in { } brackets in equation (6.61). It is

$$\frac{\mu}{\omega^4}\sum_{\vec{k}} |\varphi_{\vec{k}}|^2 k_3^2 = \frac{\mu}{\omega^4} \frac{V}{(2\pi)^3} 2\pi \frac{1}{V}\left(\frac{e}{\varepsilon_0 \mu}\right)^2 \int_0^{k_M} dk\, k^2 \frac{1}{k^2}\int_{-1}^1 d\lambda(k^2\lambda^2). \tag{6.70}$$

In equation (6.70) we have used the result that $k_3 = k \cos \theta$, and have made the substitution $\lambda = \cos \theta$ where θ is the polar angle. The final result, from equations (6.61) and (6.70), is

$$(m^{*\prime} - m) = \frac{n_0}{3\mu} \left(\frac{e}{\varepsilon_0 \omega^2} \right)^2.$$ (6.71)

The reader should check that the dimensionality of the ratio of the calculated expressions E_s' to $(m^{*\prime} - m)$ in equations (6.69) and (6.71) has the required dimensionality of $(\text{speed})^2$.

In summary, we have derived the hamiltonian for an electron with its self-induced polarization field in a dielectric solid, equation (6.17) with equations (6.30) and (6.31). We have analysed the hamiltonian equations of motion to obtain equations of motion for the polarization field, equation (6.34) with equations (6.35) and (6.2), and for the electron, equation (6.33). The electron's motion, $\vec{r}_{el} = \vec{v}t$ leads to a solution for the polarization field, equation (6.36). Substituting both into the hamiltonian gives the total energy in terms of *electron* momentum $\vec{p}$, equation (6.54), or in terms of *polaron* velocity $\vec{v}$, namely

$$H = \left\{ \frac{m^{*\prime} v^2}{2} + E_s'(\vec{v}) \right\},$$ (6.72)

where $E_s'(\vec{v})$ is given in equation (6.62). In terms of the electron momentum, the electron's mass m can be renormalized giving m^*, equation (6.57), to take account of the contribution of the polarization field to the kinetic energy associated with the electron's motion, leaving a momentum-dependent potential, equation (6.59). In lowest order, i.e. for small momentum, this potential gives a momentum-independent energy shift, but higher-order corrections bring in momentum dependence. In terms of the polaron velocity $\vec{v}$, the electron's mass m is renormalized to $m^{*\prime}$, equations (6.61)–(6.63), with equation (6.72). In this case, the kinetic energy term $(m^{*\prime} v^2)/2$ is clearly identified as characteristic of the polaron quasiparticle, both in terms of its effective mass $m^{*\prime}$ and its velocity $\vec{v}$. The remaining term in the energy, E_s', equations (6.63) and (6.62), is identified as a velocity-dependent potential or energy shift. When one takes account of the discrete atomic nature of the dielectric solid, one can explicitly evaluate the mass renormalization $(m^{*\prime} - m)$, equation (6.71), and the lowest order, velocity-independent energy shift $(\approx E_s')$, equation (6.69).

The polaron theory presented in this chapter is entirely classical. When values of μ, ω and n_0 for typical materials, such as NaCl and KCl, are inserted into equation (6.71) for the mass renormalization a large number is obtained (see exercise 6.1). This is not in agreement with either the quantum theory of the polaron, or with experiment. The theory represented here, however, has the merit of illustrating in a particularly simple way the self-energy and mass renormalization effects that are characteristic of particle–field interaction. The polaron problem is nearly as old as quantum theory itself: Mott and Gurney (1940, Chapter III, section 5.3), describe Landau's work dated 1933 [Landau (1933)]. It still represents a very lively field of theoretical and experimental investigation. The recent textbook by Marder (2000) gives a good introduction to both aspects of the subject.

On the experimental side, Marder presents two ratios: in his notation (m^*/m), where m^* is band mass due to the periodic potential of the crystal lattice, and m is the free electron mass; and (m^*_{pol}/m), where m^*_{pol} is the polaron effective mass, corresponding to our quantity m^* in equation (6.57): see also equation (6.67) and the intervening discussion. From Marder's notation, the polaron effective mass in units of band mass is

$$\left(\frac{m^*_{pol}}{m^*}\right) = \frac{\left(\dfrac{m^*_{pol}}{m}\right)}{\left(\dfrac{m^*}{m}\right)}. \tag{6.73}$$

From our equation (6.57), the corresponding quantity, (m^*/m) in our notation, would always be less than unity. By contrast, the experimental results given in Marder's table 22.1 have $(m^*_{pol}/m^*) > 1$ in all but two cases, only one of which, CdF_2, is probably beyond experimental error. For the alkali halides, the ratio is of the order of two, while for common semiconductors it is close to unity.

The quantum theory of the polaron has been dominated by Fröhlich's effective hamiltonian [Fröhlich (1963, equation (4.24), p. 20)]. The total energy in this formulation, in appropriate units, depends on a material-dependent coupling constant, denoted α_p in Marder. Several approximate derivations of the effective mass ratio of equation (6.73) have been developed. In weak coupling [see for example Fröhlich (1963, equation (5.13), p. 24)], with $\alpha_p \ll 1$:

$$\left(\frac{m^*_{pol}}{m^*}\right) = \left(1 + \frac{\alpha_p}{6}\right), \tag{6.74}$$

in Marder's notation. From Marder's Table 22.1, $\alpha_p \ll 1$ for most of the semiconductor materials listed, and for these, equation (6.74) gives excellent agreement with experiment. The intermediate coupling theory of Pines (1963) extends the weak coupling result to values of $\alpha_p \lesssim 3$, as follows:

$$\left(\frac{m^*_{pol}}{m^*}\right) = \left\{\frac{1}{\left(1 + \dfrac{\alpha_p}{6}\right)} + \frac{0.02\alpha_p^2}{\left(1 + \dfrac{\alpha_p}{6}\right)^2}\right\}^{-1}. \tag{6.75}$$

When the values of α_p for the alkali halides are substituted into equation (6.75), we obtain the results shown in table 6.1. We see that while both theory

Table 6.1. Effective mass ratios (m^*_{pol}/m^*), equation (6.73), for polarons in some alkali halides, experimental and as calculated from intermediate coupling theory, equation (6.75), where α_p is Fröhlich's polaron coupling constant; based on table 22.1 of Marder (2000).

Crystal	α_p	(m^*_{pol}/m^*)	
		Experiment	Theory
KI	2.51	1.66	1.30
KBr	3.14	1.79	1.35
RbI	3.16	1.96	1.35
KCl	3.45	2.12	1.37
CsI	3.67	2.29	1.38
RbCl	3.84	2.38	1.39

and experiment have the same trend as α_p, the discrepancy between experiment and theory is 22% for the smallest value of α_p, namely 2.51, and the discrepancy rises fast, to 42% at the largest value of α_p, namely 3.84. We conclude that in the intermediate coupling range, the theory is not very accurate. Feynman (1955) has applied his path integral method to the polaron theory, obtaining analytical results that are expected to be valid for all coupling strengths, subject to the limitation of a gaussian wave function for the electron. Application of Feynman's results to a polaronic picture of the diffuse excited state of the F center (an electron bound to an anion vacancy) in BaF_2, where α_p has been estimated at 4.63, leads to a promising result in comparison with the optical excitation energy [Vail *et al.* (2002)]. This is discussed further in Chapter 11, especially section 11.8.

6.4   Polaron in a magnetic field: quantization

Let us now consider a polaron in a magnetic field, and suppose that we want to get a quantum-mechanical description. Then we want a hamiltonian operator, in general a functional of $\vec{p}$ and $\vec{r}_{el}$, the canonical momenta and generalized coordinates of the electron, where $\vec{p}$ and $\vec{r}_{el}$ are operators whose components satisfy the canonical commutation rules:

$$[p_\alpha, p_\beta] = 0, \qquad (6.76)$$

$$[r_{el,\alpha}, r_{el,\beta}] = 0, \qquad (6.77)$$

$$[p_\alpha, r_{el,\beta}] = -i\hbar\delta_{\alpha\beta}. \qquad (6.78)$$

Since the usual approach is to start with a classical hamiltonian and then introduce the operator properties of equations (6.76)–(6.78), we would naturally consider the hamiltonian of equation (6.54), which consists of the bare electron's kinetic energy and an effective, momentum-dependent potential, or else equation (6.59), which consists of a polaron kinetic energy incorporating a renormalized effective mass m^*, and a somewhat different effective potential, still, however, in terms of electron momentum $\vec{p}$.

We now turn on a magnetic field, with magnetic induction field $\vec{B}(\vec{r})$, expressible in terms of a vector potential $\vec{A}(\vec{r})$ by

$$\vec{B} = (\vec{\nabla} \times \vec{A}). \tag{6.79}$$

Classically, we know that the hamiltonian is modified by the replacement

$$\vec{p} \to [\vec{p} - e\vec{A}(\vec{r}_{\text{el}})], \tag{6.80}$$

where, in the latter, $\vec{p}$ is the canonical momentum, not the particle's mechanical momentum $m\vec{v}$ [see e.g. Goldstein (1980), section 7.9]. The kinetic energy term, for example in equation (6.59), becomes

$$\frac{1}{2m^*} p^2 \to \frac{1}{2m^*} [\vec{p} - e\vec{A}(\vec{r}_{\text{el}})]^2. \tag{6.81}$$

Quantization of such an expression has been found not to present any new problem of interpretation, so long as the ordering of terms is maintained as follows:

$$(\vec{p} - e\vec{A}) \cdot (\vec{p} - e\vec{A}) = [p^2 - e(\vec{p} \cdot \vec{A} + \vec{A} \cdot \vec{p}) + e^2 A^2]. \tag{6.82}$$

In equation (6.82), since $\vec{p}$ and $\vec{r}_{\text{el}}$ do not commute,

$$\vec{p} \cdot \vec{A}(\vec{r}_{\text{el}}) \neq \vec{A}(\vec{r}_{\text{el}}) \cdot \vec{p}. \tag{6.83}$$

Schafroth (1958) has discussed the ambiguity that arises when one tries to understand the magnetic properties of a system of particles (in the present case, one particle) described by an effective potential that has momentum dependence of order higher than quadratic. In equation (6.59) we have a term

$$\left[1 - \left(\frac{k_3}{m\omega}\right)^2 p^2\right]^{-2}. \tag{6.84}$$

The meaning of such a quantum-mechanical operator is in terms of the power series expansion, whose lowest-order term beyond quadratic is

$$(p^2)^2 = p^4. \tag{6.85}$$

There is no fundamental criterion favoring one form of p^4 over another. In the absence of magnetic field, they are identical in view of the commutation relations (6.76), i.e.

$$p_\alpha p_\alpha p_\beta p_\beta = p_\alpha p_\beta p_\alpha p_\beta, \tag{6.86}$$

where Einstein summation over cartesian components is used. Furthermore they are both gauge covariant and hermitian [Schafroth (1960)]. However, when we turn on a magnetic field according to equation (6.80) we are led to consider

$$(\pi_\alpha \pi_\alpha \pi_\beta \pi_\beta - \pi_\alpha \pi_\beta \pi_\alpha \pi_\beta), \tag{6.87}$$

where

$$\pi_\alpha = [p_\alpha - eA_\alpha(\vec{r}_{el})]. \tag{6.88}$$

From the commutation relations, equations (6.76)–(6.78), we deduce

$$[p_\alpha, A_\beta(\vec{r}_{el})] = -i\hbar \partial_\alpha A_\beta, \tag{6.89}$$

whence

$$[\pi_\alpha, \pi_\beta] = i\hbar e(\partial_\alpha A_\beta - \partial_\beta A_\alpha). \tag{6.90}$$

With equation (6.90), the expression in equation (6.87) reduces to

$$
\begin{aligned}
(\pi_\alpha \pi_\alpha \pi_\beta \pi_\beta &- \pi_\alpha \pi_\beta \pi_\alpha \pi_\beta) \\
&= -\pi_\alpha(-i\hbar e)(\partial_\alpha A_\beta - \partial_\beta A_\alpha)\pi_\beta \\
&= (ie\hbar)\pi_\alpha\{\pi_\beta(\partial_\alpha A_\beta - \partial_\beta A_\alpha) + i\hbar \partial_\beta(\partial_\alpha A_\beta - \partial_\beta A_\alpha)\} \\
&= \{(ie\hbar)\pi_\alpha \pi_\beta(\partial_\alpha A_\beta - \partial_\beta A_\alpha) - e\hbar^2 \pi_\alpha \partial_\beta(\partial_\alpha A_\beta - \partial_\beta A_\alpha)\}. \tag{6.91}
\end{aligned}
$$

It is clear that all terms with $\alpha = \beta$ are zero. Performing the summation over all other terms we obtain, from the right-hand side of equation (6.91),

$$
\begin{aligned}
ie\hbar\{&\pi_1 \pi_2 B_3 - \pi_1 \pi_3 B_2 - \pi_2 \pi_1 B_3 + \pi_2 \pi_3 B_1 + \pi_3 \pi_1 B_2 - \pi_3 \pi_2 B_1\} \\
&-e\hbar^2\{\pi_1 \partial_2 B_3 - \pi_1 \partial_3 B_2 - \pi_2 \partial_1 B_3 + \pi_2 \partial_3 B_1 + \pi_3 \partial_1 B_2 - \pi_3 \partial_2 B_1\} \\
= ie\hbar\{&(i\hbar e B_3)B_3 + (i\hbar e B_2)B_2 + (i\hbar e B_1)B_1\} \\
&-e\hbar^2\{(\pi_1 \partial_2 - \pi_2 \partial_1)B_3 + (\pi_3 \partial_1 - \pi_1 \partial_3)B_2 + (\pi_2 \partial_3 - \pi_3 \partial_2)B_1\} \\
= -e\hbar^2\{&eB^2 + (\vec{\pi} \times \vec{\nabla}) \cdot \vec{B}\} \\
= -e\hbar^2\{&eB^2 + \vec{\pi} \cdot (\vec{\nabla} \times \vec{B})\}. \tag{6.92}
\end{aligned}
$$

In obtaining equation (6.92) we have used the generalization of equation (6.89) along with equations (6.90) and (6.79). We recall that $\vec{\pi}$ is the particle's mechanical momentum $m\vec{v}$. From the Maxwell equation,

$$\vec{\nabla} \times \vec{B} = \mu \vec{J} \tag{6.93}$$

for a time-independent system, where μ is the magnetic permeability of the medium, and $\vec{J}$ is the electrical current density which is the source of the magnetic field. The spatial variables in equation (6.92) are those of the electron $\vec{r}_{el}$: see equation (6.88). Thus if, as is usually the case, the source currents of the magnetic field lie outside the sample, as we shall assume here, then $\vec{J}(\vec{r}_{el}) = 0$. In this case we see, from equations (6.91)–(6.93) that the difference between two hamiltonian terms derived from p^4, with non-zero magnetic field, is $(-e^2\hbar^2 B^2)$. This is non-zero, and involves a measurable physical quantity. The two terms represent two distinct physical situations with distinct physical properties, and both equally conform to the quantization algorithm. There is thus an ambiguity associated with this algorithm when it is applied to a hamiltonian, or an effective hamiltonian as here for the polaron, that contains momentum beyond second order. Schafroth (1958) gives other examples, from p^4 and p^6 terms, and discusses the problem more fully in relation to gauge covariance [Schafroth (1960), especially in sections 13c, d and e]. The latter is highly recommended to the reader. To eliminate such ambiguities we must rely on experimental evidence that excludes some magnetic effects but exhibits others. From the purely theoretical viewpoint, one must introduce the magnetic field and quantization at the level of equation (6.17), prior to eliminating the field variables.

Chapter 7

Atomistic quantum theory of solids

7.1 Introduction

The preceding six chapters have addressed physical properties of solids viewed as continuous media. However, many important properties of technological materials depend upon details of the atomic structure that cannot be adequately represented by continuum models, or even by classical discrete atoms. An obvious example is optical properties due to chemical impurities. More generally, however, it is interesting to ask to what extent arbitrary properties of a solid material can be related directly, and rigorously, to the nuclei and electrons of which the material is ultimately composed. The nuclei specify the chemical composition of the solid and, thereby, the crystal and defect structure under given thermodynamic conditions; whence also both equilibrium and dynamical properties and processes. In this chapter we shall illustrate how the thermodynamic equation of state of a solid is related to the electrically neutral collection of nuclei and electrons of which it is made up. We shall establish a formal framework so that the reader can see how, by improving the initial model, and by adopting one or another set of systematic approximations in the mathematical treatment, one can simulate an extremely wide range of phenomena. The approach will be based on elementary concepts of quantum mechanics and statistical thermodynamics. In later chapters it will become evident that, given presently available computing power, implementation of large parts of this agenda for specific materials and properties is a practical undertaking. Much of this chapter closely follows Maradudin (1974).

7.2 The hamiltonian of a solid

We consider a solid to be a quantum-mechanical system consisting of N electrons and N_1 nuclei. Then Schrödinger's equation for stationary energy eigenstates is

$$H|E_\lambda\rangle = E_\lambda|E_\lambda\rangle. \tag{7.1}$$

We write the hamiltonian H in configuration space as the sum of three terms: nuclear, electronic, and nucleus–electron interaction:

$$H = (H_n + H_e + H_{ne}). \qquad (7.2)$$

Let nuclear coordinates be denoted $\vec{R}_J$, $J = 1, 2, \ldots, N_1$, collectively denoted $\underline{R} = (\vec{R}_1, \vec{R}_2, \ldots, \vec{R}_{N_1})$, and electron coordinates $\underline{r}_j = (\vec{r}_j, s_j)$, where $\vec{r}_j$ is position coordinates and s_j is spin, $j = 1, 2, \ldots, N$, collectively denoted $\underline{r} = (\underline{r}_1, \underline{r}_2, \ldots, \underline{r}_N)$. Then nuclear and electronic energies each have kinetic and potential parts,

$$H_n = (T_n + V_n), \qquad H_e = (T_e + V_e), \qquad (7.3a, b)$$

where the nuclear and electronic kinetic energy parts are, respectively,

$$T_n = \sum_{J=1}^{N_1} \left(-\frac{\hbar^2}{2M_J} \right) \nabla_J^2, \qquad T_e = \sum_{j=1}^{N} \left(-\frac{\hbar^2}{2m} \right) \nabla_j^2. \qquad (7.4a, b)$$

In equations (7.4a,b), M_J are nuclear masses, m is the electron's mass, $\vec{\nabla}_J$ is gradient with respect to $\vec{R}_J$, and $\vec{\nabla}_j$ is gradient with respect to $\vec{r}_j$.

For the potential energies, we adopt a very simple model, in which the nuclei and electrons are non-relativistic charged point masses. Spin-dependent effects, apart from those arising from the Pauli principle (see Chapter 12), can therefore not be described by this model. Neglecting gravitational and magnetic effects, potential energies are then exclusively electrostatic, as follows:

$$V_n(\underline{R}) = \frac{e^2}{4\pi\varepsilon_0} \cdot \frac{1}{2} {\sum_{J,J'}}' \frac{Z_J Z_{J'}}{|\vec{R}_J - \vec{R}_{J'}|}, \qquad (7.5a)$$

$$V_e(\underline{r}) = \frac{e^2}{4\pi\varepsilon_0} \cdot \frac{1}{2} {\sum_{j,j'}}' |\vec{r}_j - \vec{r}_{j'}|^{-1}. \qquad (7.5b)$$

Similarly, the nucleus–electron interaction, equation (7.2), is

$$V_{ne}(\underline{R}, \underline{r}) = -\frac{e^2}{4\pi\varepsilon_0} {\sum_{J,j}}' Z_J |\vec{r}_j - \vec{R}_J|^{-1}. \qquad (7.6)$$

For equations (7.5a) and (7.6), Z_J is the charge of the Jth nucleus, in units of $e > 0$. In equations (7.5) and (7.6), ε_0 is the permittivity of free space.

7.3 Nuclear dynamics: the adiabatic approximation

The adiabatic approximation has a hierarchy of levels [see for example, Vail (1987), Stoneham (1975), Born and Huang (1954)]. Here we adopt the simplest, namely the average field approximation. Denote the energy eigenstate $|E_\lambda\rangle$, equation (7.1), by $\Psi_\lambda(\underline{r}, \underline{R})$ in position representation, and then denote it by

$$|E_\lambda\rangle \equiv |\Psi_\lambda\rangle. \qquad (7.7)$$

Now in a crystalline solid, atomic sites are identifiable: denote them collectively by $\underline{R}_0 = (\vec{R}_{10}, \ldots, \vec{R}_{N_10})$. In fact, $\underline{R}_0$ defines the crystal structure. The average field approximation is then

$$\Psi_\lambda(\underline{r}, \underline{R}) \approx \psi_\lambda(\underline{r}, \underline{R}_0) \cdot \varphi_\lambda(\underline{R}). \qquad (7.8)$$

The implication of equation (7.8) is that the electronic motion is determined by the crystal structure, approximately, independent of the nuclear dynamics. A well-known exception to this approximation is superconductivity, where the electronic states depend crucially on the dynamical interaction of electrons and nuclei, the so-called electron–phonon interaction (see section 14.2).

Within the form of the adiabatic approximation given by equation (7.8), we can examine the nuclear motions by averaging over the electronic distribution represented by $\psi_\lambda(\underline{r}, \underline{R}_0)$ in equation (7.8). Thus from equations (7.1), (7.7) and (7.8),

$$\langle\psi_\lambda|H|\Psi_\lambda\rangle = E_\lambda|\varphi_\lambda\rangle, \qquad (7.9)$$

assuming ψ_λ (and φ_λ) are normalized. In position representation,

$$\langle\psi_\lambda|H|\Psi_\lambda\rangle = \left\{\int d\tau_{\underline{r}}\, \psi_\lambda^*(\underline{r}, \underline{R}_0) \cdot H(\underline{r}, \underline{R}) \cdot \psi_\lambda(\underline{r}, \underline{R}_0)\right\}\varphi_\lambda(\underline{R}). \qquad (7.10)$$

From equations (7.9) and (7.10) we can identify the quantity in { } brackets as an effective nuclear hamiltonian; it depends only on $\underline{R}$, and is the expectation value of the total hamiltonian H with respect to the electronic state $|\psi_\lambda\rangle$. Explicitly,

$$\{T_n + W(\underline{R}; \psi_\lambda)\}\varphi_\lambda(\underline{R}) = E_\lambda\varphi_\lambda(\underline{R}) \qquad (7.11)$$

where

$$W(\underline{R}; \psi_\lambda) = \{V_n(\underline{R}) + \langle\psi_\lambda|H_e + V_{ne}|\psi_\lambda\rangle\}. \qquad (7.12)$$

We have used, in equation (7.11), the fact that T_n and V_n are independent of $\underline{r}$, and ψ_λ is normalized. In equation (7.12) we see that the effective nuclear potential W consists of two parts, $V_n(\underline{R})$, which is purely repulsive [see equation (7.5a)], and an electronic term that depends on the electronic state $|\psi_\lambda\rangle$, which accounts for the interatomic binding of the crystal. For absolute zero, $T = 0$, the crystal structure, defined by $\underline{R}_0$ (for a given electronic state) is determined by the equilibrium condition

$$\left(\frac{\partial W}{\partial \underline{R}}\right)_{\underline{R} = \underline{R}_0} = 0. \qquad (7.13)$$

This corresponds to a model of the system in which the nuclei are viewed as classical particles, collectively in their lowest state of total energy. Note that in equation (7.13), ψ_λ is independent of $\underline{R}$.

7.4 The harmonic approximation

We now take account of the fact that the nuclei spend most of the time close to the crystalline sites. [There are, of course, exceptions to this, such as so-called anharmonic or quantum crystals, and also very-low atomic number impurities such as hydrogen.] We therefore write

$$\vec{R}_J = (\vec{R}_{J0} + \vec{u}_J), \qquad J = 1, 2, \ldots, N_1, \tag{7.14}$$

and adopt the approximation that $\vec{u}_J$ are small. Note that $\psi_\lambda(\underline{r}, \underline{R}_0)$ is independent of $\vec{u}_J$. The discrete variable $\vec{u}_J$ here corresponds to the distortion field $\vec{u}(\vec{r})$ in the continuum mechanics of Chapters 1 to 6. We introduce a matrix notation, in which $\underline{u}$ is a column matrix, with elements $u_{J\alpha}$, $J = 1, 2, \ldots, N_1$ with $\alpha = 1, 2, 3$ corresponding to cartesian components. Let us now expand $W(\underline{R}; \psi_\lambda)$, equation (7.12), to second order in $\underline{u}$: this is the harmonic approximation:

$$W \approx \{W^{(0)} + \underline{W}^{(1)T} \cdot \underline{u} + \tfrac{1}{2}\underline{u}^T \cdot \underline{W}^{(2)} \cdot \underline{u}\}. \tag{7.15}$$

In equation (7.15) superscripts indicate order of differentiation, in the following sense:

$$W^{(0)} = W(\underline{R}_0, \psi_\lambda) \tag{7.16}$$

$$W_k^{(1)} = \left(\frac{\partial W}{\partial u_{J\alpha(k)}}\right)_{u=0} \tag{7.17}$$

where $J\alpha(k)$ is the kth element of $\underline{u}$, and

$$W_{k,k'}^{(2)} = \left(\frac{\partial^2 W}{\partial u_{J\alpha(k)} \, \partial u_{J'\alpha'(k')}}\right)_{u=0}. \tag{7.18}$$

Thus $W^{(1)}$ is a column matrix and $\underline{W}^{(2)}$ is a square matrix. Superscript T indicates transpose, in equation (7.15). Since, from equation (7.14), $\partial/\partial\underline{R} = \partial/\partial\underline{u}$, the equilibrium condition equation (7.13) gives $\underline{W}^{(1)} = 0$. This, with equations (7.11) and (7.15), gives the Schrödinger equation for the nuclei:

$$\{T_n + \tfrac{1}{2}\underline{u}^T \cdot \underline{W}^{(2)}(\psi_\lambda) \cdot \underline{u}\}\varphi_\lambda = [E_\lambda - W^{(0)}(\psi_\lambda)]\varphi_\lambda. \tag{7.19}$$

In equation (7.19) we have indicated that the force constant matrix $\underline{W}^{(2)}$ depends on the electronic state $|\psi_\lambda\rangle$, as do the total energy eigenvalue E_λ and the nuclear state $|\varphi_\lambda\rangle$.

7.5   Phonons

The nuclear Schrödinger equation (7.19) can be written in position representation as

$$\left\{ \sum_{J\alpha} \frac{P_{J\alpha}^2}{2M_J} + \frac{1}{2} \sum_{J\alpha,J'\alpha'} W^{(2)}_{J\alpha,J'\alpha'} u_{J\alpha} u_{J'\alpha'} \right\} \varphi_\lambda = E'_\lambda \varphi_\lambda \qquad (7.20)$$

where

$$P_{J\alpha} = \left(-i\hbar \frac{\partial}{\partial u_{J\alpha}} \right), \qquad (7.21a)$$

and

$$E'_\lambda = [E_\lambda - W^{(0)}(\psi_\lambda)]. \qquad (7.12b)$$

We now proceed to show that, under very general conditions, equation (7.20) is equivalent to a system of non-interacting linear simple harmonic oscillators [see Maradudin (1974), Peierls (1955)].

7.5.1   Periodic boundary conditions for bulk properties

We begin with the Born–von Karmann boundary conditions. For this, we consider a macroscopic sample of crystal to consist of a large number of smaller macroscopic components, each with N_1 nuclei. All such components that are not too near the surfaces of the crystal will be very similar: approximately identical. Almost all such components will be surrounded by similar components, all of which have properties that are representative of the bulk material. We then say that such a component, subject to periodic boundary conditions, will represent the bulk behaviour of the material.

For simplicity we now consider a monatomic crystal with one atom per primitive unit cell. For simplicity of illustration, consider the case where the crystal's primitive translation vectors are orthogonal, e.g. orthorhombic. Then

$$\vec{R}_{J0} = \sum_{\alpha=1}^{3} a_\alpha \hat{\varepsilon}_\alpha \nu_{J\alpha} \qquad (7.22)$$

where $(a_\alpha \hat{\varepsilon}_\alpha)$ are primitive translation vectors, $\hat{\varepsilon}_\alpha$ ($\alpha = 1, 2, 3$) are orthonormal basis vectors in real space, and $\nu_{J\alpha}$ are integers. We shall Fourier transform the nuclear Schrödinger equation (7.20). We have noted that the vector $\vec{u}_J$ is the distortion field of the crystal, having meaning only at atomic sites J. In fact, for this discrete atomistic model of the crystal, spatial positions are determined by J.

In such a space, consider the basis functions

$$\eta_{\vec{k}}(J) = N_1^{-1/2} \cdot \exp(i\vec{k} \cdot \vec{R}_{J0}). \qquad (7.23)$$

These basis functions must satisfy the Born–von Karmann periodic boundary conditions. Let the periodically repeated volume be

$$V = \prod_{\alpha=1}^{3}(N_{1\alpha}a_\alpha), \qquad (7.24)$$

where

$$N_1 = \prod_{\alpha=1}^{3} N_{1\alpha}. \qquad (7.25)$$

Equation (7.24) says that there are $N_{1\alpha}$ atoms (primitive unit cells) in the α direction in V, and equation (7.25) says that there are N_1 atoms in V. Then the periodic boundary conditions are

$$u_{J+N_{1\alpha},\alpha'} = u_{J,\alpha'}, \qquad \alpha = 1,2,3, \text{ for given } \alpha'. \qquad (7.26)$$

The Fourier series for $\vec{u}_J$ is

$$\vec{u}_J = (N_1)^{-1/2} \sum_{\vec{k}} \vec{q}_{\vec{k}} \exp(i\vec{k}\cdot\vec{R}_{J0}). \qquad (7.27)$$

Substituting this into equation (7.26),

$$\sum_{\vec{k}} \vec{q}_{\vec{k}} \exp(i\vec{k}\cdot\vec{R}_{J+N_{1\alpha},0}) = \sum_{\vec{k}} \vec{q}_{\vec{k}} \exp(i\vec{k}\cdot\vec{R}_{J0}). \qquad (7.28)$$

Now,

$$\vec{R}_{J+N_{1\alpha},0} = (\vec{R}_{J0} + N_{1\alpha}a_\alpha\hat{\varepsilon}_\alpha). \qquad (7.29)$$

Thus, for arbitrary $\vec{q}_{\vec{k}}$ in equation (7.28), we require

$$\exp(i\vec{k}\cdot\hat{\varepsilon}_\alpha N_{1\alpha}a_\alpha) = 1$$

or

$$k_\alpha = \left(\frac{2\pi}{N_{1\alpha}a_\alpha}\right)n_\alpha, \qquad n_\alpha = 0,\pm1,\pm2,\dots \qquad (7.30)$$

Consider two values of n_α differing by $N_{1\alpha}$, e.g.

$$k'_\alpha = \left(\frac{2\pi}{N_{1\alpha}a_\alpha}\right)(n_\alpha + N_{1\alpha}) = \left(k_\alpha + \frac{2\pi}{a_\alpha}\right).$$

We then have

$$\exp(i\vec{k}'\cdot\vec{R}_{J0}) = \exp(i\vec{k}\cdot\vec{R}_{J0})\cdot\exp\left\{i\left(\sum_\alpha \frac{2\pi}{a_\alpha}\cdot R_{J0,\alpha}\right)\right\}.$$

But from equation (7.22)

$$\exp\left\{i\left(\sum_\alpha \frac{2\pi}{a_\alpha} \cdot R_{J0,\alpha}\right)\right\} = \exp\left\{i\left(\sum_\alpha \frac{2\pi}{a_\alpha} \cdot a_\alpha \nu_{J\alpha}\right)\right\}$$

$$= \exp\left(2\pi i \sum_\alpha \nu_{J\alpha}\right) = 1,$$

since $\nu_{J\alpha}$ are integers. It follows that

$$\exp(i\vec{k}' \cdot \vec{R}_{J0}) = \exp(i\vec{k} \cdot \vec{R}_{J0}).$$

Thus the set of integers n_α in equation (7.30) for k_α may be restricted to $N_{1\alpha}$ consecutive values:

$$k_\alpha = \left(\frac{2\pi}{N_{1\alpha}a_\alpha}\right)n_\alpha, \qquad n_\alpha = 0, 1, 2, \ldots, (N_1 - 1). \qquad (7.31)$$

We can now show that the basis functions $\eta_{\vec{k}}(J)$ are an orthonormal set, in the sense

$$\sum_J \eta_{\vec{k}}^*(J)\eta_{\vec{k}'}(J) = \delta_{\vec{k},\vec{k}'}. \qquad (7.32)$$

Clearly, for $\vec{k}' = \vec{k}$, we have from equation (7.23)

$$\sum_{J=1}^{N_1} |\eta_{\vec{k}}(J)|^2 = N_1^{-1} \sum_{J=1}^{N_1} (1) = 1.$$

For $\vec{k}' \neq \vec{k}$ in the left-hand side of equation (7.32) we have

$$N_1^{-1} \sum_J \exp[-i(\vec{k} - \vec{k}') \cdot \vec{R}_{J0}]. \qquad (7.33)$$

In summing over atomic sites J, we have $\nu_{J\alpha} = 1, 2, \ldots, N_{1\alpha}$ in equation (7.22) for $\vec{R}_{J0}$. Let

$$(k_\alpha - k'_\alpha) = \frac{2\pi}{N_{1\alpha}a_\alpha} \cdot (n_\alpha - n'_\alpha)$$

from equation (7.31), and with $(n_\alpha - n'_\alpha) = n''_\alpha$, an integer,

$$(\vec{k} - \vec{k}') \cdot \vec{R}_{J0} = \sum_\alpha (k_\alpha - k'_\alpha)R_{J\alpha,0}$$

$$= \sum_\alpha \frac{2\pi}{N_{1\alpha}a_\alpha} \cdot n''_\alpha \cdot a_\alpha \nu_{J\alpha}$$

$$= \sum_\alpha \frac{2\pi}{N_{1\alpha}} \cdot n''_\alpha \cdot \nu_{J\alpha}. \qquad (7.34)$$

Thus, from equation (7.33) we have, with equation (7.34),

$$N_1^{-1} \sum_J \prod_\alpha [\exp(-2\pi i n_\alpha''/N_{1\alpha})]^{\nu_{J\alpha}} = N_1^{-1} \prod_\alpha \sum_{\nu_{J\alpha}=1}^{N_{1\alpha}} x_\alpha^{\nu_{J\alpha}} \tag{7.35}$$

where

$$x_\alpha = [\exp(-2\pi i n_\alpha''/N_{1\alpha})]. \tag{7.36}$$

Now

$$\sum_{\nu_{J\alpha}=1}^{N_{1\alpha}} x_\alpha^{\nu_{J\alpha}} = \frac{(1 - x_\alpha^{N_{1\alpha}})}{(1 - x_\alpha)}.$$

But from equation (7.36),

$$x_\alpha^{N_{1\alpha}} = \exp(-2\pi i n_\alpha'') = 1,$$

so $(1 - x_\alpha^{N_{1\alpha}}) = 0$ and so therefore

$$N_1^{-1} \sum_J \exp[-i(\vec{k} - \vec{k}') \cdot \vec{R}_J] = 0, \qquad \text{if } \vec{k} \neq \vec{k}'.$$

This completes the proof of the orthonormality of the basis set $\eta_{\vec{k}}(J)$, equation (7.32).

Finally, we mention the completeness of the basis set $\eta_{\vec{k}}(J)$, equation (7.23). If they are a complete set, it will follow from the Fourier expansion, equation (7.27), that for arbitrary distortion field $\vec{u}_J$,

$$\vec{q}_{\vec{k}} = (N_1)^{-1/2} \sum_J \vec{u}_J \exp(-i\vec{k} \cdot \vec{R}_{J0}). \tag{7.37}$$

We leave this as an exercise, given orthonormality, equation (7.32).

7.5.2 The dynamical matrix of the crystal

We now return to the equation of motion, equation (7.20). From translational invariance in the crystal, all properties of the system that depend on two sites, including the elements of the force constant matrix $\underline{\underline{W}}^{(2)}$, depend only on the relative positions of the two sites, i.e.

$$W_{J\alpha,J'\alpha'}^{(2)} = W_{0\alpha,(J'-J)\alpha'}^{(2)}.$$

We denote the latter in a new notation by

$$W_{\alpha,\alpha'}(J' - J) \equiv W_{J\alpha,J'\alpha'}^{(2)}. \tag{7.38}$$

For simplicity of illustration, consider a monatomic crystal. Equation (7.20) now becomes, with $M_J = M$ (all J),

$$\left\{\frac{1}{2M}\sum_{J\alpha} P_{J\alpha}^2 + \frac{1}{2}\sum_{J,J',\alpha,\alpha'} W_{\alpha,\alpha'}(J-J')u_{J\alpha}u_{J'\alpha'}\right\}\varphi_\lambda = E'_\lambda\varphi_\lambda. \tag{7.39}$$

The elements of the distortion field $u_{J\alpha}$ are canonical coordinates of the system, $3N_1$ in number. Their canonical momenta $P_{J\alpha}$ are also a field, with space dependence J. The canonical quantum-mechanical commutation rules are

$$[P_{J\alpha}, u_{J'\alpha'}] = -i\hbar\delta_{J,J'}\delta_{\alpha,\alpha'}. \tag{7.40}$$

The Fourier series and Fourier transform of $\vec{P}_J$ are, respectively,

$$\vec{P}_J = N_1^{-1/2}\sum_{\vec{k}} \vec{p}_{\vec{k}}\exp(i\vec{k}\cdot\vec{R}_{J0}), \tag{7.41}$$

and

$$\vec{p}_{\vec{k}} = N_1^{-1/2}\sum_{J} \vec{P}_J\exp(-i\vec{k}\cdot\vec{R}_{J0}), \tag{7.42}$$

corresponding to equations (7.27) and (7.37) for $\vec{u}_J$, where $\vec{k}$ is restricted as in equation (7.31).

The Fourier coefficients $\vec{q}_{\vec{k}}$, equation (7.27), are a new set of canonical coordinates, $3N_1$ in number: see equation (7.31). The canonical commutation rules, equation (7.40), in terms of $\vec{q}_{\vec{k}}$ and $\vec{p}_{\vec{k}}$, equations (7.27) and (7.41), become

$$\begin{aligned}
[p_{\vec{k}\alpha}, q_{\vec{k}'\alpha'}] &= N_1^{-1}\sum_{JJ'}[P_{J\alpha}, u_{J'\alpha'}]\exp[i(\vec{k}\cdot\vec{R}_{J0} + \vec{k}'\cdot\vec{R}_{J'0})] \\
&= -i\hbar N_1^{-1}\sum_{J,J'}\delta_{JJ'}\delta_{\alpha\alpha'}\exp[i(\vec{k}\cdot\vec{R}_{J0} + \vec{k}'\cdot\vec{R}_{J'0})] \\
&= -i\hbar\delta_{\alpha\alpha'}\cdot N_1^{-1}\sum_{J}\exp[i(\vec{k}+\vec{k}')\cdot\vec{R}_{J0}] \\
&= -i\hbar\delta_{\alpha\alpha'}\sum_{J}\eta^*_{-\vec{k}}(J)\eta_{\vec{k}'}(J) \quad \text{from equation (7.23)} \\
&= -i\hbar\delta_{\alpha\alpha'}\cdot\delta_{-\vec{k},\vec{k}'} \quad \text{from equation (7.32).}
\end{aligned} \tag{7.43}$$

Thus $p_{-\vec{k}\alpha}$ (not $p_{\vec{k}\alpha}$) is the canonical momentum for $q_{\vec{k}\alpha}$.

From equation (7.39), the effective nuclear hamiltonian H is

$$H = \left\{\frac{1}{2}M\sum_{J\alpha} P_{J\alpha}^2 + \frac{1}{2}\sum_{J,J',\alpha\alpha'} W_{\alpha\alpha'}(J-J')u_{J\alpha}u_{J'\alpha'}\right\}. \tag{7.44}$$

In terms of the new canonical variables, $q_{\vec{k}\alpha}$ and $p_{-\vec{k}\alpha}$, substitution of equations (7.27) and (7.41) into equation (7.44) gives

$$H = N_1^{-1}\left\{ \frac{1}{2M} \sum_{J,\alpha} \sum_{\vec{k},\vec{k}'} p_{\vec{k}\alpha} p_{\vec{k}'\alpha} \exp[i(\vec{k}+\vec{k}')\cdot\vec{R}_{J0}] \right.$$

$$\left. + \frac{1}{2} \sum_{J,J',\alpha,\alpha'} W_{\alpha\alpha'}(J-J') \sum_{\vec{k},\vec{k}'} q_{\vec{k}\alpha} q_{\vec{k}'\alpha'} \exp[i(\vec{k}\cdot\vec{R}_{J0}+\vec{k}'\cdot\vec{R}_{J'0})] \right\}$$

$$= \left\{ \frac{1}{2M} \sum_{\vec{k},\vec{k}',\alpha} p_{\vec{k}\alpha} p_{\vec{k}'\alpha} \delta_{\vec{k},-\vec{k}'} + \frac{1}{2} N_1^{-1} \sum_{J,J'',\alpha,\alpha'} W_{\alpha\alpha'}(J'') \sum_{\vec{k},\vec{k}'} q_{\vec{k}\alpha} q_{\vec{k}'\alpha'} \right.$$

$$\left. \times \exp[i(\vec{k}+\vec{k}')\cdot\vec{R}_{J0}] \exp(-i\vec{k}'\cdot\vec{R}_{J''0}) \right\}$$

$$= \left\{ \frac{1}{2M} \sum_{\vec{k}\alpha} p_{\vec{k}\alpha} p_{-\vec{k}\alpha} + \frac{1}{2} \sum_{\vec{k}\vec{k}',\alpha\alpha'} W_{-\vec{k}',\alpha\alpha'} q_{\vec{k}\alpha} q_{\vec{k}'\alpha'} \delta_{\vec{k}',-\vec{k}} \right\}$$

$$= \left\{ \frac{1}{2M} \sum_{\vec{k}\alpha} p_{\vec{k}\alpha} p_{-\vec{k}\alpha} + \frac{1}{2} \sum_{\vec{k},\alpha,\alpha'} W_{\vec{k},\alpha\alpha'} q_{\vec{k}\alpha} q_{-\vec{k}\alpha'} \right\}. \tag{7.45}$$

In equation (7.45), we have introduced the dynamical matrix, in the notation

$$W_{\vec{k},\alpha\alpha'} = \sum_J W_{\alpha\alpha'}(J) \exp(i\vec{k}\cdot\vec{R}_{J0}). \tag{7.46}$$

This differs by a factor $N_1^{1/2}$ from the Fourier transform of $W_{\alpha\alpha'}(J)$, as we have previously defined Fourier transforms: see equations (7.37) and (7.42).

So far, the vectors $\vec{k}$ have been defined relative to a given coordinate system, defined by $\hat{\varepsilon}_\alpha$; see equation (7.22). We now adopt a different convention. For each $\vec{k}$ vector, $W_{\vec{k},\alpha\alpha'}$ is a 3×3 matrix. By rotation to principal axes, $W_{\vec{k},\alpha\alpha'}$ becomes diagonal. In that case, $W_{\vec{k},\alpha\alpha'}$ has only three elements,

$$W_{\vec{k},\alpha\alpha'} = W_{\vec{k},\alpha\alpha}\cdot\delta_{\alpha\alpha'},$$

We denote

$$W_{k,\alpha\alpha} \equiv W_{\vec{k},\alpha}. \tag{7.47}$$

This does not affect the $\vec{k}$-dependence of $W_{\vec{k},\alpha\alpha'}$, equation (7.46), which is borne only by scalar quantities $(\vec{k}\cdot\vec{R}_{J0})$. From equation (7.45) we now have

$$H = \sum_{\vec{k},\alpha}\left\{ \frac{1}{2M} p_{\vec{k}\alpha} p_{-\vec{k}\alpha} + \frac{1}{2} W_{\vec{k}\alpha} q_{\vec{k}\alpha} q_{-\vec{k}\alpha} \right\}. \tag{7.48}$$

We re-emphasize that now $W_{\vec{k},\alpha}$ is the diagonal element of $W_{\vec{k},\alpha\alpha'}$, equations (7.46), (7.38) and (7.18), in the principal axes coordinate system of $W_{\vec{k},\alpha\alpha'}$.

7.5.3 The normal modes of crystal vibration

We proceed to introduce a further canonical transformation which will cast equation (7.48) in the form of a set of independent simple harmonic oscillators. For this we consider real variables. We note that, from equation (7.37), since $\vec{u}_J$ are real displacements,

$$q^*_{\vec{k}\alpha} = q_{-\vec{k}\alpha}. \tag{7.49}$$

Thus consider

$$Q^{(1)}_{\vec{k}\alpha} = 2^{-1/2}(q_{\vec{k}\alpha} + q_{-\vec{k}\alpha}) = 2^{1/2}\,\mathrm{Re}(q_{\vec{k}\alpha}) \tag{7.50}$$

and

$$Q^{(2)}_{\vec{k}\alpha} = -\mathrm{i}2^{-1/2}(q_{\vec{k}\alpha} - q_{-\vec{k}\alpha}) = 2^{1/2}\,\mathrm{Im}(q_{\vec{k}\alpha}). \tag{7.51}$$

Similarly, from equation (7.42), since $\vec{P}_J$ are real momenta,

$$p_{-\vec{k}\alpha} = p^*_{\vec{k}\alpha}. \tag{7.52}$$

We therefore construct

$$P^{(1)}_{\vec{k}\alpha} = 2^{-1/2}(p_{\vec{k}\alpha} + p_{-\vec{k}\alpha}) = 2^{1/2}\,\mathrm{Re}(p_{\vec{k}\alpha}), \tag{7.53}$$

$$P^{(2)}_{\vec{k}\alpha} = \mathrm{i}2^{-1/2}(p_{\vec{k}\alpha} - p_{-\vec{k}\alpha}) = -2^{1/2}\,\mathrm{Im}(p_{\vec{k}\alpha}). \tag{7.54}$$

We note that $Q^{(1)}_{\vec{k}\alpha}$ and $P^{(1)}_{\vec{k}\alpha}$ are of even parity, and $Q^{(2)}_{\vec{k}\alpha}$ and $P^{(2)}_{\vec{k}\alpha}$ are of odd parity, as functions of $\vec{k}$.

If we invert relations (7.50) to (7.54), we have

$$q_{\vec{k}\alpha} = 2^{-1/2}(Q^{(1)}_{\vec{k}\alpha} + \mathrm{i}Q^{(2)}_{\vec{k}\alpha}) \tag{7.55}$$

$$p_{\vec{k}\alpha} = 2^{-1/2}(P^{(1)}_{\vec{k}\alpha} - \mathrm{i}P^{(2)}_{\vec{k}\alpha}). \tag{7.56}$$

Substituting these into the hamiltonian, equation (7.48), we obtain

$$H = \frac{1}{2}\sum_{j=1,2}\sum_{\vec{k}\alpha}\left\{\frac{1}{2M}\cdot P^{(j)^2}_{\vec{k}\alpha} + \frac{1}{2}W_{\vec{k},\alpha}Q^{(j)^2}_{\vec{k}\alpha}\right\}. \tag{7.57}$$

Consider the special case where the crystal has a center of symmetry. Then, in that case,

$$W_{-\vec{k},\alpha} = W_{\vec{k},\alpha}, \tag{7.58}$$

and because $Q^{(j)}_{\vec{k}\alpha}$ and $P^{(j)}_{\vec{k}\alpha}$ have definite parity,

$$H^{(j)}_{\vec{k}\alpha} \equiv \left\{\frac{1}{2M}\cdot P^{(j)^2}_{\vec{k}\alpha} + \frac{1}{2}W_{\vec{k}\alpha}Q^{(j)^2}_{\vec{k}\alpha}\right\} = H^{(j)}_{-\vec{k},\alpha}. \tag{7.59}$$

From the commutation rules, equation (7.43), for $\vec{p}_{\vec{k}}$ and $\vec{q}_{\vec{k}}$, we find those for $P_{\vec{k}}^{(j)}$ and $Q_{\vec{k}}^{(j)}$, $j = 1$, 2, from equations (7.50), (7.51), (7.53) and (7.54),

$$[P_{\vec{k}\alpha}^{(1)}, Q_{\vec{k}'\alpha'}^{(1)}] = \tfrac{1}{2}(-i\hbar)\{2\delta_{\vec{k}\vec{k}'}\delta_{\alpha\alpha'} + 2\delta_{\vec{k}',-\vec{k}}\delta_{\alpha\alpha'}\}. \tag{7.60}$$

Thus, if for the moment we assume that $\vec{k}' \neq -\vec{k}$, we have canonical commutation rules,

$$[P_{\vec{k}\alpha}^{(1)}, Q_{\vec{k}'\alpha'}^{(1)}] = -i\hbar\delta_{\vec{k},\vec{k}'} \cdot \delta_{\alpha\alpha'}. \tag{7.61}$$

Similarly, we obtain

$$[P_{\vec{k}\alpha}^{(2)}, Q_{\vec{k}'\alpha'}^{(2)}] = \tfrac{1}{2}(-i\hbar)\{2\delta_{\vec{k}\vec{k}'}\delta_{\alpha\alpha'} - 2\delta_{\vec{k}',-\vec{k}}\delta_{\alpha\alpha'}\} \tag{7.62}$$

or, if $\vec{k} \neq -\vec{k}'$,

$$[P_{\vec{k}\alpha}^{(2)}, Q_{\vec{k}'\alpha'}^{(2)}] = -i\hbar\delta_{\vec{k},\vec{k}'} \cdot \delta_{\alpha\alpha'}. \tag{7.63}$$

Furthermore,

$$[P_{\vec{k}\alpha}^{(i)}, Q_{\vec{k}'\alpha'}^{(j)}] = 0 \qquad \text{for } i \neq j. \tag{7.64}$$

From equations (7.61) and (7.63), we conclude that, $\vec{P}_{\vec{k}}^{(1)}$ and $\vec{Q}_{\vec{k}}^{(1)}$ are a canonical pair, and $\vec{P}_{\vec{k}}^{(2)}$, $\vec{Q}_{\vec{k}}^{(2)}$ are a canonical pair, both subject to the given restrictions on $\vec{k}$ and $\vec{k}'$, namely $\vec{k} \neq -\vec{k}'$.

Now consider the hamiltonian in the form of equation (7.57). Consider $j = 1$. Because of the symmetry relation equation (7.59), we can limit the sum over $\vec{k}$ to a single hemisphere, say $k_z > 0$, and restrict $\vec{k}$ to a semicircle for $k_z = 0$, and introduce a factor 2. Then the restriction $\vec{k}' \neq -\vec{k}$, under which equation (7.61) is valid, is satisfied. Similarly, for $j = 2$, limit the $\vec{k}$-sum to $k_z < 0$ and the other semicircle for $k_z = 0$ and multiply by 2, validating equation (7.63). Thus $j = 1, 2$ refer to $k_z > 0$ and $k_z < 0$ respectively, limited to mutually exclusive semicircles for $k_z = 0$. The overall factor $\tfrac{1}{2}$ now cancels in equation (7.57), leaving

$$H = \sum_{\vec{k},\alpha}\left\{\frac{1}{2M} \cdot P_{\vec{k}\alpha}^2 + \frac{1}{2}W_{\vec{k}\alpha}Q_{\vec{k}\alpha}^2\right\}, \tag{7.65}$$

with the sum over $\vec{k}$ unrestricted. In equation (7.65) $P_{\vec{k}\alpha}$ and $Q_{\vec{k}\alpha}$ correspond to $P_{\vec{k}\alpha}^{(j)}$ and $Q_{\vec{k}\alpha}^{(j)}$ with $j = 1$ or 2 in the respective regions of $\vec{k}$.

For bound states of the crystal, the force constants $W_{\vec{k}\alpha}$ in equation (7.65) will be real and positive. Thus introduce angular frequencies $\omega_{\vec{k}\alpha}$:

$$\omega_{\vec{k}\alpha}^2 = M^{-1}W_{\vec{k}\alpha}. \tag{7.66}$$

We take it to be known that the harmonic oscillator hamiltonian

$$H_{\vec{k}\alpha} = \left\{\frac{1}{2M}P_{\vec{k}\alpha}^2 + \frac{1}{2}M\omega_{\vec{k}\alpha}^2 Q_{\vec{k}\alpha}^2\right\} \tag{7.67}$$

has an eigenvalue spectrum

$$E_{\vec{k}\alpha} = (n_{\vec{k}\alpha} + \tfrac{1}{2})\hbar\omega_{\vec{k}\alpha}, \tag{7.68}$$

where $n_{\vec{k}\alpha} = 0, 1, 2, \ldots$ We denote the corresponding single-oscillator eigenstates $\chi_{n_{\vec{k}\alpha}}(Q_{\vec{k}\alpha})$. Furthermore, the Schrödinger equation for the nuclear motions, equation (7.20), now has the form

$$\left\{ \sum_{\vec{k},\alpha} H_{\vec{k}\alpha} \right\} \varphi_\lambda = [E_\lambda - W^{(0)}(\psi_\lambda)]\varphi_\lambda, \tag{7.69}$$

from equation (7.21b), where φ_λ is separable into $3N_1$ single-oscillator eigenstates:

$$\varphi_\lambda(\underline{Q}) = \prod_{\vec{k}\alpha} \chi_{n_{\vec{k}\alpha}}(Q_{\vec{k}\alpha}). \tag{7.70}$$

In equation (7.70) $\underline{Q}$ stands for the set of oscillator coordinates $\{Q_{\vec{k}\alpha}\}$, which are related to the nuclear displacements $\{\vec{u}_J\}$ through equations (7.50) and (7.51), and (7.37). In particular, from equation (7.37) we see that the normal modes of the crystal are collective modes, $Q_{\vec{k}\alpha}$, depending on all atomic sites J, for given $\vec{k}\alpha$.

7.5.4   Electrons and phonons: total energy

From equations (7.69) and (7.68) we obtain the total energy E_λ of the crystal for a given electronic state $\psi_\lambda(\underline{r})$ and a given phonon state $\{n_{\vec{k}\alpha}\}$, the latter representing the level of excitation of all of the $3N_1$ independent harmonic modes of oscillation of the crystal, or in alternative terminology, the number of phonons of each normal mode angular frequency $\omega_{\vec{k}\alpha}$. Specifically,

$$E_\lambda \equiv E_{\lambda,\{n\}} = \left\{ W^{(0)}(\psi_\lambda) + \sum_{\vec{k}\alpha}(n_{\vec{k}\alpha} + \tfrac{1}{2})\hbar\omega_{\vec{k}\alpha} \right\}, \tag{7.71}$$

where with $E_{\lambda,\{n\}}$ we indicate the phonon distribution $n_{\vec{k}\alpha}$, for all $\vec{k}\alpha$, by $\{n\}$. In equation (7.71), from equations (7.16), (7.12), (7.6) and (7.5a),

$$W^{(0)}(\psi_\lambda) = \{V_n(\underline{R}_0) + \langle\psi_\lambda|H_e + V_{ne}(\underline{R}_0)|\psi_\lambda\rangle\}. \tag{7.72}$$

In equation (7.71), we note that $\omega_{\vec{k}\alpha}$ is a function of the electronic state,

$$\omega_{\vec{k}\alpha} \equiv \omega_{\vec{k}\alpha}(\psi_\lambda); \tag{7.73}$$

see equations (7.66), (7.47), (7.46), (7.38), (7.18) and (7.12). Thus, even in the case where no phonons are present, i.e. $n_{\vec{k}\alpha} = 0$ for all $\vec{k}\alpha$, we have

$$E_\lambda = \left\{ W^{(0)}(\psi_\lambda) + \frac{1}{2}\sum_{\vec{k}\alpha} \hbar\omega_{\vec{k}\alpha}(\psi_\lambda) \right\}. \tag{7.74}$$

The second term in equation (7.74) is the zero-point energy of the phonon field, and depends on the electronic state. This is understandable, since the interatomic forces in the crystal depend upon the electronic configuration.

We can now discuss the determination of the electronic state. In principle, one could variationally estimate the many-electron wave function ψ_λ from equation (7.71) for a given level $\{n_{\vec{k}\alpha}\}$ of phonon excitation. However, since the phonon state is not stationary, but participates thermodynamically in the crystal's properties, such calculations are not worthwhile. Electronic states near absolute zero may be estimated variationally from equation (7.74), with $n_{\vec{k}\alpha} = 0$ for all $\vec{k}\alpha$. Even this is seldom done, though it is practicable. Usually the phonon zero-point energy is neglected, and the variational method is applied only to $W^{(0)}(\psi_\lambda)$, the so-called static lattice approximation. This corresponds to the picture of the nuclei as classical particles: see the comment following equation (7.13). Thus consider

$$\frac{\delta}{\delta\psi_\lambda} W^{(0)} = 0, \qquad \text{subject to normalization, } \langle\psi_\lambda|\psi_\lambda\rangle = 1, \qquad (7.75)$$

where

$$W^{(0)} = \langle\psi_\lambda|H_s|\psi_\lambda\rangle, \qquad (7.76)$$

and the static lattice hamiltonian H_s is

$$H_s = \left[\frac{e^2}{4\pi\varepsilon_0} \cdot \frac{1}{2}{\sum_{J,J'}}' Z_J Z_{J'} |\vec{R}_{J0} - \vec{R}_{J'0}|^{-1} \right.$$
$$\left. + \sum_{j=1}^{N}\left\{ -\frac{\hbar^2}{2m}\nabla_j^2 - \frac{e^2}{4\pi\varepsilon_0}\sum_J Z_J|\vec{r}_j - \vec{R}_{J0}|^{-1} + \frac{e^2}{4\pi\varepsilon_0}\frac{1}{2}\sum_{j'}|\vec{r}_j - \vec{r}'_j|^{-1} \right\} \right].$$
$$(7.77)$$

See equations (7.5a), (7.3b), (7.4b), (7.5b) and (7.6). Methodologies for implementing equations (7.75)–(7.77) for the electronic states of a crystal are well developed [Pisani *et al.* (1988), Kunz (1982)]. We take up this subject of the electronic state later, in Chapters 12 and 14.

7.6 Statistical thermodynamics of a solid

The basic result of statistical thermodynamics, based on the Gibbs canonical ensemble, is

$$F = -k_B T \ln Z \qquad (7.78)$$

where F is the Helmholtz free energy, k_B is Boltzmann's constant, T is Kelvin temperature, and Z is the partition function

$$Z = \sum_l \exp(-\beta E_l), \qquad \beta = (k_B T)^{-1}, \qquad (7.79)$$

where E_l are eigenvalues of H, the quantum-mechanical hamiltonian of the system [see e.g. Schrödinger (1952)].

This result relates to equilibrium thermodynamics through the following well known relationships,

$$F = (U - TS), \tag{7.80}$$

where U is internal energy, and S is entropy; and where the combined first and second laws of thermodynamics is

$$dU = (T\,dS - p\,dV) \tag{7.81}$$

where (p, V) are (pressure, volume). From equations (7.80) and (7.81),

$$dF = (-S\,dT - p\,dV), \tag{7.82}$$

so the equation of state is

$$p = -\left(\frac{\partial F}{\partial V}\right)_{\mathrm{T}}. \tag{7.83}$$

7.6.1   Partition function of the crystal

We now examine the partition function Z, equation (7.79), in terms of our crystalline solid. In equation (7.71), we have the energy eigenstates labelled by the electronic state λ and the phonon distribution $\{n\}$. Thus, we have

$$E_l = E_{\lambda,\{n\}}. \tag{7.84}$$

In this chapter, we are concentrating on nuclear dynamical properties, rather than electronic properties. Thus, in equation (7.71), suppose that $\lambda = 0$ is the electronic ground state (static lattice approximation, equations (7.75), (7.76)), and that electronic excited states are considerably higher. Then in equation (7.79) with equation (7.84), the sum over λ will be dominated by the term $\lambda = 0$. We then have

$$Z \approx \sum_{\{n\}} \exp(-\beta E_{0,\{n\}}) = Z_{\mathrm{el}} \cdot Z_{\mathrm{ph}}, \tag{7.85}$$

where

$$Z_{\mathrm{el}} = \exp\{-\beta W^{(0)}(\psi_0)\} \tag{7.86}$$

and

$$\begin{aligned}
Z_{\mathrm{ph}} &= \sum_{\{n\}} \exp\left\{-\beta \sum_{\vec{k}\alpha}(n_{\vec{k}\alpha} + \tfrac{1}{2})\hbar\omega_{\vec{k}\alpha}\right\} \\
&= \sum_{\{n\}} \prod_{\vec{k}\alpha} \exp\{-\beta(n_{\vec{k}\alpha} + \tfrac{1}{2})\hbar\omega_{\vec{k}\alpha}\}. \tag{7.87}
\end{aligned}$$

We note that there are only $3N_1$ factors in $\prod_{\vec{k}\alpha}$; see equation (7.31), whereas the possible range of values each $n_{\vec{k}\alpha}$ is infinite, in principle: $n_{\vec{k}\alpha} = 0, 1, 2, \ldots$.

In practice, of course, an infinitely high level of excitation of a normal mode would imply infinitely large nuclear excursions, leading to mechanical breakdown of the crystal. The following results are therefore only approximately valid, to the extent that infinitely high values of $n_{\vec{k}\alpha}$ in equation (7.87) contribute negligibly because the exponents are negative.

In equation (7.87), the sum over $\{n\}$ means that each of the $n_{\vec{k}\alpha}$ may take any value ≥ 0, integer. Thus, equation (7.87) is of the form

$$Z_{\text{ph}} = \sum_{\{n\}} \prod_{j=1}^{3N_1} \exp\{-\beta\hbar\omega_j(n_j + \tfrac{1}{2})\}$$

$$= \sum_{n_1=0}^{\infty} \sum_{n_2} \cdots \sum_{n_{3N_1}} \prod_j \exp\{-\beta\hbar\omega_j(n_j + \tfrac{1}{2})\}$$

$$= \prod_j \sum_{n_j=0}^{\infty} \exp\{-\beta\hbar\omega_j(n_j + \tfrac{1}{2})\}. \tag{7.88}$$

Now, consider

$$\sum_{n=0}^{\infty} \exp\{-\beta\hbar\omega(n + \tfrac{1}{2})\} = \exp(-\beta\hbar\omega/2) \sum_{n=0}^{\infty} \{\exp(-\beta\hbar\omega)\}^n$$

$$= \exp(-\beta\hbar\omega/2) \cdot \{1 - \exp(-\beta\hbar\omega)\}^{-1}$$

$$= \{\exp(\beta\hbar\omega/2) - \exp(-\beta\hbar\omega/2)\}^{-1}$$

$$= \tfrac{1}{2} \operatorname{csch}(\beta\hbar\omega/2). \tag{7.89}$$

Combining equations (7.88) and (7.89),

$$Z_{\text{ph}} = \prod_{j=1}^{3N_1} \tfrac{1}{2}\operatorname{csch}(\beta\hbar\omega_j/2), \qquad j = \vec{k}\alpha. \tag{7.90}$$

7.6.2   Equation of state of the crystal

Returning now to statistical thermodynamics, we have, in equation (7.83),

$$p = -\left(\frac{\partial F}{\partial V}\right)_{\text{T}}, \tag{7.91}$$

where, from equations (7.78) and (7.90),

$$F = -\frac{1}{\beta}\ln Z = -\frac{1}{\beta} \cdot \sum_j \ln\{\tfrac{1}{2}\operatorname{csch}(\beta\hbar\omega_j/2)\}$$

$$= \frac{1}{\beta}\sum_j \ln\{2\sinh(\beta\hbar\omega_j/2)\}. \tag{7.92}$$

Thus from equation (7.91)

$$
\begin{aligned}
p &= -\frac{1}{\beta} \cdot \sum_j \frac{\partial}{\partial V}\{\ln[2\sinh(\beta\hbar\omega_j/2)]\}|_T \\
&= -\frac{1}{\beta} \cdot \sum_j \frac{2\cosh(\beta\hbar\omega_j/2)}{2\sinh(\beta\hbar\omega_j/2)} \cdot \frac{\beta\hbar}{2}\left(\frac{\partial\omega_j}{\partial V}\right)_T \\
&= -\frac{\hbar}{2}\sum_j \coth(\beta\hbar\omega_j/2)\cdot\left(\frac{\partial\omega_j}{\partial V}\right)_T.
\end{aligned}
$$

Thus

$$
p = -\frac{\hbar}{2}\sum_{\vec{k}\alpha}\coth(\hbar\omega_{\vec{k}\alpha}/2k_B T)\left(\frac{\partial\omega_{\vec{k}\alpha}}{\partial V}\right)_T. \tag{7.93}
$$

The equation of state, equation (7.93), is of the form $p = p(V,T)$, where V is the volume of the crystal. Let us try to identify the volume (V) dependence. The temperature (T) dependence is already explicit. The V-dependence resides in $\{\omega_{\vec{k}\alpha}\}$. From equations (7.66), (7.47), (7.46), (7.38) and (7.18),

$$
\omega_{\vec{k}\alpha}^2 = M^{-1}\sum_J\left(\frac{\partial^2 W}{\partial u_{0\alpha}\,\partial u_{J\alpha}}\right)_{u=0}\exp(-i\vec{k}\cdot\vec{R}_{J0}). \tag{7.94}
$$

From equations (7.22) and (7.30), we note that $(\vec{k}\cdot\vec{R}_{J0})$ is independent of a_α, and therefore independent of V. The equation of state gives us the variation of volume with pressure (at given T) for a given sample of material. Thus we must consider N_1, the number of atoms, to be fixed in the present case. For simplicity of illustration consider a simple cubic crystal, for which $a_\alpha = a$, and

$$
V = N_1 a^3 \tag{7.95}
$$

whence

$$
\frac{\partial}{\partial V} = \left(\frac{\partial a}{\mathrm{d}V}\right)\cdot\frac{\partial}{\partial a},
$$

and

$$
\left(\frac{\partial a}{\partial V}\right) = (3N_1 a^2)^{-1}.
$$

Thus in equation (7.93), the equation of state, we encounter

$$
\left(\frac{\partial\omega_{\vec{k}\alpha}}{\partial V}\right)_T = (3N_1 a^2)^{-1}\left(\frac{\mathrm{d}\omega_{\vec{k}\alpha}}{\mathrm{d}a}\right).
$$

The volume dependence in the equation of state then becomes a question of how $\omega_{\vec{k}\alpha}$ depends on a. Referring to equation (7.94), it becomes in turn, a question of how W, equation (7.12), or rather its second derivative in the equilibrium configuration $\underline{R}_0$ of the crystal, depends on a. We shall not pursue the analytical details of this back through V_n and V_{ne}, equations (7.5a) and (7.6) nor, even more difficult, through $\psi_\lambda(\underline{r}, \underline{R}_0)$ [see equation (7.8) and equations (7.75)–(7.77)]. Suffice it to say that, within the present formulation, such a process is possible, both in principle and in practice.

7.6.3   Thermodynamic internal energy of the crystal; phonons as bosons

Let us now consider the internal energy U, from equation (7.80):

$$U = (F + TS), \tag{7.96}$$

From equation (7.82),

$$S = -\left(\frac{\partial F}{\partial T}\right)_V. \tag{7.97}$$

From equation (7.92), with $\beta = (k_B T)^{-1}$, we find

$$S = k_B \sum_j \{(\beta\hbar\omega_j/2)\coth(\beta\hbar\omega_j/2) - \ln[2\sinh(\beta\hbar\omega_j/2)]\}. \tag{7.98}$$

Thus, from equation (7.96) with equations (7.92) and (7.98),

$$U = \sum_j \left(\frac{\hbar\omega_j}{2}\right)\coth(\beta\hbar\omega_j/2). \tag{7.99}$$

Let us write the internal energy in terms of the distribution of phonons among normal modes, as a function of temperature,

$$U = \sum_j \hbar\omega_j(\bar{n}_j + \tfrac{1}{2}), \tag{7.100}$$

where $\bar{n}_j \equiv \bar{n}_{\vec{k}\alpha}(T)$ is the mean number of phonons in mode $j \equiv \vec{k}\alpha$, at temperature T. Then from equations (7.99) and (7.100)

$$\hbar\frac{\omega_j}{2}\coth(\beta\hbar\omega_j/2) = \hbar\omega_j(\bar{n}_j + \tfrac{1}{2})$$

or

$$\bar{n}_j = \tfrac{1}{2}\{\coth(\beta\hbar\omega_j/2) - 1\} = [\exp(\beta\hbar\omega_j) - 1]^{-1}. \tag{7.101}$$

This is recognized as the distribution function for Bose statistics [see for example Huang (1967), sections 9.5 and A.1], and leads us to identify phonons, the quantum excitations of the normal modes of a crystal, as bosons.

7.7 Summary

Beginning with a collection of N_1 nuclei, each with charge Z, and N electrons, we have used the principles of quantum mechanics to arrive at a description of a macroscopic solid crystal. If the solid is electrically neutral, N is equal to (ZN_1). When nuclear excursions are limited to the harmonic approximation, equation (7.15), and the electronic system is confined to a single quantum state, a useful approximation is introduced which allows determination of the electronic state, equations (7.75)–(7.77), the nuclear quantum dynamics, equation (7.65), and the crystal structure, equation (7.13). These three features must, in general, be determined with mutual self-consistency. The nuclear dynamics in this case consists of simple harmonic collective motions, called phonons. The quantum statistical thermodynamics of this set of oscillators leads to the explicit thermodynamic equation of state, equations (7.93) and (7.94). Examination of the thermodynamic internal energy leads to the identification of phonons as bosons.

The chapter as a whole illustrates how, for the equation of state,

$$p = f(V, T),$$

one can obtain the function f in terms of the fundamental parameters of the atomic system, namely the nuclear charges Z and masses M.

The current state of the art of computational modelling and simulation is such that much of the *tour de force* formulated above is actually being carried out. The result is that we are in the early stages of a period in which computer modelling and simulation can be used in the search for solid state and molecular structures with specified properties, leaving actual fabrication to last, in cases where that happens to be the most efficient approach.

Chapter 8

Phonons

8.1 Introduction

In the preceding chapter, we showed that, under commonly prevalent conditions, a crystal behaves dynamically like a set of independent harmonic oscillators [see equation (7.65)]. The quantized form of these oscillators are phonons. The generalized coordinates of the oscillators are, in general, not atomic coordinates, but rather collective coordinates involving all the atoms of the crystal [see equations (7.37), (7.50) and (7.51)]. The normal mode frequencies $\omega_{\vec{k}\alpha}$, equation (7.66), are obtainable classically if the force constants are known, and we have shown [equations (7.12), (7.18), (7.38), (7.46) and (7.48)] how to derive them under the approximations of Chapter 7 from the quantum-mechanical state of the electrons in the crystal.

In this chapter, we shall solve for the normal mode frequencies of several very simple model crystals. In that way, we shall see some specific features of phonons that carry over in some sense to more realistic crystalline systems. Our models will be one-dimensional, and the force constants will be limited to nearest-neighbor interaction.

The results of this chapter are well-presented in many other works, but they are repeated here so that this important subject is not left in the very general, and intuitively unappealing, form of equations (7.67)–(7.70). Notable references are Born and Huang (1954, section II.5), and Ashcroft and Mermin (1976, Chapter 22).

In section 8.2 we discuss the monatomic linear chain, revealing the sharp qualitative distinction between the dynamical behavior of a continuous medium (Chapter 2) and a medium of discrete atoms. In section 8.3 we discuss the diatomic linear chain, illustrating the distinct natures of optical and acoustic branches of the phonon spectrum. In section 8.4 we briefly discuss the localized mode in the crystal associated with a point defect.

8.2   Monatomic linear chain

Consider a model in which an infinite set of identical classical atoms of mass M is constrained to lie along a straight line, the x-axis. Let interatomic forces be harmonic (spring-like) with force constant K, limited to nearest-neighbor interaction. Let atomic equilibrium positions be $x_j = ja$ where j is an integer or zero, defining an equilibrium separation distance a. Let $u_j(t)$ be small atomic displacements in oscillations of this 'crystal', and let us apply periodic boundary conditions to a large set of N such atoms:

$$u_j(t) = u_{j+N}(t). \tag{8.1}$$

From Newton's second law of motion, we have, for atom j,

$$M\frac{d^2x_j}{dt^2} = M\frac{d^2u_j}{dt^2} = K[(u_{j+1} - u_j) - (u_j - u_{j-1})], \qquad j = 0, 1, 2, \ldots, (N-1). \tag{8.2}$$

In the linear chain, atomic positions $x_j = ja$ are discrete, determined by the integers j. Thus the continuum analogue of the position variable j in the discrete linear chain is x. Correspondingly, the discrete variable $u_j(t)$ has the continuum analogue $u(x, t)$,

$$u_j(t) \rightarrow u(x, t). \tag{8.3}$$

Thus, in equation (8.2),

$$(u_{j+1} - u_j) \rightarrow [u(x + \Delta x, t) - u(x, t)], \tag{8.4}$$

where

$$\Delta x = a. \tag{8.5}$$

Similarly,

$$(u_j - u_{j-1}) \rightarrow [u(x, t) - u(x - \Delta x, t)]. \tag{8.6}$$

From equations (8.4)–(8.6)

$$[(u_{j+1} - u_j) - (u_j - u_{j-1})] \rightarrow \left\{\frac{\partial}{\partial x}[u(x, t)] - \frac{\partial}{\partial x}[u(x - \Delta x, t)]\right\}\Delta x, \tag{8.7}$$

when $\Delta x = a$ is very small, in macroscopic terms. Thus

$$[(u_{j+1} - u_j) - (u_j - u_{j-1})] \rightarrow \frac{\partial^2}{\partial x^2}u(x, t) \cdot a^2. \tag{8.8}$$

Similarly equation (8.2) corresponds to

$$M\frac{\partial^2 u}{\partial t^2} = (Ka^2)\frac{\partial^2 u}{\partial x^2}. \tag{8.9}$$

If we identify the constant $(Ka^2/M) = v^2$, we have, from equation (8.9),

$$\left(\frac{\partial^2 u}{\partial x^2} - \frac{1}{v^2}\frac{\partial^2 u}{\partial t^2}\right) = 0. \tag{8.10}$$

where v has the dimensionality of speed. In fact, equation (8.10) is the one-dimensional form of the equation of motion for waves in a macroscopic continuum that we obtained in equation (2.41).

We now return to the discrete atomic case, equation (8.2), and seek a normal mode solution, in which all atoms have the same angular frequency ω:

$$u_j(t) = u_j(0)\,e^{-i\omega t}. \tag{8.11}$$

Substituting from equation (8.11) into equation (8.2) we have

$$-M\omega^2 u_j(0) = -K[2u_j(0) - u_{j+1}(0) - u_{j-1}(0)]. \tag{8.12}$$

We now consider the phase relationship between consecutive atoms on the linear chain in their vibrations. Let

$$u_{j+1}(0) = e^{i\xi}u_j(0). \tag{8.13}$$

Substituting from equation (8.13) into equation (8.12) we have

$$M\omega^2 = K(2 - e^{i\xi} - e^{-i\xi})$$

$$= 2K[1 - \cos(\xi)] = 4K\sin^2\left(\frac{\xi}{2}\right). \tag{8.14}$$

We see from equation (8.14) that, for ω independent of j, the phase ξ in equation (8.13) must also be independent of j. We now apply the periodic boundary condition of equation (8.1), with equations (8.11) and (8.13):

$$u_{j+N}(0) = u_j(0)\,e^{iN\xi} = u_j(0). \tag{8.15}$$

This requires

$$N\xi = 2\pi n, \qquad \text{with } n = 0, \pm1, \pm2, \ldots, \tag{8.16}$$

or

$$\xi \equiv \xi_n = \frac{2\pi n}{N}. \tag{8.17}$$

Now from equation (8.13) we have

$$u_j(0) = e^{ij\xi}u_0(0). \tag{8.18}$$

We now introduce the wave number k_n, as in Chapter 2 [see equation (2.45)]

$$k_n = \frac{2\pi n}{Na} = \frac{\xi_n}{a}. \tag{8.19}$$

Along with the notation $x_j = ja$, equation (8.19) enables us to see the normal mode, from equations (8.11) and (8.18), as a longitudinal travelling wave:

$$u_{j,n}(t) = u_0(0) \exp[\mathrm{i}(j\xi_n - \omega_n t)]$$
$$= u_0(0) \exp[\mathrm{i}(k_n \cdot x_j - \omega_n t)]. \tag{8.20}$$

From the periodic boundary condition of equation (8.1), we see that the normal modes $u_{j,n}(t)$, equation (8.20), are only distinct for a set of N consecutive values of n. Without loss of generality we can take N to be an odd number, and we can then limit n to the following range

$$-\tfrac{1}{2}(N - 1) \leq n \leq \tfrac{1}{2}(N - 1), \tag{8.21}$$

which now applies in equation (8.16). This says, *inter alia*, that there are only N normal modes for a segment of linear chain that contains N atoms; i.e. there are exactly as many modes as there are degrees of freedom in the mechanical system (because each atom in the one-dimensional system has only one degree of freedom). From equation (8.14) with equation (8.19) we find the normal mode frequencies to be

$$\omega_n = \omega(k_n) = 2\left(\frac{K}{M}\right)^{1/2} \left|\sin\left(\frac{k_n a}{2}\right)\right| = 2\left(\frac{K}{M}\right)^{1/2} \left|\sin\left(\frac{n\pi}{N}\right)\right|. \tag{8.22}$$

We note first that

$$\omega(k_n) = \omega(-k_n),$$

and that, for small $k_n > 0$,

$$\omega_n \approx \left(\frac{Ka^2}{M}\right)^{1/2} \cdot k_n = v \cdot k_n, \tag{8.23}$$

having used equations (8.9)–(8.10). The approximately linear dispersion relation, equation (8.23), for small k_n is the same as that for a continuous medium, equation (2.46). Indeed, the speed v in equation (8.23) is the direct analogue of the longitudinal speed of wave propagation in a continuous medium. We can see this as follows. We have

$$v = \left(\frac{Ka^2}{M}\right)^{1/2} = \left[\frac{Ka}{(M/a)}\right]^{1/2}. \tag{8.24}$$

Now for the linear chain, (M/a) is the mass density, corresponding to ρ_0 in equation (2.42) for v_{L}:

$$v_{\mathrm{L}} = \left[\frac{(\lambda + 2\mu)}{\rho_0}\right]^{1/2}. \tag{8.25}$$

Also, from the summary following equation (1.63), the numerator of equation (8.25), right hand side, is

$$(\lambda + 2\mu) = c_{11}, \tag{8.26}$$

the longitudinal elastic constant. Now the one-dimensional definition of c_{11}, equation (8.26), comes from equation (1.48),

$$\sigma_1 = c_{11}\varepsilon_1, \tag{8.27}$$

where σ_1 and ε_1 are stress and strain respectively in Voigt notation. Stress is defined as force per unit area: see equation (1.34a) and discussion. In one dimension, where there is no cross-sectional area, the stress in the linear continuum must be replaced by force. Strain is fractional deformation: see equation (1.10) and discussion. Thus, here, from equation (8.27),

$$c_{11} = \frac{\sigma_1}{\varepsilon_1}. \tag{8.28}$$

In words, c_{11} is force per unit fractional deformation. But the corresponding numerator in equation (8.24) is

$$Ka = \frac{K}{(a^{-1})},$$

the force per unit deformation within a primitive unit cell of the linear crystal, divided by a^{-1}, or force per unit fractional deformation, in direct correspondence with equation (8.28). The process of relating bulk properties of a crystal, defined for a continuous medium, to the details of a classical atomistic model is illustrated extensively for a realistic three-dimension system in Chapter 9.

The fact that the speed of wave propagation in the linear chain is the same as that which one obtains by viewing the chain as an approximate linear continuum means that the dispersion relation

$$\omega = v \cdot k \tag{8.29}$$

from equation (2.46) for the linear continuum coincides with that for the linear chain, equation (8.22), at small k: see equation (8.23). This is illustrated in figure 8.1. The reader should show, from equation (2.45), that even if we apply periodic boundary conditions to the linear continuum, the values of k

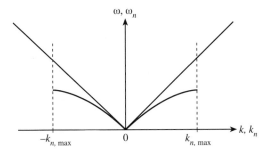

Figure 8.1 Dispersion relation ω versus k for a linear elastic continuum (straight line), and ω_n versus k_n for a monatomic linear chain of discrete atoms [see equation (8.22)].

are unrestricted in magnitude; that is, the constraint of equation (8.21) does not apply. Now the meaning of k (or k_n for the linear chain of discrete atoms) is that of a wave number, in the following sense:

$$k = \frac{2\pi}{\lambda}, \tag{8.30}$$

where λ is a wavelength. The shortest wavelength that can be identified in a discrete linear chain is two atomic spacings,

$$\lambda_{\min} = 2a, \tag{8.31}$$

so the largest wave number $k_{n,\max}$ is

$$k_{n,\max} = \frac{\pi}{a}. \tag{8.32}$$

Now, according to equations (8.21) and (8.19),

$$k_{n,\max} = \left(\frac{2\pi}{aN}\right) \cdot \frac{1}{2}(N - 1). \tag{8.33}$$

In the limit $N \to \infty$, this becomes

$$k_{n,\max} = \frac{\pi}{a}, \tag{8.34}$$

in agreement with equation (8.32). By contrast, a linear medium that is literally continuous can sustain waves of all wavelength, down to $\lambda = 0$. This means that $k_{\max}$ is infinite. We further note, in figure 8.1, that $\omega(k_n)$ (or $\omega(k)$) are degenerate as between $\pm k_n$ (or $\pm k$).

A feature of our normal modes worth noting has to do with the density of states (or modes) $g(\omega)$ as a function of ω; that is, $g(\omega)$ is the number of states per unit range of values of ω, evaluated at a particular value of ω. It is easily deduced, qualitatively, from figure 8.1, and is shown for both discrete chain and continuum in figure 8.2. For the more realistic model, with discrete atoms, the singularity in $g(\omega)$ at $\omega_{n,\max}$ is indicative of the van Hove singularities [van Hove (1953); see also Wannier (1959, Chapter 3), and Ashcroft and Mermin (1976, Chapter 23)] that are so interesting in three-dimensional systems.

The normal modes given in equation (8.20) are complex. To represent the real atomic displacements, we must use the real or imaginary part of $u_{j,n}(t)$, namely,

$$\mathrm{Re}[u_{j,n}(t)] = u_0(0) \cos(k_n x_j - \omega_n t),$$

$$\mathrm{Im}[u_{j,n}(t)] = u_0(0) \sin(k_n x_j - \omega_n t). \tag{8.35}$$

In an infinite linear chain, the two are indistinguishable, and both therefore represent the same normal mode. Let us consider the two extremities of the dispersion relation, figure 8.1. At $k_n = 0$ (infinite wavelength), all atoms j

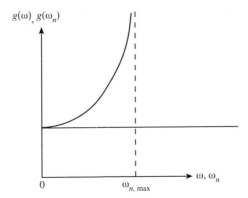

Figure 8.2 Density of phonon states $g(\omega)$ versus ω for a linear elastic continuum (straight line), and $g(\omega_n)$ versus ω_n for a linear chain of discrete atoms: see figure 8.1.

oscillate in phase:

$$\operatorname{Re}[u_{j,n=0}(t)] = u_0(0)\cos(\omega_0 t). \tag{8.36}$$

In this case there is no actual wave: the wavelength is infinite. At $k_n = k_{n,\max}$, equation (8.34), or $\bar{n} = n_{\max}$:

$$\operatorname{Re}[u_{j,n=\bar{n}}(t)] = u_0(0)\cos\left(\frac{\pi}{a}\cdot x_j - \omega_{\bar{n}}t\right)$$

$$= u_0(0)\cos(j\pi - \omega_{\bar{n}}t)$$

$$= u_0(0)\cos(\omega_{\bar{n}}t - j\pi). \tag{8.37}$$

Thus, in this mode, successive atoms on the linear chain are 180° out of phase with each other at any instant of time, corresponding to the shortest wavelength, as in equation (8.31). This is, in fact, a standing wave, for

$$\cos(\omega_{\bar{n}}t - \pi j) = [\cos(\omega_{\bar{n}}t)\cos(\pi j) + \sin(\omega_{\bar{n}}t)\sin(\pi j)]$$

$$= (-1)^j \cdot \cos(\omega_{\bar{n}}t). \tag{8.38}$$

Thus, in equation (8.37), with equation (8.38), all ions have zero displacement when

$$(\omega_{\bar{n}}t) = (m + \tfrac{1}{2})\pi, \qquad m = 0, 1, 2, \dots, \tag{8.39}$$

i.e. all at the same times, and similarly their extremal displacements, $\pm u_0(0)$, all occur at the same times. For all other cases, n not equal to either zero or $n_{\max}$, the normal modes are travelling waves.

We feel that, from the simple example in the foregoing discussion, the nature of phonon modes and spectra can be more easily visualized than in the general theory of Chapter 7.

8.3   Diatomic linear chain

The diatomic linear chain represents the case of a crystal whose basis consists of two distinct chemical species. As in section 8.2, we limit interatomic interactions to nearest neighbors, and distinguish the two species by their masses M and m. Consider the one-dimensional Bravais lattice

$$r_l = al, \qquad l = 0, \pm 1, \pm 2, \dots, \tag{8.40}$$

where a is the length of a primitive unit cell. Let the equilibrium positions of atoms of masses M and m respectively be $X_l^{(0)}$ and $x_l^{(0)}$,

$$X_l^{(0)} = al, \tag{8.41}$$

$$x_l^{(0)} = a(l + \tfrac{1}{2}). \tag{8.42}$$

For the vibrating 'crystal', the time-dependent atomic positions are, respectively,

$$X_l(t) = [X_l^{(0)} + U_l(t)], \tag{8.43}$$

$$x_l(t) = [x_l^{(0)} + u_l(t)], \tag{8.44}$$

where U_l and u_l are small displacements, within the harmonic approximation. We apply periodic boundary conditions to a region of N consecutive primitive unit cells:

$$U_{l+N}(t) = U_l(t), \qquad u_{l+N}(t) = u_l(t). \tag{8.45}$$

Then, with nearest-neighbor interatomic force constants K, we have the equations of motion,

$$M \frac{d^2 U_l}{dt^2} = K[(u_l - U_l) - (U_l - u_{l-1})], \tag{8.46}$$

$$m \frac{d^2 u_l}{dt^2} = K[(U_{l+1} - u_l) - (u_l - U_l)]. \tag{8.47}$$

Now, as in section 8.2, equation (8.12), we seek normal modes:

$$U_l(t) = U_l(0)\, e^{-i\omega t}, \tag{8.48}$$

$$u_l(t) = u_l(0)\, e^{-i\omega t}. \tag{8.49}$$

Substitution of equations (8.48) and (8.49) into equations (8.46) and (8.47) gives

$$-M\omega^2 U_l(0) = -K[2U_l(0) - u_l(0) - u_{l-1}(0)], \tag{8.50}$$

$$-m\omega^2 u_l(0) = -K[2u_l(0) - U_{l+1} - U_l]. \tag{8.51}$$

Again, as in section 8.2, equation (8.13), we seek the phase relationships amongst the atoms in consecutive primitive unit cells:

$$U_{l+1}(0) = e^{i\xi} U_l(0),$$ (8.52)

$$u_l(0) = e^{i\xi} u_{l-1}(0).$$ (8.53)

Substituting equations (8.52) and (8.53) into equations (8.50) and (8.51), we have

$$(M\omega^2 - 2K)U_l(0) + K(1 + e^{-i\xi})u_l(0) = 0,$$ (8.54)

$$(m\omega^2 - 2K)u_l(0) + K(e^{i\xi} + 1)U_l(0) = 0.$$ (8.55)

The necessary and sufficient condition for a solution of equations (8.54) and (8.55) is

$$\{(M\omega^2 - 2K)(m\omega^2 - 2K) - K^2(1 + e^{-i\xi})(1 + e^{i\xi})\} = 0.$$ (8.56)

Now,

$$(1 + e^{-i\xi})(1 + e^{i\xi}) = e^{-i\xi/2}[2\cos(\xi/2)] \cdot e^{i\xi/2}[2\cos(\xi/2)]$$

$$= 4\cos^2(\xi/2).$$ (8.57)

Thus, equation (8.56) reduces to

$$\{(Mm)(\omega^2)^2 - 2K(M+m)(\omega^2) + 4K^2[1 - \cos^2(\xi/2)]\} = 0,$$ (8.58)

whose solution is

$$\omega^2 = \frac{2K(M+m) \pm [4K^2(M+m)^2 - 16K^2 Mm \sin^2(\xi/2)]^{1/2}}{(2Mm)}$$

$$= K\left\{\frac{1}{\mu} \pm \left[\frac{1}{\mu^2} - \frac{4}{Mm}\sin^2(\xi/2)\right]^{1/2}\right\},$$ (8.59)

where

$$\mu = \frac{Mm}{(M+m)}$$ (8.60)

is the reduced mass of the primitive unit cell. From the periodic boundary condition, equations (8.45), along with the assumed phase relations of equations (8.52) and (8.53), we have

$$U_{l+N}(0) = e^{iN\xi} U_l(0) = U_l(0),$$ (8.61)

$$u_{l+N}(0) = e^{iN\xi} u_l(0) = u_l(0).$$ (8.62)

This is the single condition, as in equations (8.16) and (8.17),

$$\xi \equiv \xi_n = \frac{2\pi}{N}n, \qquad n = 0, \pm 1, \pm 2, \ldots$$ (8.63)

Thus, introducing k_n as in equation (8.19), we have from equations (8.52), (8.53), (8.41) and (8.42), the travelling wave modes for arbitrary n:

$$U_{l,n}(t) = U_0(0) \exp[i(k_n \cdot X_l^{(0)} - \omega_n t)], \tag{8.64}$$

$$u_{l,n}(t) = u_0(0) \exp[i(k_n \cdot x_l^{(0)} - \omega_n t)]. \tag{8.65}$$

The dispersion relations, equation (8.59), now become

$$\omega_n^2 \equiv \omega^2(k_n) = K\left\{\frac{1}{\mu} \pm \left[\frac{1}{\mu^2} - \frac{4}{mM}\sin^2(k_n a/2)\right]^{1/2}\right\}. \tag{8.66}$$

We note first that the phonon spectrum for this diatomic linear chain has two branches, corresponding to the ($\pm$) signs in equation (8.66). Let us examine these two branches, first at $k_n = 0$, i.e. at $n = 0$:

$$\omega_{n=0}^{(+)} = \left(\frac{2K}{\mu}\right)^{1/2}, \qquad \omega_{n=0}^{(-)} = 0. \tag{8.67}$$

For small n, we note that

$$\sin(k_n a/2) \approx (k_n a/2);$$

so we have

$$\left[1 - \frac{4\mu^2}{Mm}\sin^2(k_n a/2)\right]^{1/2} \approx \left\{1 - \frac{1}{2}\frac{4\mu^2}{Mm}\left(\frac{k_n a}{2}\right)^2\right\} = \left\{1 - \frac{\mu^2}{2Mm}k_n^2 a^2\right\}. \tag{8.68}$$

Thus, equation (8.66) with equation (8.68) gives

$$\omega^{(-)}(k_n) \approx \left(\frac{K\mu}{2Mm}\right)^{1/2}(k_n a) = \left[\frac{Ka^2}{2(M+m)}\right]^{1/2}(k_n). \tag{8.69}$$

This shows that $\omega^{(-)}(k_n)$ rises linearly from $k_n = 0$, qualitatively the same as the monatomic linear chain's dispersion relation, equation (8.23), which was shown to be analogous to longitudinal vibration, or sound waves, in a continuous medium. For this reason, $\omega^{(-)}$ is called the *acoustic branch*. These matters, equations (8.67) and (8.69), are illustrated in figure 8.3, along with the short-wave behavior to be discussed below.

Consider now the maximum value of k_n, in a symmetrical region, as discussed for equation (8.21). With N an odd number, we have from equations (8.63) and (8.19),

$$k_{n,\max} = \frac{2\pi}{Na}\frac{1}{2}(N-1) \approx \frac{\pi}{a}, \tag{8.70}$$

exactly as for the linear chain, bearing in mind now, however, that a is the primitive unit cell length, not the interatomic spacing, which is $(a/2)$. In

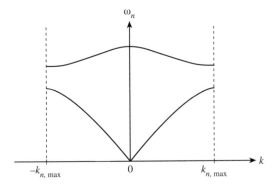

Figure 8.3 Dispersion relation for a diatomic linear chain, ω_n versus k_n, showing acoustic and optical branches, lower and upper curves respectively.

equation (8.66), we now have

$$\sin^2(k_{n,\max}a/2) = 1. \tag{8.71}$$

Thus,

$$\omega_{n,\max}^{(+)} = \left(\frac{K}{\mu}\right)^{1/2}\left\{1 + \left[1 - \frac{4\mu^2}{Mm}\right]^{1/2}\right\}^{1/2} \approx \left(\frac{2K}{m}\right)^{1/2}, \tag{8.72}$$

having assumed $M > m$, and having used equation (8.60). Similarly we obtain

$$\omega_{n,\max}^{(-)} = \left(\frac{2K}{M}\right)^{1/2} < \omega_{n,\max}^{(+)}. \tag{8.73}$$

If we compare equation (8.72) with the long-wave case, $k_n = 0$, for $\omega_n^{(+)}$, equation (8.67), we see

$$\omega_{n=0}^{(+)} = \left[\frac{2K}{m}\left(\frac{m+M}{M}\right)\right]^{1/2} > \omega_{n,\max}^{(+)}. \tag{8.74}$$

The branch $\omega_n^{(+)}$ is called the *optical branch*, corresponding to higher frequencies than the acoustic branch. This is all illustrated in figure (8.3).

Let us now examine the normal mode displacements. From equations (8.54) and (8.55),

$$(M\omega_n^2 - 2K)U_l(0) + 2K\exp\left(-ik_na/2\right)\cos(k_na/2)u_l(0) = 0, \tag{8.75}$$

$$U_l(0) = -\frac{2K}{(M\omega^2 - 2K)} \cdot \exp(-ik_na/2)\cos(k_na/2)u_l(0). \tag{8.76}$$

At $k_n = 0$, from equation (8.67) we have for the optical branch

$$[M\omega_0^{(+)^2} - 2K] = 2K\left[\frac{(M+m)}{m} - 1\right] = \frac{2M}{m}K. \qquad (8.77)$$

Thus,

$$U_l(0) = -\frac{m}{M}u_l(0). \qquad (8.78)$$

Thus, in the long-wave optical mode, the two atoms in all primitive unit cells are displaced in opposite directions at any instant, with the center of mass fixed,

$$\{MU_l(0) + mu_l(0)\} = 0, \qquad (8.79)$$

from equation (8.78), and all unit cells are in phase: $\xi_n = 0$ for $n = 0$, in equation (8.63). This is therefore a standing wave of wavelength a. For the acoustic branch at $k_n = 0$, we have $\omega_0^{(-)} = 0$ from equation (8.69) so, from equation (8.76),

$$U_l(0) = u_l(0), \qquad (8.80)$$

and both ions in all primitive cells have the same displacement: the 'crystal' moves as a rigid structure with infinite wavelength, as in the monatomic case. These results are illustrated in figure 8.4(a).

Next let us consider the short-wave length limit, $k_n = k_{n,max} = (\pi/a)$, equation (8.70). Then the cosine in equation (8.76) is zero, and from equation (8.72) the denominator $(M\omega^2 - 2K)$ is not zero for $M > m$. We therefore have for the short-wave optical mode

$$U_l(0) = 0. \qquad (8.81)$$

The phase shift ξ_n, equation (8.63), with $n = (N-1)/2 \approx N/2$ is π: the lighter atoms of mass m are 180° out of phase in successive primitive unit cells, so this standing wave has wavelength $(2a)$. For the short-wave acoustic mode, we have $\omega_n = \omega_{n,max}^{(-)}$, equation (8.73). Then since $(M\omega - 2K) = 0$, we must revert to equation (8.75), to get

$$u_l(0) = 0. \qquad (8.82)$$

Strictly speaking, equation (8.75) is satisfied in this case with $k_{n,max} = (\pi/a)$ in the cosine, without requiring equation (8.82), but for k_n slightly less than $k_{n,max}$, equation (8.82) is approximately valid. Thus, complementary to the short-wave optical case, equation (8.81), the short-wave acoustic mode has the light atoms of mass m fixed while the heavier atoms of mass M oscillate 180° out of phase in successive primitive unit cells in a standing wave of wavelength $(2a)$. These results are illustrated in figure 8.4(b).

In summary, with this very simple model we have been able to show the qualitative nature of the dispersion relation, figure 8.3, along with the

(a)

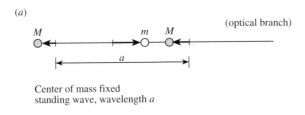

(optical branch)

Center of mass fixed
standing wave, wavelength a

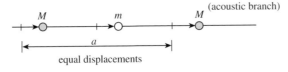

equal displacements

$k = 0$: all unit cells in phase

(b)

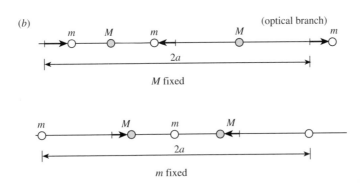

$k = k_{max}$: standing waves wavelength $2a$

Figure 8.4 (*a*) Normal modes of diatomic linear chain: long wavelength limit ($k = 0$); see equations (8.79) and (8.80). (*b*) Normal modes of diatomic linear chain: short wavelength limit ($k = k_{max}$); see equations (8.81) and (8.82).

relatively simple pattern of oscillations for the four limiting cases, low- and high-frequency limits of optical and acoustic branches.

8.4 Localized mode of a point defect

Point defects are an important feature of crystals, both for practical and theoretical reasons. They are discussed extensively in the present work in Chapters 10 and 11. Here we shall adapt the methods of the previous two sections to study the effect of a point defect on the vibrational property of a monatomic linear chain. Specifically, we consider a mass defect of mass *m* at the origin in the linear chain of section 8.2, keeping the interatomic force constant unchanged, even for interaction of the defect with the host 'crystal'. Then equation (8.2) is unchanged,

$$M \frac{\mathrm{d}^2 u_j}{\mathrm{d}t^2} = K[(u_{j+1} - u_j) - (u_j - u_{j-1})], \tag{8.83}$$

except for the range of values of j:

$$j = 1, 2, \ldots, (N-1). \tag{8.84}$$

For the defect atom, we have

$$m \frac{\mathrm{d}^2 u_0}{\mathrm{d}t^2} = K[(u_1 - u_0) - (u_0 - u_{-1})]. \tag{8.85}$$

In the present case, periodic boundary conditions mean that we are considering an infinite 'crystal' with a highly dilute (N very large) periodic array of point defects. Again we seek normal modes

$$u_j(t) = u_j(0)\, \mathrm{e}^{-\mathrm{i}\omega t}, \tag{8.86}$$

as in equation (8.12).

Substitution from equations (8.86) into equations (8.83) and (8.85) gives

$$(M\omega^2 - 2K)u_j + K(u_{j+1} + u_{j-1}) = 0, \tag{8.87}$$

$$(m\omega^2 - 2K)u_0 + K(u_1 + u_{-1}) = 0. \tag{8.88}$$

We also consider the question of phase relations among the oscillating atoms, as in equation (8.13),

$$u_{j+1}(0) = \mathrm{e}^{\mathrm{i}\xi} u_j(0). \tag{8.89}$$

We must recognize, however, that travelling waves of the type shown in equation (8.20) will not be stable in this 'lattice': they would be scattered by the point defect at the origin. Instead, we shall look for a normal mode centered on the defect, such that the phase differences will be symmetrical about the origin:

$$u_{-j-1}(0) = \mathrm{e}^{\mathrm{i}\xi} u_{-j}(0). \tag{8.90}$$

With equations (8.89) and (8.90), the first is meant to apply to atoms to the right of the origin and the second to atoms to the left, with both applying to $j = 0$. With periodic boundary conditions, we are entitled to recast the constraint of equation (8.84) as follows:

$$j = -\tfrac{1}{2}(N-1), \ldots, -2, -1, +1, 2, \ldots, \tfrac{1}{2}(N-1), \tag{8.91}$$

to conform with equation (8.90). Now consider equation (8.87) with $j = 1$,

$$[(M\omega^2 - 2K)u_1 + K(u_2 + u_0)] = \{(M\omega^2 - 2K)\,e^{i\xi} + K(e^{2i\xi} + 1)\}u_0 = 0,$$
$$(8.92)$$

and from equation (8.88)

$$[(m\omega^2 - 2K)u_0 + K(u_1 + u_{-1})] = \{(m\omega^2 - 2K) + K(e^{i\xi} + e^{i\xi})\}u_0 = 0,$$
$$(8.93)$$

where we have used equations (8.89) and (8.90). From equation (8.92), we require

$$\omega^2 = \frac{K}{M}(2 - e^{i\xi} - e^{-i\xi}),$$
$$(8.94)$$

and similarly from equation (8.93)

$$\omega^2 = \frac{K}{m}(2 - 2\,e^{i\xi}).$$
$$(8.95)$$

If a single mode exists with the symmetry of equations (8.89) and (8.90), then equations (8.94) and (8.95) must be compatible; i.e. the phase ξ must be such that both give the same value of ω. Thus, if we denote

$$e^{i\xi} = x,$$
$$(8.96)$$

then from equations (8.94) and (8.95) we have

$$\frac{1}{M}\left(2 - x - \frac{1}{x}\right) = \frac{2}{m}(1 - x)$$

or

$$2(-m + M)x + (m - 2M)x^2 + m = 0,$$

whence

$$\left\{x^2 - \frac{2(M - m)x}{(2M - m)} - \frac{m}{(2M - m)}\right\} = 0.$$
$$(8.97)$$

The solutions of equation (8.97) are simply

$$x = 1 \quad \text{or} \quad x = \left[\frac{-m}{(2M - m)}\right].$$
$$(8.98)$$

We ignore the solution $x = 1$, which would simply give $u_j(0) = u_0(0)$ for all j: oscillation of the whole 'crystal' rigidly, which is the long-wave limit, with $\omega = 0$: see equation (8.95), with $e^{i\xi} = 1$.

For the symmetrical mode, we see from equations (8.96) and (8.98) that

$$e^{i\xi} = \left[\frac{-m}{(2M - m)}\right]$$

or

$$i\xi = \left\{ \ln(-1) + \ln\left[\frac{m}{(2M-m)}\right] \right\}$$

$$= \left\{ i\pi - \ln\left(\frac{2M}{m} - 1\right) \right\},$$

whence

$$\xi = \left\{ \pi + i\ln\left(\frac{2M}{m} - 1\right) \right\}. \tag{8.99}$$

We rewrite equation (8.99) as

$$e^{i\xi} = -e^{-\alpha}, \tag{8.100}$$

where we have used $\ln(1/y) = (-\ln y)$, and we have defined

$$\alpha = \ln\left(\frac{2M}{m} - 1\right). \tag{8.101}$$

From equation (8.101) we see that α is real, as is $e^{i\xi}$, equation (8.100). The real part of the phase shift ξ in equation (8.99) is π radians, or 180°, leaving successive atoms for $|j| > 0$ out of phase by this amount, oscillating in opposite directions: see equations (8.89) and (8.90). The imaginary part of the phase gives a real factor which together with $u_j(0)$ determines the amplitude of oscillation of the atoms, relative to that of the central defect atom

$$u_{\pm j}(0) = e^{-j\alpha}u_0(0). \tag{8.102}$$

We can rewrite this as

$$u_{\pm j}(0) = e^{-\alpha' x_j}u_0(0), \tag{8.103}$$

where

$$\alpha' = \frac{\alpha}{a}, \qquad x_j = ja. \tag{8.104}$$

Since α is dimensionless, α' has dimensionality (length)$^{-1}$. Denote α' as

$$\alpha' = \frac{\alpha}{a} = \frac{1}{R}, \qquad R = \left(\frac{a}{\alpha}\right). \tag{8.105}$$

Then R is the range of the exponential decay of the oscillation's amplitude in equation (8.103)

$$u_{\pm j}(0) = e^{-x_j/R}u_0(0). \tag{8.106}$$

The range R is the distance within which the wave amplitude is reduced to $e^{-1} = 0.368$ of its central value. Since this mode is exponentially localized in space, it is called a *local mode*. We remark that this exponentially damped wave has some limited similarity with the Rayleigh surface waves described in Chapter 4, where also the waves acquired an imaginary part to their phases as a result of breaking the translational invariance of the medium.

Finally, let us determine the local mode frequency, from equations (8.95), (8.100) and (8.101)

$$\omega^2 = \frac{2K}{m}(1 - e^{i\xi}),$$

whence

$$\omega = \left[\frac{2K}{m}(1 + e^{-\alpha})\right]^{1/2} = \left[\frac{2K}{m}\left(1 + \left(\frac{2M}{m} - 1\right)^{-1}\right)\right]^{1/2}$$

$$= 2\left(\frac{K}{M}\right)^{1/2}\left[\frac{M^2}{m(2M - m)}\right]^{1/2}. \tag{8.107}$$

The factor $2(K/M)^{1/2}$ in equation (8.107) represents the maximum frequency of the lattice when there is no point defect: see equations (8.34) and (8.22). The other factor is

$$\left[\frac{M^2}{m(2M - m)}\right]^{1/2}. \tag{8.108}$$

This is greater than unity for $M > m$, with m the defect mass, and M the host atom mass. To prove this, assume the contrary, and show that it leads to a contradiction. Thus the local mode frequency lies above the perfect 'crystal' spectrum, figure 8.1. A useful reference for this section, and much else, is Kittel (1953, Chapter 5, especially pages 156–158). A more general discussion of localized phonon modes is given by Maradudin (1963).

Chapter 9

Classical atomistic modelling of crystals

9.1 Introduction

In Chapter 7 we gave a quantum mechanical framework for the theory of solids. We noted, in section 7.1, the central role played by the atomic nuclei in determining the structure and properties of a solid. Furthermore, we have commented, in section 1.1, on the fact that under 'terrestrial' conditions, solids are composed of identifiable atoms. The purpose of this chapter is to conceptually bridge the gap between the atomistic description of a solid, represented in Chapter 7, and the continuum description represented in Chapter 1. To that end we shall first, in section 9.2, describe qualitatively how a *classical* atomistic model may be inferred from the quantum-mechanical formulation of Chapter 7. Also in section 9.2 we shall specify the classical atomistic shell model for insulating crystals. The remainder of this chapter will then be largely devoted to derivations from the shell model of various bulk properties that are defined in terms of the continuum model of a solid. Thus in section 9.3 we derive the cohesive energy of the crystal. In section 9.4 we use this to derive the elastic constants: see section 1.4.1, especially equation (1.48). In section 9.5 we derive the dielectric and piezoelectric constants. The content of sections 9.3–9.5 is based on the report by Harding (1982).

9.2 The shell model for insulating crystals

In Chapter 7 we introduced a form of the adiabatic approximation which we called the average field approximation: see equation (7.8). In this approximation, the energy of a solid could be written as the sum of nuclear kinetic energy T_n, equation (7.4a), and an effective potential energy W, equation (7.12); see also equation (7.11). The effective potential energy of the crystal includes the kinetic energy T_e of the electrons in the electronic part H_e of the system's hamiltonian: see equations (7.12), (7.3b) and (7.4b). It depends on the many-electron state of the crystal $\psi_\lambda(\underline{r}, \underline{R}_0)$, where $\underline{r}$ is the collection of electron coordinates, and $\underline{R_0}$ is the collection of nuclear equilibrium positions, here considered for the infinite perfect crystal. From equations (7.12), (7.5a) and (7.6), the effective potential energy has the form

$$W = W(\underline{R}, \psi_\lambda(\underline{R_0})), \tag{9.1}$$

where the r-dependence of ψ_λ is not shown because it is integrated out in equation (7.12).

In this chapter we shall consider a crystal that is in equilibrium with some external stress. In that case, the equilibrium configuration $\underline{R}$ will be different from the perfect-crystal configuration $\underline{R_0}$, and the electronic state ψ_λ will be determined by $\underline{R}$. This is distinct from the situation discussed in Chapter 7, where $\underline{R}$ is a set of dynamical variables: see equations (7.12) and (7.14). In the present case, we therefore have, in place of equation (9.1),

$$W = (\underline{R}, \psi_\lambda(\underline{R})). \tag{9.2}$$

In this effective potential, the combination of the atomic nuclei, represented by the variables $\underline{R}$ standing alone in equation (9.2), with the $\underline{R}$-dependent argument $\psi_\lambda(\underline{R})$, constitutes a potential due to a set of interacting atoms. The specific configurations of the individual atoms are determined by details of the electronic distribution defined by $\psi_\lambda(\underline{R})$. The spherically sym-metrical part of ψ_λ about a particular nucleus at $\vec{R}_J$, combined with the nuclear charge Z_J, might be simply modelled by an atomic core charge. The remainder of the electronic distribution about $\vec{R}_J$ contributes further charge to the atom, and electric multipole moments of dipole, quadrupole, and higher orders.

The preceding discussion has been presented because, although we shall not carry it further, it can form the basis of a derivation of the atomistic structure and the multipole moments of atoms in a crystal described in terms of nuclear positions $\underline{R}$ and an electronic many-body wave function ψ_λ, as in equation (9.2).

From the above, it is clear that a simple classical model of a crystal can be based on ions with charges Q_J, core charges $(Q_J - Y_J)$, and associated electric multipole moments. For highly ionic crystals, the perfect, undistorted crystal can be represented to good approximation, for several basic proper-ties, by ions with point charges Q_J. Under static or dynamic deformation the ions acquire dipole moments. This can be represented by breaking the ionic charge into two parts, a core of charge $(Q_J - Y_J)$ and a so-called *shell* charge, Y_J. If the deformation displaces the shell charge Y_J by a finite amount $\vec{u}_J$ relative to the core, then the ion in fact acquires multipole moments above dipole order as well, but for small $\vec{u}_J$, the dipole effect dominates.

The earliest atomistic model of an ionic crystal neglected ionic polarizability, and assumed that the rigid ions could be represented by point charges Q_J only. It was understood that the Coulomb attractions between positive and negative ions (cations and anions) were in fact balanced by short-range interatomic repulsive forces of quantum-mechanical origin due to the electronic structure of the ions. The balance between Coulomb attraction and short-range repulsion accounted for the specific lattice spacing of the crystal. A simple form of nearest-neighbor anion–cation repulsion, inspired by quantum-mechanical analysis, was determined by Born and Mayer (1932). The Born–Mayer potential $V_{JJ'}$ is

$$V_{JJ'}(r) = B_{JJ'} \exp(-r/\rho_{JJ'}), \qquad (9.3)$$

where r is the interionic distance. For the rocksalt crystal structure, for example, the parameters $B_{JJ'}$ and $\rho_{JJ'}$ in equation (9.3) can be determined by fitting the calculated lattice spacing and bulk modulus to the experimental values: see Born and Huang (1954, Chapter I, section 3). This model of a crystal has been extensively reviewed by Tosi (1968). The model in this form does not represent lattice dynamic or dielectric properties very well. The attempt to improve the results by including ionic polarizability in the form of free-ion polarizability [e.g. Tessman *et al.* (1953)] was never very successful. Over time, it came to be understood that ionic polarization depended not on only the local electric field in the crystal, but also on the polarizing effect of one ion pressing upon another, an aspect of the short-range repulsion.

Thereafter, the shell model as we have described it was developed, first by Dick and Overhauser (1958), in which both the shell–core coupling and the short-range interionic forces were taken to be harmonic. This was clearly satisfactory when the application of the model was limited to harmonic effects. It has been found that the best modelling is obtained if the short-range interionic forces act between *shells*. The effectiveness of the model in this form was demonstrated by the work of Brockhouse and coworkers [Woods *et al.* (1960)], who by fitting the parameters to bulk properties were able to reproduce experimental phonon dispersion relations for a variety of alkali halides with impressive accuracy. Somewhat later [see, for example, Lidiard and Norgett (1972)] the short-range interionic potentials were applied in the anharmonic Buckingham form, as follows:

$$V_{JJ'}(r) = \{B_{JJ'} \exp(-r/\rho_{JJ'}) - C_{JJ'}r^{-6}\}. \qquad (9.4)$$

In equation (9.4) we see not only the Born–Mayer repulsion of equation (9.3), but the so-called van der Waals attraction, $(C_{JJ'}r^{-6})$, which is also of quantum mechanical origin. Sometimes a further term is added to $V_{JJ'}$, of the form $(D_{JJ'}r^{-n})$ when n is in the range 8 to 14. This adds a hard core of repulsion to the otherwise finite value of the Born–Mayer term at $r = 0$. It is possible to represent the short-range interionic potential even more accurately by using a spline fit, rather than the simple analytical forms mentioned earlier, and to introduce angle-dependent three-body forces for partly covalent materials [Leslie (1981)]. Furthermore, quadrupolar deformation of the ions has been incorporated in some calculations [Jacobs *et al.*

(1980), Bilz (1985)]. By around 1980, the shell model had been applied successfully in a wide range of systems, resulting in major compilations of data to describe both phonon dispersion relations [Bilz and Kress (1979)] and point defect properties [Stoneham (1981)]. Applications of the shell model continue to proliferate in collating and predicting both the crystal properties and defect processes of an ever-growing range of systems. The current state of the art in computational modelling at this level, and in flexible, user-friendly software, is represented by the work of Gale (1997).

For the purposes of the present work, we adopt a relatively simple, but highly effective form of the shell model. The ionic total charges Q_J are taken to be given integers (units of $|e|$, where e is the electron's charge). Shell charges are denoted Y_J, so that core charges are $(Q_J - Y_J)$. The core and shell of a given ion, of species J, are harmonically coupled with a coupling constant K_J which is assumed to include the corresponding core–shell Coulomb interaction. Short-range interionic potentials $V_{JJ'}$ are taken to be of the Buckingham form, equation (9.4). Thus, for a binary solid, for example, with two ionic species $J = 1$, 2, we have in general two shell charges Y_1, Y_2; two shell-core force constants K_1, K_2; and three sets of Buckingham potential parameters: B_{11}, ρ_{11}, C_{11}; B_{22}, ρ_{22}, C_{22}; and B_{12}, ρ_{12}, C_{12}. This is a total of 13 parameters. Ideally, they are determined by fitting calculated results of the model to a corresponding number, or greater number, of experimental bulk properties. The accuracy of the fit that is obtainable is one measure of the appropriateness of the model for the particular material. The business of fitting shell-model parameters to bulk properties is somewhat of an art, best learned from an expert practitioner. The subject of computational modelling of insulators using the shell model has been discussed extensively in a book edited by Catlow and Mackrodt (1982), especially in Chapters 1 and 10.

9.3 Cohesive energy of a crystal

We begin with the total energy of a perfect infinite shell-model crystal. We allow for small displacements of cores and shells from their equilibrium positions. We consider only stationary configurations of this classical system: the static-crystal approximation. Let $\vec{r}_l$ be a Bravais lattice site for the crystal. Let basis ions of the crystal be labelled by an index n, and let τ label cores ($\tau = 1$) and shells ($\tau = 2$). Then let the position of core or shell τ of basis ion n in primitive unit cell l, measured relative to $\vec{r}_l$, be $\vec{r}_n^\tau(l)$.

There are three sets of potentials that contribute to the total potential energy: Coulomb, interionic shell–shell short range, and core–shell short-range potentials. Denote the position of a core or shell by $\vec{X}_n^\tau(l)$:

$$\vec{X}_n^\tau(l) = [\vec{r}_l + \vec{r}_n^\tau(l)]. \tag{9.5}$$

Denote the distance between two elements (cores and shells) by $X_{nn'}^{\tau\tau'}(l, l')$:

$$X_{nn'}^{\tau\tau'}(l, l') = |\vec{X}_n^{\tau}(l) - \vec{X}_{n'}^{\tau'}(l')|. \qquad (9.6a)$$

Then the total Coulomb energy E_C of the crystal is

$$E_C = \frac{1}{(4\pi\varepsilon_0)} \frac{1}{2} \sideset{}{''}\sum_{\substack{ln\tau \\ l'n'\tau'}} \frac{q_n^{\tau} q_{n'}^{\tau'}}{X_{nn'}^{\tau\tau'}(l, l')}, \qquad (9.6b)$$

in SI units, where q_n^{τ} is the charge of an element, and 'double prime' on the summation means omit the terms $(ln\tau) \equiv (l'n'\tau')$ *and* omit terms $l = l'$, $n = n'$, $\tau \neq \tau'$. The latter restriction eliminates Coulomb interaction between the core and shell of a given ion: recall that this interaction is assumed to be included in the core–shell harmonic interaction.

Let the core–shell interaction v_n for the ionic species of basis ion type n be

$$v^{(n)}(u) = \tfrac{1}{2} K_n u^2, \qquad (9.7)$$

where u is the core–shell separation distance, and K_n is the corresponding harmonic force constant. Then we can write the core–shell interactions as follows:

$$v_{nn'}^{\tau\tau'}(l, l') = \delta_{l,l'}\delta_{n,n'}, \delta_{\tau,1}\delta_{\tau',2} \tfrac{1}{2} K_n [X_{nn'}^{\tau\tau'}(l, l')]^2. \qquad (9.8)$$

The Krönecker deltas in equation (9.8) ensure that this interaction is only between the core and shell of a single ion. Similarly the short-range shell–shell interionic potentials, exemplified by equation (9.4), can be written

$$V_{nn'}^{\tau\tau'}(l, l') = \delta_{\tau,2}\delta_{\tau',2} V_{nn'}(X_{nn'}^{\tau\tau'}(l, l')). \qquad (9.9)$$

The Krönecker deltas in equation (9.9) ensure that the interaction is between shells only ($\tau = 2$, $\tau' = 2$). We can now combine the short-range intraionic potentials $v_{nn'}^{\tau\tau'}(l, l')$ and the short-range interionic potentials $V_{nn'}^{\tau\tau'}(l, l')$ as follows. Define

$$\Phi_{nn'}^{\tau\tau'}(l, l') = [v_{nn'}^{\tau\tau'}(l, l') + V_{nn'}^{\tau\tau'}(l, l')]. \qquad (9.10)$$

Then the total short-range energy of the shell-model crystal is

$$E_s = \frac{1}{2} \sideset{}{'}\sum_{ln\tau, l'n'\tau'} \Phi_{nn'}^{\tau\tau'}(l, l'), \qquad (9.11)$$

where 'prime' on the summation means omit the terms $(ln\tau) \equiv (l'n'\tau')$.

Now combining equations (9.6) and (9.11) we have the total energy E of the crystal:

$$E = \frac{1}{2} \sideset{}{'}\sum_{ln\tau, l'n'\tau'} \left[\frac{q_n^{\tau} q_{n'}^{\tau'}}{4\pi\varepsilon_0 X_{nn'}^{\tau\tau'}(l, l')} + \Phi_{nn'}^{\tau\tau'}(l, l') \right]. \qquad (9.12)$$

In equation (9.12), the 'prime' on the summation incorporates the constraints of both equations (9.6) and (9.11). Note that equation (9.12) is of the form

$$E = \sum_{l'} E_{l'}, \tag{9.13}$$

where

$$E_{l'} = \frac{1}{2} \sum_{l n \tau, n' \tau'}' \left[\frac{q_n^\tau q_{n'}^{\tau'}}{4\pi\varepsilon_0 X_{nn'}^{\tau\tau'}(l, l')} + \Phi_{nn'}^{\tau\tau'}(l, l') \right]. \tag{9.14}$$

Now consider the case of homogeneous strain of the whole infinite perfect crystal. It corresponds to having the configurations of all unit cells identical, provided the unit cells are very small compared with macroscopic dimensions. We shall assume here that all *primitive* unit cells are identical. This is the modelling equivalent of the state of a finite perfect crystal when a bulk property is being measured experimentally. In equation (9.13), $E_{l'}$ is the total energy of primitive unit cell l' interacting with the rest of the infinite crystal, plus the internal interaction energy of that unit cell. But since all unit cells are identical, the values of $E_{l'}$ for all values of l' are equal. In particular, they are equal to E_0, ($l' = 0$). Thus we have

$$E = NE_0, \tag{9.15}$$

where N is the number (infinite) of unit cells. We define the *cohesive energy* E_{coh} by

$$E_{\mathrm{coh}} = E_0, \tag{9.16}$$

where

$$E_{\mathrm{coh}} = \frac{1}{2} \sum_{l n \tau, n' \tau'}' \left[\frac{q_n^\tau q_{n'}^{\tau'}}{4\pi\varepsilon_0 X_{nn'}^{\tau\tau'}(l, 0)} + \Phi_{nn'}^{\tau\tau'}(l, 0) \right]. \tag{9.17}$$

Thus, although the total energy is infinite, the cohesive energy is not.

9.4   Elastic constants

We refer to Chapter 1, where the concepts of stress, strain and elastic constants are discussed in detail. Under a small strain (harmonic approximation, linear elasticity theory) we have the following relationship, equation (1.52),

$$W = \frac{1}{2} \sum_{n,m} c_{nm} \varepsilon_m \varepsilon_n. \tag{9.18}$$

In equation (9.18), W is the energy per unit volume, and ε_n and c_{nm} are strain and elastic constants respectively, in Voigt notation: see equations (1.44) and (1.48). Now the cohesive energy E_{coh}, equation (9.17), is the energy per primitive unit cell. Thus if V is the volume of the primitive unit cell, we have

$$W = \frac{E_{\text{coh}}}{V}, \tag{9.19}$$

where W is given by equation (9.18). The elastic constants c_{nm}, equation (9.18) are determined from

$$c_{nm} = \frac{\partial^2 W}{\partial \varepsilon_n \, \partial \varepsilon_m} = \frac{1}{V} \frac{\partial^2 E_{\text{coh}}}{\partial \varepsilon_n \, \partial \varepsilon_m}. \tag{9.20}$$

Our task now is to relate the macroscopic strain ε_n to the core and shell displacements which are the arguments of E_{coh}: see equation (9.17).

Referring again to Chapter 1, we recall that, in the continuum picture of the solid, a small segment of material spanned by the vector $\mathrm{d}x_j$ ($j = 1, 2, 3$) is deformed under small strain ε_{ij}, in the absence of rotation, into a segment spanned by $\mathrm{d}x_i'$ ($i = 1, 2, 3$) according to the relation given in equation (1.8):

$$\mathrm{d}x_i' = (\mathrm{d}x_i + \varepsilon_{ij} \, \mathrm{d}x_j), \tag{9.21}$$

where ε_{ij} is the strain tensor, with the Einstein summation convention in equation (9.21). In terms of the matrix notation of equation (1.8), this is

$$\underline{\mathrm{d}r'} = (\underline{I} + \underline{\varepsilon}) \cdot \underline{\mathrm{d}r}, \tag{9.22}$$

where $\underline{I}$ is the identity matrix with elements $I_{ij} = \delta_{ij}$. For a homogeneous strain, where $\underline{\varepsilon}$ is independent of position in the solid, equation (9.22) integrates trivially to

$$\underline{r'} = (\underline{I} + \underline{\varepsilon}) \cdot \underline{r} \tag{9.23}$$

To help us remember that the strain is small, we replace $\underline{\varepsilon}$ by $\underline{\underline{\Delta\varepsilon}}$, so that equation (9.23) becomes

$$\underline{r'} = (\underline{I} + \underline{\underline{\Delta\varepsilon}}) \cdot \underline{r}. \tag{9.24}$$

We begin to apply the homogeneous strain $\underline{\underline{\Delta\varepsilon}}$ to the shell model crystal by applying it, as in equation (9.24) to positions of all elements (cores and shells); i.e. we replace $\underline{r}$ in equation (9.24) by $\underline{X}_n^{\tau}(l)$: see equation (9.5). Having done so, we note that the result will not be completely correct. The reason is that applying equation (9.24) to the Bravais lattice positions $\underline{r}_l$ will be correct, but applying it to shells and cores at $\underline{r}_n^{\tau}(l)$ within the unit cell will not, in general, leave them in equilibrium. There must be a further set of displacements $\underline{\delta r}_n^{\tau}(l)$, the same for all l as for $l = 0$, which will represent the deformation of the unit cell by the strain $\underline{\underline{\Delta\varepsilon}}$. We further simplify the notation as follows. Let

$$\underline{X}_{n\tau l} \equiv \underline{X}_n^{\tau}(l); \tag{9.25}$$

see equation (9.5). Then we have, for the effect of strain,

$$\underline{X}'_{n\tau l} = [\underline{X}_{n\tau l} + \underline{\underline{\Delta\varepsilon}} \cdot \underline{X}_{n\tau l} + \underline{\delta r}^\tau_n(0)]. \tag{9.26}$$

The equilibrium condition for $\underline{\delta r}^\tau_n(0)$ will become clear later. We emphasize that the matrix notation beginning with equation (9.22) and ending with equation (9.26) deals with matrices of dimension three, corresponding to the three cartesian components of vectors and tensors in 3-space.

The cohesive energy, E_{coh}, equation (9.17), now becomes a function of $\underline{\underline{\Delta\varepsilon}}$ and $\underline{\delta r}^\tau_n(0)$ under strain: see equations (9.6)–(9.10), with $\vec{X}^\tau_n(l) \equiv \vec{X}_{n\tau l}$ replaced by $\vec{X}'_{n\tau l}$, equation (9.26). We now revert to the Voigt notation of equation (1.44), used in equation (9.20). We introduce the column matrix $\underline{\Delta\varepsilon}$, having six elements. We also introduce the column matrix $\underline{\delta r}$, constructed from the elements of $\delta\vec{r}^\tau_n(0)$, which has [2 × (number of ions in the basis) × 3] elements: 2 for $\tau = 1, 2$; 3 for the cartesian components. We then introduce a deformation vector (column matrix) $\underline{\delta}$, as follows:

$$\underline{\delta} = \left(\frac{\Delta\varepsilon}{\delta r}\right). \tag{9.27}$$

In linear elasticity, all the strains are assumed to be small, so all of the elements of $\underline{\delta}$ must be small. Thus, in harmonic approximation, we can write E_{coh} as

$$E_{coh}(\underline{\delta}) \approx \{E_{coh}(\underline{0}) + \underline{g}^T \cdot \underline{\delta} + \tfrac{1}{2}\underline{\delta}^T \cdot \underline{\underline{W}} \cdot \underline{\delta}\}, \tag{9.28}$$

where $\underline{g}^T$ is the transpose of the column vector $\underline{g}$, whose elements are first derivatives of $E_{coh}(\underline{\delta})$ with respect to elements of $\underline{\delta}$. Similarly, $\underline{\underline{W}}$ in equation (9.28) is a square matrix of the same dimensionality as $\underline{\delta}$, whose elements are second derivatives of E_{coh} with respect to elements of $\underline{\delta}$. The derivatives in $\underline{g}$ and in $\underline{\underline{W}}$ are evaluated at zero strain $\underline{\delta} = 0$, as follows:

$$g_m = \left(\frac{\partial E_{coh}}{\partial \delta_m}\right)_{\underline{\delta}=0}, \tag{9.29}$$

$$W_{mn} = \left(\frac{\partial^2 E_{coh}}{\partial \delta_m \, \partial \delta_n}\right)_{\underline{\delta}=0}. \tag{9.30}$$

For the equilibrium configuration of the crystal in the absence of strain, we require

$$\left(\frac{\partial E_{coh}}{\partial \delta_m}\right)_{\underline{\delta}=0} = 0 = g_m. \tag{9.31}$$

This determines the equilibrium configuration $\underline{X}_{n\tau l}$, equation (9.26). The cohesive energy, equation (9.28) then becomes

$$E_{coh}(\underline{\delta}) \approx \{E_{coh}(\underline{0}) + \tfrac{1}{2}\underline{\delta}^T \cdot \underline{\underline{W}} \cdot \underline{\delta}\}. \tag{9.32}$$

We now address the question of the relationship between the bulk homogeneous strain $\underline{\Delta\varepsilon}$ and the internal strain $\underline{\delta r}$ in the unit cell: see equation (9.27). To this end, let us write all matrices in terms of submatrices spanned by the elements of $\underline{\Delta\varepsilon}$ and of $\underline{\delta r}$, as in equation (9.27) for $\underline{\delta}$. Then, in equation (9.32), $\underline{\underline{W}}$ takes the form

$$\underline{\underline{W}} = \begin{pmatrix} \underline{\underline{W}}_{\varepsilon\varepsilon} & \underline{\underline{W}}_{\varepsilon r} \\ \underline{\underline{W}}_{r\varepsilon} & \underline{\underline{W}}_{rr} \end{pmatrix}. \tag{9.33}$$

In equation (9.33), the subscripts ε and r indicate dimensionalities of $\underline{\Delta\varepsilon}$ and of $\underline{\delta r}$ respectively. Note that $\underline{\underline{W}}_{r\varepsilon} = \underline{\underline{W}}_{\varepsilon r}^{\mathrm{T}}$: see equation (9.30). Thus, combining equations (9.28), (9.27) and (9.33), we have

$$E_{\mathrm{coh}}(\underline{\Delta\varepsilon}, \underline{\delta r}) = \left\{ E_{\mathrm{coh}}(0) + \frac{1}{2}\left[(\underline{\Delta\varepsilon}^{\mathrm{T}}, \underline{\delta r}^{\mathrm{T}}) \cdot \begin{pmatrix} \underline{\underline{W}}_{\varepsilon\varepsilon} & \underline{\underline{W}}_{\varepsilon r} \\ \underline{\underline{W}}_{r\varepsilon} & \underline{\underline{W}}_{rr} \end{pmatrix} \cdot \begin{pmatrix} \underline{\Delta\varepsilon} \\ \underline{\delta r} \end{pmatrix} \right] \right\}$$

$$= E_{\mathrm{coh}}(0) + \tfrac{1}{2}\{ \underline{\Delta\varepsilon}^{\mathrm{T}} \cdot \underline{\underline{W}}_{\varepsilon\varepsilon} \cdot \underline{\Delta\varepsilon} + \underline{\Delta\varepsilon}^{\mathrm{T}} \cdot \underline{\underline{W}}_{\varepsilon r} \cdot \underline{\delta r}.$$

$$+ \underline{\delta r}^{\mathrm{T}} \cdot \underline{\underline{W}}_{r\varepsilon} \cdot \underline{\delta\varepsilon} + \underline{\delta r}^{\mathrm{T}} \cdot \underline{\underline{W}}_{rr} \cdot \underline{\delta r} \}. \tag{9.34}$$

The equilibrium configuration within the unit cell, for given bulk strain $\underline{\Delta\varepsilon}$, is given by

$$\left[\frac{\partial E_{\mathrm{coh}}}{\partial(\underline{\delta r})} \right]_\varepsilon = 0. \tag{9.35}$$

From equation (9.34), this becomes

$$(\underline{\Delta\varepsilon}^{\mathrm{T}} \cdot \underline{\underline{W}}_{\varepsilon r} + \underline{\delta r}^{\mathrm{T}} \cdot \underline{\underline{W}}_{rr}) = 0. \tag{9.36}$$

From equation (9.36) we can solve for $\underline{\delta r}$ as a function of $\underline{\Delta\varepsilon}$:

$$\underline{\delta r} = (-\underline{\underline{W}}_{rr}^{-1} \cdot \underline{\underline{W}}_{r\varepsilon} \cdot \underline{\Delta\varepsilon}). \tag{9.37}$$

This is the deformation generated in the unit cell by the strain $\underline{\Delta\varepsilon}$.

Finally, we can obtain the internal energy density W in terms of Voigt strains $\underline{\Delta\varepsilon}$ in the form of equation (9.18), by substituting from equation (9.37) for $\underline{\delta r}$ into equation (9.34):

$$E_{\mathrm{coh}}(\underline{\Delta\varepsilon}) = \{ E_{\mathrm{coh}}(0) + \tfrac{1}{2}\underline{\Delta\varepsilon}^{\mathrm{T}} \cdot [\underline{\underline{W}}_{\varepsilon\varepsilon} - (\underline{\underline{W}}_{\varepsilon r} \cdot \underline{\underline{W}}_{rr}^{-1} \cdot \underline{\underline{W}}_{r\varepsilon})] \cdot \underline{\Delta\varepsilon} \}. \tag{9.38}$$

We return to equation (9.20) to obtain the explicit formula for the bulk elastic constants (in Voigt notation) in terms of the shell model parameters:

$$c_{nm} = \frac{1}{V} \frac{\partial^2(E_{\mathrm{coh}})}{\partial(\Delta\varepsilon_n)\partial(\Delta\varepsilon_m)}$$

$$= \frac{1}{V} \{ \underline{\underline{W}}_{\varepsilon\varepsilon} - (\underline{\underline{W}}_{\varepsilon r} \cdot \underline{\underline{W}}_{rr}^{-1} \cdot \underline{\underline{W}}_{r\varepsilon}) \}_{nm}, \tag{9.39}$$

where in equation (9.39) the subscripts *nm* refer to the matrix element of the matrix in { } brackets. The explicit dependence on the shell-model parameters can be traced back through equations (9.33), (9.30), (9.27), (9.26), (9.25), (9.17), (9.10) and (9.8). Thus, for a given shell model we can calculate the elastic constants.

9.5   Dielectric and piezoelectric constants

We now consider the role of static, uniform electric fields in relation to bulk properties of a crystal. Suppose that the crystal is subjected to a uniform electric field of external origin, $\vec{E}_{\text{ext}}$, for example by being placed between the plates of a capacitor. Then

$$\vec{E}_{\text{ext}} = \frac{1}{\varepsilon_0}\vec{D} \qquad (9.40)$$

where the electric displacement $\vec{D}$ is determined from the free charge density ρ on the capacitor plates from Gauss's law,

$$\vec{\nabla}\cdot\vec{D} = \rho. \qquad (9.41)$$

We use SI units, and in equation (9.40) ε_0 is the *permittivity of free space*. In terms of the shell model, the resultant electric field inside the dielectric will act on all shells and cores of the crystal, through their charges q_n^T, displacing them from their field-free positions, and producing a polarizing effect and a deformation of the crystal. For an isotropic dielectric continuum the polarization is described in terms of a *polarization* vector $\vec{P}$, representing the electric dipole moment per unit volume. In a linear dielectric, $\vec{P}$ is proportional to the *internal* electric field which we denote $\vec{E}$. Then

$$\vec{P} = \chi\vec{E} \qquad (9.42)$$

where χ is the *electric susceptibility*. The internal electric field arises from the combination of external free charge and polarization of the medium. Specifically, if the dielectric has no free charge, then in the dielectric,

$$\varepsilon_0\vec{E} = (\vec{D} - \vec{P}) \qquad (9.43)$$

where $\vec{D}$ is still given by equation (9.40). We can rewrite equation (9.43), with equation (9.42),

$$\vec{D} = (\varepsilon_0\vec{E} + \vec{P}) = (\varepsilon_0 + \chi)\vec{E} = K\varepsilon_0\vec{E}, \qquad (9.44)$$

where we have introduced the *dielectric constant K*,

$$K = \left(1 + \frac{\chi}{\varepsilon_0}\right). \qquad (9.45)$$

A good reference for the preceding discussion is Reitz *et al.* (1979).

Actually, equations (9.42)–(9.45) represent an approximation in which the polarizing effect occurs without bulk strain. There are two kinds of strain effects observed for dielectric media. One is *electrostriction*, in which the material expands under an applied field $\vec{E}_{\text{ext}}$, equation (9.40), regardless of the field's direction: the strain is *quadratic* in the field. The other strain-related effect is *piezoelectric*. The piezoelectric effect occurs only for crystals that do not have a center of symmetry: see for example Nye (1957), section 4.1. For such crystals the relationship between strain and field is linear so that, in a simple geometry, the field in one direction will compress the crystal, and in the opposite direction it will expand it. The so-called *direct piezoelectric effect* involves electric polarization induced by an applied strain. In the so-called *converse piezoelectric effect*, an applied electric field induces a linear strain.

The results of the preceding section on elastic constants are evidently valid for non-piezoelectric crystals in the absence of external electric fields. Let us now consider a piezoelectric material subject to a uniform external field $\vec{E}_{\text{ext}}$, equation (9.40), and to specified external stresses. As we have indicated, the resultant internal electric field, along with the applied stress, will induce shell-model core and shell displacements. The cohesive energy of the crystal, equation (9.34), will now be modified by the interaction energy of shell–model point charges q_n^{τ} with the internal electric field $\vec{E}$. We consider only the case of weak fields, where the piezoelectric effect dominates over electrostriction. Core and shell displacements $\delta\vec{r}_n^{\tau}(0)$, equation (9.26), due to the weak field and applied stress will also be small. In that case, their electric dipole moment $[q_n^{\tau}\delta\vec{r}_n^{\tau}(0)]$ is the dominant multipole moment, apart from their charge, which is included in E_c, equation (9.6). The interaction energy of this dipole moment with the electric field is

$$\left\{ -\sum_{n,\tau} q_n^{\tau}\delta\vec{r}_n^{\tau}(0)\cdot\vec{E} \right\}. \tag{9.46}$$

We wish to express the quantity in equation (9.46) in terms of the column matrix $\underline{\delta r}$, introduced prior to equation (9.27). Expression (9.46) is clearly the inner product of the vector $\underline{\delta r}$ with another vector whose elements are also determined by n, τ and α, where α labels cartesian components, in this case components of $\vec{E}$, and where n, τ comes from the charge q_n^{τ}. We denote this vector $(\underline{qE})$. Then

$$-\sum_{n,\tau} q_n^{\tau}\delta\vec{r}_n^{\tau}(0)\cdot\vec{E} = -(\underline{qE})^{\text{T}}\cdot\underline{\delta r} \tag{9.47}$$

We add this to the cohesive energy in harmonic approximation, equation (9.28),

$$E_{\text{coh}}(\underline{\delta},\vec{E}) = \{E_{\text{coh}}(0) + \underline{g}^{\text{T}}\cdot\underline{\delta} + \tfrac{1}{2}\underline{\delta}^{\text{T}}\cdot\underline{\underline{W}}\cdot\underline{\delta} - (\underline{qE})^{\text{T}}\cdot\underline{\delta r}\}. \tag{9.48}$$

We note that if the expansion of equation (9.48) is relative to the perfect undistorted crystal configuration as in equations (9.28)–(9.30), then in equation (9.48), $E_{coh}(0)$ refers to $\vec{E} = 0$, and the equilibrium condition is

$$\left[\frac{\partial E_{coh}(\underline{\delta}, \vec{E})}{\partial(\underline{\delta})}\right]_{\underline{\delta}=0, \vec{E}=0} = 0 = \underline{g}. \tag{9.49}$$

The term in $\underline{g}$ is therefore zero in E_{coh}, equation (9.48). We can then proceed to the expression analogous to equation (9.34) in terms of the sub-matrices of $\underline{\underline{W}}$ and $\underline{\delta}$, equations (9.27) and (9.33) respectively,

$$E_{coh}(\underline{\delta}, \vec{E}) = \{E_{coh}(0) + \tfrac{1}{2}[\Delta\varepsilon^{\mathrm{T}} \cdot \underline{\underline{W}}_{\varepsilon\varepsilon} \cdot \Delta\varepsilon$$
$$+ 2\underline{\delta r}^{\mathrm{T}} \cdot \underline{\underline{W}}_{r\varepsilon} \cdot \Delta\varepsilon + \underline{\delta r}^{\mathrm{T}} \cdot \underline{\underline{W}}_{rr} \cdot \underline{\delta r}] - (\underline{qE})^{\mathrm{T}} \cdot \underline{\delta r}\}. \tag{9.50}$$

In equation (9.50) we have introduced $(\Delta\varepsilon)$ for the net strain due to applied stress and electric field.

Analogous to equation (9.35), we now determine the core and shell displacements in terms of the induced strain $\Delta\varepsilon$ and the electric field $\vec{E}$. We require, from equation (9.50), equilibrium condition

$$\left(\frac{\partial E_{coh}}{\partial(\underline{\delta r})}\right)_{\Delta\varepsilon, \vec{E}} = [\underline{\underline{W}}_{r\varepsilon} \cdot \Delta\varepsilon + \underline{\underline{W}}_{rr} \cdot \underline{\delta r} - (\underline{qE})] = 0. \tag{9.51}$$

The solution of equation (9.51) for $\underline{\delta r}$ is

$$\underline{\delta r} = -\underline{\underline{W}}_{rr}^{-1} \cdot [\underline{\underline{W}}_{r\varepsilon} \cdot \Delta\varepsilon - (\underline{qE})]. \tag{9.52}$$

We now express the electric displacement $\vec{D}$, equation (9.43), in terms of the shell model. From that we shall see that the dielectric and piezoelectric constants can be determined. First we require the polarization:

$$\vec{P} = \frac{1}{V}\sum_{n,\tau} q_n^\tau \delta\vec{r}_n^\tau(0). \tag{9.53}$$

Then, from equation (9.43) with equation (9.53),

$$\vec{D} = \left\{\varepsilon_0 \vec{E} + \frac{1}{V}\sum_{n,\tau} q_n^\tau \delta\vec{r}_n^\tau(0)\right\}. \tag{9.54}$$

We wish to introduce $\underline{\delta r}$ into equation (9.54) from equation (9.52) in order to have $\vec{D}$ as a function of $\vec{E}$ and $\Delta\varepsilon$. For this purpose, consider equation (9.54) in cartesian component form:

$$D^\alpha = \left\{\varepsilon_0 E^\alpha + \frac{1}{V}\sum_{n,\tau} q_n^\tau \delta r_n^{\tau,\alpha}(0)\right\}, \qquad \alpha = 1, 2, 3. \tag{9.55}$$

Let us write matrices $\underline{\delta r}$, $\underline{\underline{W}}_{rr}$ and $\underline{\underline{W}}_{r\varepsilon}$ in equation (9.52) in terms of cartesian component submatrices as follows:

$$\underline{\delta r} = \begin{pmatrix} \underline{\delta r}^1 \\ \underline{\delta r}^2 \\ \underline{\delta r}^3 \end{pmatrix}, \qquad \underline{\underline{W}}_{rr} = \begin{pmatrix} \underline{\underline{W}}_{rr}^{11} & \underline{\underline{W}}_{rr}^{12} & \underline{\underline{W}}_{rr}^{13} \\ \underline{\underline{W}}_{rr}^{21} & \underline{\underline{W}}_{rr}^{22} & \underline{\underline{W}}_{rr}^{23} \\ \underline{\underline{W}}_{rr}^{31} & \underline{\underline{W}}_{rr}^{32} & \underline{\underline{W}}_{rr}^{33} \end{pmatrix}, \qquad \underline{\underline{W}}_{r\varepsilon} = \begin{pmatrix} \underline{\underline{W}}_{r\varepsilon}^1 \\ \underline{\underline{W}}_{r\varepsilon}^2 \\ \underline{\underline{W}}_{r\varepsilon}^3 \end{pmatrix},$$

$$(9.56)$$

where superscripts 1, 2, 3 label the corresponding cartesian component, denoted by α in equation (9.55). With this notation, matrices have dimensionality determined by the union of the dimensionalities of n and τ, where the dimensionality of τ is two (cores and shells), and the dimensionality of n is the number of atoms in the crystal basis. The last term in equation (9.55) is now

$$\frac{1}{V}(\underline{q}^{\mathrm{T}} \cdot \underline{\delta r}^\alpha). \tag{9.57}$$

When we substitute for $\underline{\delta r}^\alpha$ from equation (9.52) into equation (9.55) with equation (9.57) we obtain

$$D^\alpha = \left\{ \varepsilon_0 E^\alpha - \frac{1}{V} \underline{q}^{\mathrm{T}} \cdot (\underline{\underline{W}}_{rr}^{-1})^{\alpha\beta} \cdot [\underline{\underline{W}}_{r\varepsilon}^\beta \cdot \underline{\Delta\varepsilon} - \underline{q}E^\beta] \right\}. \tag{9.58}$$

In equation (9.58), the Einstein summation convention applies to the repeated index β. We rewrite equation (9.58) as

$$D^\alpha = \left\{ \left[\varepsilon_0 \delta^{\alpha\beta} + \frac{1}{V} \underline{q}^{\mathrm{T}} \cdot (\underline{\underline{W}}_{rr}^{-1})^{\alpha\beta} \cdot \underline{q} \right] E^\beta - \frac{1}{V} \underline{q}^{\mathrm{T}} \cdot (\underline{\underline{W}}_{rr}^{-1})^{\alpha\beta} \cdot \underline{\underline{W}}_{r\varepsilon}^\beta \cdot \underline{\Delta\varepsilon} \right\}. \tag{9.59}$$

Referring to equations (9.44) and (9.59), we now see that the dielectric constant K, and the electric susceptibility χ, are second rank tensors, in general. From equation (9.59) we see

$$K^{\alpha\beta} = \frac{1}{\varepsilon_0} \left(\frac{\partial D^\alpha}{\partial E_\beta} \right)_{\Delta\varepsilon} = \left[\delta^{\alpha\beta} + \frac{1}{\varepsilon_0 V} \underline{q}^{\mathrm{T}} \cdot (\underline{\underline{W}}_{rr}^{-1})^{\alpha\beta} \cdot \underline{q} \right]. \tag{9.60}$$

Explicitly, this is the static dielectric constant for constant strain, expressed in terms of the shell model of the crystal.

A form of direct piezoelectric constant is also derived from equation (9.59):

$$\underline{\lambda}^\alpha = \left(\frac{\partial D^\alpha}{\partial(\Delta\varepsilon)} \right)_{\bar{E}} = \left[-\frac{1}{V} \underline{q}^{\mathrm{T}} \cdot (\underline{\underline{W}}_{rr}^{-1})^{\alpha\beta} \cdot \underline{\underline{W}}_{r\varepsilon}^\beta \right]. \tag{9.61}$$

This is a direct piezoelectric constant at constant field, expressed in terms of the shell model. It is not the conventional one, which is given by

$$\underline{d}^\alpha = \left(\frac{\partial D^\alpha}{\partial(\Delta\sigma)} \right)_{\bar{E}}, \tag{9.62}$$

where $\underline{\Delta\sigma}$ is the *stress* in Voigt notation [see Nye (1957), Chapter X, section 3]. In equation (9.61) we have defined a piezoelectric constant in terms of *strain* $\underline{\Delta\varepsilon}$. We can obtain $\underline{d}^\alpha$ from equation (9.61) by using the elastic constants:

$$\underline{\Delta\sigma} = \underline{c} \cdot \underline{\Delta\varepsilon}; \qquad (9.63)$$

see equation (1.48). Then, from equations (9.61) and (9.63),

$$\underline{\lambda}^\alpha = \left(\frac{\partial(\underline{\Delta\sigma})}{\partial(\underline{\Delta\varepsilon})}\right)_{\vec{E}}^{T} \cdot \left(\frac{\partial D^\alpha}{\partial(\underline{\Delta\sigma})}\right)_{\vec{E}} = \underline{c} \cdot \left(\frac{\partial D^\alpha}{\partial(\underline{\Delta\sigma})}\right)_{\vec{E}}, \qquad (9.64)$$

where the elastic constants must be determined at the given fixed value of $\vec{E}$. Equation (9.64) with equation (9.62) gives

$$\underline{d}^\alpha = \underline{c}^{-1} \cdot \underline{\lambda}^\alpha. \qquad (9.65)$$

Referring to equations (9.61) and (9.62), if we express each element of the Voigt strain $(\underline{\Delta\varepsilon})$ or stress $(\underline{\Delta\sigma})$ in terms of the corresponding elements of the cartesian tensor strain $(\underline{\Delta\varepsilon})$ and stress $(\underline{\Delta\sigma})$ then, for example in equation (9.61), we have

$$\lambda^{\alpha\beta\gamma} = \left(\frac{\partial D^\alpha}{\partial(\Delta\varepsilon_{\beta\gamma})}\right)_{\vec{E}} = \left[-\frac{1}{V}q^{T} \cdot (\underline{W}_{rr}^{-1})^{\alpha\beta'} \cdot \underline{W}_{r\varepsilon}^{\beta'}\right]_{\beta\gamma}, \qquad (9.66)$$

which displays explicitly the piezoelectric constant as a third rank cartesian tensor. Furthermore, we can consider the case $\vec{E} = 0$. Then, from equation (9.43),

$$\vec{D} = \vec{P} \qquad (\vec{E} = 0). \qquad (9.67)$$

In that case, equations (9.61) and (9.62) become

$$\lambda^\alpha = \left(\frac{\partial P^\alpha}{\partial(\underline{\Delta\varepsilon})}\right)_{\vec{E}=0}, \qquad d^\alpha = \left(\frac{\partial P^\alpha}{\partial(\underline{\Delta\sigma})}\right)_{\vec{E}=0}. \qquad (9.68)$$

The latter of equations (9.68) gives us the formula that we intuitively associate with the direct piezoelectric effect: the rate of change of polarization with respect to applied stress at zero electric field.

To summarize, the main purpose of this chapter is to show explicitly how to relate experimental bulk measurements such as elastic, dielectric and piezoelectric constants for a crystal to the parameters of a shell model for the crystal. By combining formulae such as equations (9.39), (9.60) and (9.61) with equations (9.33), (9.30), (9.10), (9.9) and (9.4), we obtain explicit relationships. Such relationships are used in computer programs that fit or calculate physical properties with shell models.

Chapter 10

Classical atomic diffusion in solids

10.1 Introduction

In the previous chapter we gave considerable detail about how an atomistic model of a crystal can be developed and used to analyse the bulk properties. We spoke only about perfect crystals, with periodically repeated unit cells. A large fraction of the subject of the materials science of solids is based upon imperfect crystals that contain chemical impurities or other point defects. The next chapter is devoted to a description of selected theoretical calculations for point defects in insulators. Section 11.2 presents some results for classical diffusion, and describes two fundamental defect mechanisms: vacancy diffusion and interstitial diffusion. The reader is referred to that discussion at this time. The movement of ions, atoms and molecules through a crystal is responsible for a host of properties. We mention only a very few examples at this time. A more wide-ranging discussion can be found in works devoted to solid state diffusion, such as that by Borg and Dienes (1988).

Charge transport in ionic insulators is usually dominated by ionic conduction, in which host ions of the material move under an applied electric field by vacancy or interstitial diffusion. The optical properties of ionic crystals can often be altered by additive coloration, in which the surface of the crystal is exposed to a gas of the crystal's cation species. The excess of positive ions diffuses into the crystal, creating anion vacancies that establish overall electrical neutrality in the crystal by trapping excess electrons, forming F centers (see sections 11.2.2, 11.3, 11.6 and 11.8). The F centers have optical characteristics different from those of the host crystal. The formation or dissociation of point defect complexes (sections 11.2.2 and 11.3), and aggregation of impurities at dislocations or grain boundaries involve atomic diffusion in solids. Solid state chemical reactions involving impurities or interdiffusion are controlled by diffusion kinetics.

We shall see that equilibrium concentrations of point defects are determined by the temperature of the crystal, as are diffusion rates. Atomic diffusion occurs by the conversion of phonon energy into the activation energy for atomic site-to-site displacement. The distribution of phonon energy, of course, depends on the temperature, as discussed in section 7.6.

Not all solid state diffusion can be described classically. The subject of quantum diffusion is introduced briefly in section 11.9. It represents the case where the essentially quantum-mechanical nature of diffusing atoms cannot be ignored. In terms of the particle–wave dualism of the quantum mechanics of particles and fields, the particle-like characteristics dominate in classical diffusion and the wave-like characteristics dominate, or are at least not negligible, in quantum diffusion.

In section 10.2 we derive the diffusion equation and apply it to the case of a slab source of the diffusing species in the material. In section 10.3 the latter problem is discussed, not on the basis of the diffusion equation, but in terms of atomic diffusion as a random walk process. In section 10.4 the principles of statistical thermodynamics are applied in discussing the equilibrium concentration of point defects in a solid as a function of temperature. In section 10.5, the Vineyard relation is derived, to demonstrate the temperature dependence of diffusion.

Throughout this chapter, our presentation closely follows that of Borg and Dienes (1988).

10.2 The diffusion equation

10.2.1 Derivation

Consider atomic point defects of a given species diffusing in one dimension, by successive jumps between atomic planes in an infinite solid, where the planes are perpendicular to the direction of diffusion. Assume that the temperature is constant and uniform throughout the material. Consider two planes at x and $(x + \Delta x)$, where Δx is the interplanar spacing, and x is the direction of diffusion. Let $N(x)$ be the density of diffusing point defects on a plane at x. The units of $N(x)$ are defects per square meter: m^{-2}. $N(x)$ is then the planar concentration. We introduce the concept of jump frequency Γ: the average number of plane-to-plane jumps executed per unit time by a diffusing defect: units: s^{-1}. The jump frequency, or jump rate, Γ, is determined by the specific nature of the defect and of the crystal, as well as by the temperature, as discussed in section 10.5. For simplicity, let us consider only jumps between nearest-neighbor sites in the crystal, so that the jump distance is Δx.

We now consider the defect current density $J(x)$ at any instant of time in the crystal at x due to planar concentration $N(x)$ and jump frequency Γ. It has two possible directions in one dimension, relating to the fact that (except at the surface) an atom may jump in the forward or backward direction. The current density at any instant of time, for instantaneous jumps, at

$(x + (\Delta x/2))$, halfway between atomic planes at x and $(x + \Delta x)$ is, on average,

$$J\left(x + \frac{\Delta x}{2}\right) = [J_+(x) - J_-(x + \Delta x)] \qquad (10.1)$$

where $J_\pm(x)$ is the magnitude of current density originating from a plane at x, in the positive or negative x-direction respectively. Now,

$$J_\pm(x) = \tfrac{1}{2}\Gamma \cdot N(x), \qquad (10.2)$$

from the definitions of jump rate Γ and planar concentration $N(x)$ given above, where the factor $\frac{1}{2}$ takes account of the fact that on average, a forward jump and a backward jump are equally probable for a given atom. From equations (10.1) and (10.2), we have

$$J\left(+\frac{\Delta x}{2}\right) = -\tfrac{1}{2}\Gamma[N(x + \Delta x) - N(x)]. \qquad (10.3)$$

We now introduce the volume concentration, henceforth referred to simply as the *concentration*, $C(x)$:

$$C(x) = \frac{N(x)}{(\Delta x)}. \qquad (10.4)$$

Combining equations (10.3) and (10.4),

$$J\left(x + \frac{\Delta x}{2}\right) = -\tfrac{1}{2}\Gamma(\Delta x)^2 \frac{[C(x + \Delta x) - C(x)]}{\Delta x}. \qquad (10.5)$$

Up to now, we have referred to Δx as the interplanar spacing, which is small compared with macroscopic scale, but non-zero. We now consider the case where $J(x)$ and $C(x)$ are determined by macroscopic measurements, so that x may be viewed approximately as a continuous variable. We replace the factor $(\Delta x)^2$ in equation (10.5) by a^2, where a is the interplanar spacing of the crystal. Then

$$J\left(x + \frac{\Delta x}{2}\right) = -(\tfrac{1}{2}\Gamma a^2) \frac{[C(x + \Delta x) - C(x)]}{\Delta x}. \qquad (10.6)$$

Taking the limit $(\Delta x) \rightarrow$ on both sides of equation (10.6), we have

$$J(x) = -D \frac{\partial}{\partial x} C(x), \qquad (10.7)$$

where we have introduced the partial derivative because in general C will be a function of time as well as of x, as we shall see, and we have introduced the *diffusion constant* D, given in the present example by

$$D = \tfrac{1}{2}\Gamma a^2. \qquad (10.8)$$

If the diffusing species is highly dilute, so that each atom can jump in any of the three cartesian coordinate directions on average, with equal probability if the solid is isotropic, then we should have

$$D = \tfrac{1}{6}\Gamma a^2. \qquad (10.9)$$

With arbitrary boundary conditions that may not ensure uniform diffusion in the x direction, we should have the following generalization of equation (10.7) for an isotropic solid or cubic crystal:

$$\vec{J}(\vec{r}, t) = -D\vec{\nabla}C(\vec{r}, t), \qquad (10.10)$$

where we have now introduced the time explicitly into equation (10.10). For an anisotropic medium, equation (10.10) generalizes further:

$$J_i(\vec{r}, t) = -D_{ij}\partial_j C(\vec{r}, t), \qquad (10.11)$$

with the same notation as in earlier chapters for cartesian tensors, with Einstein summation convention. Thus in general the diffusion constant D_{ij} will be a second-rank tensor. Equations (10.7), (10.10) and (10.11) are forms of so-called *Fick's first law*.

In some cases, the number of point defects of a given type may not remain constant within a system. For example, a diffusion step may take an interstitial atom or ion of a given species into a vacancy of the same species. In that case the interstitial and vacancy are mutually annihilated. Another example is when electron transfer between a monovalent ion and a trivalent ion leaves two divalent ions. We exclude such situations from the present discussion, so that the number of defects, all of the same type, is constant. This constraint is in the form of a conservation rule. It means that within an arbitrary volume V of the material, the rate of change of the number of defects within V is equal to the rate at which defects enter the volume V by diffusion through the surface $S(V)$ bounding V. In the notation of the present section this can be written as

$$\frac{\mathrm{d}}{\mathrm{d}t}\int_V \mathrm{d}V\, C(\vec{r}, t) = -\int_{S(V)} \mathrm{d}S\,\hat{n}\cdot\vec{J}(\vec{r}, t), \qquad (10.12)$$

where $\hat{n}$ is a unit normal outward vector on $S(V)$. On the left-hand side of equation (10.12), $\vec{r}$ is a variable of integration throughout the fixed but arbitrary volume V in the material. Equation (10.12) may therefore be rewritten as

$$\left\{\int_V \mathrm{d}V\,\frac{\partial C}{\partial t} + \int_S \mathrm{d}S\,\hat{n}\cdot\vec{J}\right\} = 0. \qquad (10.13)$$

Gauss's theorem applies to the surface integral, whence

$$\int_V \mathrm{d}V\left\{\frac{\partial C}{\partial t} + \vec{\nabla}\cdot\vec{J}\right\} = 0. \qquad (10.14)$$

From the arbitrariness of V, equation (10.14) implies

$$\left\{\frac{\partial C}{\partial t} + \vec{\nabla} \cdot \vec{J}\right\} = 0. \tag{10.15}$$

Equations (10.12) and (10.15) are exactly equivalent: they are referred to as *continuity equations*. Continuity equations of this form apply to all manner of physical quantities that are conserved in the sense described above.

Equation (10.15) can now be combined with equation (10.10) in the case of isotropy or a cubic crystal, with the result

$$\vec{\nabla} \cdot \vec{J} = -\frac{\partial C}{\partial t} = -D\nabla^2 C \tag{10.16}$$

or

$$\frac{\partial C}{\partial t} = D\nabla^2 C. \tag{10.17}$$

Equation (10.17) is the *diffusion equation*, the equation of motion for concentration $C(\vec{r}, t)$ under the fairly general conditions described above. It is sometimes referred to as Fick's second law. As we mentioned in Chapter 3, it has the same mathematical form as the heat flow equation for the temperature $T(\vec{r}, t)$, equation (3.44), under the condition $\vec{\nabla} \cdot \vec{u} = 0$, where $\vec{u}$ is the deformation field of the solid. There are therefore whole classes of problems in heat flow and in atomic diffusion where the mathematical forms of the solution are identical, despite the physical distinctness of the two phenomena. Early in the development of quantum mechanics it was noted that Schrödinger's equation for a free particle had a similar form:

$$-\frac{\hbar^2}{2m}\nabla^2\psi = i\hbar\frac{\partial\psi}{\partial t}. \tag{10.18}$$

The fact that the 'effective diffusion constant' in this equation is pure imaginary, however, means that apparently no physical or geometric conclusions can be carried over to it from the fields of diffusion or heat transfer.

10.2.2 Planar source problem

The qualitative nature of the concentration as a function of both space and time can be displayed by a solution of the one-dimensional form of equation (10.17):

$$\frac{\partial C}{\partial t} = D\frac{\partial^2 C}{\partial x^2}. \tag{10.19}$$

Suppose that $C(x, t)$ is known explicitly at $t = 0$,

$$C(x, 0) = f_1(x), \tag{10.20}$$

with $f_1(x)$ a given function. How will the concentration profile evolve with time?

We begin with a standard approach to the solution of equation (10.19): separation of variables:

$$C(x,t) = f(x)g(t). \tag{10.21}$$

Substitution of equation (10.21) into equation (10.19) gives

$$f(x)\frac{dg}{dt} = Dg(t)\frac{d^2f}{dx^2}. \tag{10.22}$$

which we rewrite as

$$\frac{1}{Dg}\frac{dg}{dt} = \frac{1}{f}\frac{d^2f}{dx^2} = K, \tag{10.23}$$

where K is necessarily independent of both x and t, because the second equation would have it exclusively a function of x, and the first a function of t, for which the only resolution is that it is constant. From the first equation in equations (10.23),

$$\frac{dg}{dt} = DKg \rightarrow g = g_0\,e^{+DKt}. \tag{10.24}$$

If the concentration, equation (10.21), is to remain finite at long time t, the separation constant must be real and negative, from equation (10.24),

$$K = -k^2, \tag{10.25}$$

where k is real. The equation for $f(x)$, from equations (10.23) and (10.25) is

$$\frac{d^2f}{dx^2} = -k^2f, \tag{10.26}$$

whose solutions are

$$f(x) \approx e^{\pm ikx}. \tag{10.27}$$

The general solution of equation (10.19) with equation (10.21) is an arbitrary linear combination of the solutions, equations (10.24) with (10.25) and (10.27),

$$C(x,t) = \int_{-\infty}^{\infty} dk\, g_0(k)\, e^{-Dk^2 t}\, e^{ikx}, \tag{10.28}$$

where $g_0(k)$ is an arbitrary function, except that for $C(x,t)$ to be real, we must have

$$g_0(-k) = g_0^*(k), \tag{10.29}$$

where $*$ means complex conjugate.

We now apply the initial condition, equation (10.20), with equation (10.28):

$$f_1(x) = \int_{-\infty}^{\infty} dk\, g_0(k)\, e^{ikx}. \tag{10.30}$$

From Fourier's integral theorem, equation (10.30) must satisfy

$$g_0(k) = \frac{1}{2\pi} \int_{-\infty}^{\infty} dx'\, f_1(x')\, e^{-ikx'}. \tag{10.31}$$

Equation (10.29) is automatically satisfied by equation (10.31) if $f_1(x)$ is real. We substitute equation (10.31) into equation (10.28) to obtain $C(x,t)$:

$$C(x,t) = \frac{1}{2\pi} \int_{-\infty}^{\infty} dk \int_{-\infty}^{\infty} dx'\, f_1(x')\, e^{-Dk^2 t}\, e^{ik(x-x')}. \tag{10.32}$$

Consider the integral

$$I = \int_{-\infty}^{\infty} dk\, e^{-Dk^2 t}\, e^{ik(x-x')}. \tag{10.33}$$

Let

$$y^2 = Dtk^2 \tag{10.34}$$

whence

$$k = y(Dt)^{-1/2}, \tag{10.35}$$

and let

$$z = (x-x')(Dt)^{-1/2}. \tag{10.36}$$

Then equation (10.33) with equations (10.34)–(10.36) becomes

$$I = (Dt)^{-1/2} \int_{-\infty}^{\infty} dy\, e^{-y^2}\, e^{iyz}. \tag{10.37}$$

Now, from equation (10.37),

$$\frac{dI}{dz} = (Dt)^{-1/2} \int_{-\infty}^{\infty} dy\, e^{-y^2} (iy)\, e^{iyz}. \tag{10.38}$$

Evaluate the integral in equation (10.38) by parts:

$$\int_{-\infty}^{\infty} dy (y\, e^{-y^2})\, e^{iyz} = \left[e^{iyz} \left(-\frac{e^{-y^2}}{2} \right) \right]_{-\infty}^{\infty} + \int_{-\infty}^{\infty} dy \left(\frac{e^{-y^2}}{2} \right) (iz)\, e^{iyz}$$

$$= \left(\frac{iz}{2} \right) \int_{-\infty}^{\infty} dy\, e^{-y^2}\, e^{iyz}. \tag{10.39}$$

Combining equations (10.38) and (10.39), we have

$$\frac{dI}{dz} = i \left(\frac{iz}{2} \right) I. \tag{10.40}$$

We rewrite equation (10.40) and solve

$$\frac{\mathrm{d}I}{I} = -\frac{z}{2}\,\mathrm{d}z, \qquad \ln\left(\frac{I}{I_0}\right) = -\frac{z^2}{4}, \qquad I = I_0\,\mathrm{e}^{-z^2/4}. \tag{10.41}$$

From equation (10.41) with equation (10.37)

$$I_0 = I(z=0) = (Dt)^{-1/2}\int_{-\infty}^{\infty}\mathrm{d}y\,\mathrm{e}^{-y^2} = (Dt)^{-1/2}\sqrt{\pi}. \tag{10.42}$$

We introduce the change of variable, equation (10.36) into equation (10.32) for $C(x,t)$, and with equations (10.33), (10.41) and (10.42) obtain

$$C(x,t) = \frac{1}{2\sqrt{\pi}}\int_{-\infty}^{\infty}\mathrm{d}z\,f_1(x - (Dt)^{1/2}z)\,\mathrm{e}^{-z^2/4}. \tag{10.43}$$

Let us consider a simple form of the initial condition, equation (10.20), in which there is a thin planar layer of the diffusing species within an infinite solid, say at $x = 0$. This can be expressed as

$$f_1(x) = C_0\delta(x), \tag{10.44}$$

where $\delta(x)$ is the Dirac delta function. Equation (10.43) for the concentration profile $C(x,t)$ at time t becomes

$$C(x,t) = \frac{C_0}{2\sqrt{\pi}}\int_{-\infty}^{\infty}\mathrm{d}z\,\delta(x - (Dt)^{1/2}z)\,\mathrm{e}^{-z^2/4}$$

$$= C_0\,\frac{1}{2(\pi Dt)^{1/2}}\exp\left\{-\frac{x^2}{4Dt}\right\}. \tag{10.45}$$

The characteristics of this profile are, first, an amplitude that decays in time as $t^{-1/2}$ at the surface $x = 0$; second a gaussian shape $\sim\mathrm{e}^{-\alpha x^2}$ at any instant; third a spatial range of the gaussian, given by $(4Dt)^{1/2}$, that increases $\sim t^{-1/2}$, indicating the distance x at which the gaussian is e^{-1} of its maximum value, at given time t. Thus we see that as diffusion proceeds in both directions into the material from the planar layer, the concentration profile has a gaussian shape in x that diminishes in amplitude as $t^{-1/2}$ and increases in width as $t^{1/2}$.

10.3 Diffusion as a random walk

In the previous section, diffusion was viewed as a flux of particles within the solid driven by the space- and time-dependent concentration gradient, equation (10.10), subject to particle conservation, equation (10.15). In this section, we take a statistical, or probabilistic, approach. We ask, for a particle that can be expected to jump forward or backward with equal probability at each step along its way, what the probability will be of it arriving at a given distance from its starting point within the material, in n steps. This can be interpreted as the mean fraction of particles that will diffuse to this distance in n steps. The number of steps n is, of course, simply related to the time taken, through the jump frequency.

Suppose that, for a given particle, in n steps it goes a distance $x = ma$, where a is the atomic interplanar distance, and m is a positive integer. As in section 10.2.2, we consider one-dimensional diffusion. If $m < n$, it must have done m' steps backward (m' integer, positive). Then we have $(m + m')$ steps to the right, and m' steps to the left, or

$$n = (m + 2m'), \tag{10.46}$$

whence

$$m' = \tfrac{1}{2}(n - m). \tag{10.47}$$

If the particle is restricted to the region $x \geq 0$, then we must exclude paths for which, at some point along the way more backward steps have been taken than forward. This constraint is difficult to implement analytically. We therefore consider, not a semi-infinite material in $x > 0$, but an infinite one, in $-\infty < x < \infty$, as in section 10.2.2. The question then becomes one of the particle arriving at $x = \pm ma$ in n steps, with either $(m + m')$ steps to the right and m' to the left, or $(m + m')$ to the left and m' to the right, subject to equation (10.47). Then the number of distinct paths for the particle to get to $\pm m$ in n steps is just the number of ways of distributing m' backward steps amongst n steps in total; i.e. the number of combinations $_nC_{m'}$ of n steps m' at a time,

$$_nC_{m'} = \frac{n!}{m'!\,(n - m')!} = \frac{n!}{\left(\dfrac{n - m}{2}\right)!\left(\dfrac{n + m}{2}\right)!}, \tag{10.48}$$

where we have used equation (10.47) in the last step. This is, of course, the same as $_nC_{m+m'}$, given equation (10.47). Given that the probabilities of a forward and of a backward step are assumed here to be equal, whence the expression 'random walk', the probability of a specific path is $(\tfrac{1}{2})^n$. For arriving at $x = \pm ma$ in n steps, this factor is weighted by $_nC_{m'}$, equation (10.48), in arriving at the probability for a single particle, or the average distribution of a large number of particles over final states m. Thus the concentration will be proportional to $\Omega(m, n)$, where

$$\Omega(m, n) = (\tfrac{1}{2})^n \, _nC_{m'} = (\tfrac{1}{2})^n \frac{n!}{\left(\dfrac{n - m}{2}\right)!\left(\dfrac{n + m}{2}\right)!}. \tag{10.49}$$

[Note from equation (10.46) that if n is (odd/even), m is also (odd/even), so that in equation (10.49) $(n \pm m)$ is even, and so $\tfrac{1}{2}(n \pm m)$ is an integer.]

Now for the macroscopic distribution given by the concentration $C(x,t)$, the number of steps n, for macroscopic x, must be very large, since the net displacement $x = \pm ma$ must be macroscopic. We are therefore justified in using Stirling's approximation for $n!$ and for $[\frac{1}{2}(n+m)]!$ We shall also use it for $[\frac{1}{2}(n-m)]!$, on the grounds that the latter will be large for the overwhelming fraction of cases, where m and m' both will be very large. Stirling's formula is that for large n,

$$n! \approx n^n e^{-n} \sqrt{2\pi n}. \tag{10.50}$$

This result is discussed in the Appendix to this chapter. In approximating Ω, equation (10.49), we work with $\ln \Omega$,

$$\ln \Omega = n\ln(\tfrac{1}{2}) + \ln(n!) - \ln\left[\left(\frac{n-m}{2}\right)!\right] - \ln\left[\left(\frac{n+m}{2}\right)!\right], \tag{10.51}$$

and, from equation (10.50),

$$\ln(n!) \approx n\ln(n) - n + \tfrac{1}{2}\ln(2\pi n). \tag{10.52}$$

Applying equation (10.52) to $(n!)$ and to $[\frac{1}{2}(n \pm m)]!$ in equation (10.51) gives

$$\ln \Omega \approx \left\{ n\ln\left(\frac{1}{2}\right) - \frac{1}{2}\ln(2\pi) + \left(n+\frac{1}{2}\right)\ln(n) - (n+1)\ln\left(\frac{n}{2}\right) \right.$$
$$\left. -\left(\frac{n-m+1}{2}\right)\ln\left(1-\frac{m}{n}\right) - \left(\frac{n+m+1}{2}\right)\ln\left(1+\frac{m}{n}\right) \right\}. \tag{10.53}$$

If we can assume that $m \ll n$ for the overwhelming fraction of cases, then

$$\ln\left(1\pm\frac{m}{n}\right) \approx \frac{m}{n}\left\{\pm 1 - \frac{1}{2}\frac{m}{n}\right\} \approx \pm\frac{m}{n}. \tag{10.54}$$

With the approximation of equation (10.54), equation (10.53) reduces to

$$\ln \Omega \approx \left\{ -\frac{1}{2}\ln(n) - \frac{1}{2}\ln(2\pi) + \ln(2) - \frac{m^2}{2n} \right\}, \tag{10.55}$$

or

$$\ln\left[\left(\frac{n\pi}{2}\right)^{1/2}\Omega\right] = -\frac{m^2}{2n}, \tag{10.56}$$

whence

$$\Omega \approx \left(\frac{2}{n\pi}\right)^{1/2} e^{-m^2/(2n)}. \tag{10.57}$$

Suppose that there are n diffusion steps on average, per unit time. Then

$$n = \Gamma t, \tag{10.58}$$

where Γ is the step frequency. Furthermore, m net forward steps between planes with spacing a gets a diffusing particle to the following position in x:

$$ma = x. \tag{10.59}$$

Substituting from equations (10.58) and (10.59) into equation (10.57) we obtain

$$\Omega \approx \left(\frac{2}{\pi\Gamma t} \right)^{1/2} e^{-x^2/(2a^2\Gamma t)}. \tag{10.60}$$

Recall that for one-dimensional diffusion, as here, the diffusion constant D is $(\Gamma a^2/2)$, equation (10.8). Thus, equation (10.60) becomes

$$\Omega \approx \frac{a}{(\pi Dt)^{1/2}} e^{-x^2/(4Dt)}. \tag{10.61}$$

Recall that Ω is the probability of arriving at $\pm x = (ma)$; i.e. we may write $\Omega = \Omega(|x|, t)$, in equation (10.61). Then the probability $\Omega(x, t)$ of arriving at x is

$$\Omega(x, t) = \tfrac{1}{2}\Omega(|x|, t). \tag{10.62}$$

The probability density (i.e. the probability per unit length) of finding a particle at x is

$$P(x, t) = \frac{1}{a} \Omega(x, t). \tag{10.63}$$

It is left as an exercise to show that $P(x, t)$ in equation (10.63), with equations (10.62) and (10.61), is a conventionally normalized probability:

$$\int_{-\infty}^{\infty} dx\, P(x, t) = 1. \tag{10.64}$$

In the present example, all of the diffusing particles start at $x = 0$, so the initial condition is the same as that in section 10.2.2, equation (10.44), where C_0 is the initial number of particles per unit area at $x = 0$. Referring to equations (10.63) and (10.62), we deduce that the probability density $P(x, t)$ is the fractional concentration, in the sense

$$\frac{C(x, t)}{C_0} = P(x, t) = \frac{1}{2(\pi Dt)^{1/2}} e^{-x^2/(4Dt^2)}, \tag{10.65}$$

having used equation (10.61). This is the same result, equation (10.45), that we obtained by solving the diffusion equation. It is interesting how these two quite distinct mathematical approaches can be successfully applied to the same physical problem.

10.4   Equilibrium concentration of point defects

Consider a simple crystal, monatomic with a basis consisting of a single atom. At high temperature, thermal vibration ensures that some fraction of the atoms will not be at perfect crystal sites; that is, the crystal will not be in a zero-temperature, perfect crystal configuration. Suppose that as a result of some particular preparation of the material, an idealization of the atomistic configuration has only one species of point defect, namely vacancies. We now address the question of the equilibrium vacancy concentration as a function of temperature.

If there are j vacancies distributed over N atomic sites in the crystal, then the fractional vacancy concentration C_V is

$$C_V = \frac{j}{N}. \tag{10.66}$$

The number of distinct configurations g_j for given j is

$$g_j = \frac{N!}{j!(N-j)!}. \tag{10.67}$$

We shall assume that the concentration at thermal equilibrium is small; that is, the distribution of vacancies is highly diffuse. In that case, vacancy–vacancy interaction energies will be small. The reasonableness of this assumption will be considered after the fact; i.e. after we have used equilibrium statistical thermodynamics to estimate the concentration as a function of temperature.

The partition function Z for our system is

$$Z = \sum_i d_i \, e^{-E_i/(k_B T)}, \tag{10.68}$$

where k_B is the Boltzmann constant and d_i is the degeneracy of the energy eigenvalue E_i for the whole system. Let us write Z, equation (10.68), as a sum of contributions, each from the set of states with a given number, say j, of vacancies:

$$Z = \sum_j Z_j. \tag{10.69}$$

For those states of the system with j vacancies in a highly diffuse distribution, i.e. $j \ll N$, the average total energy may be written as

$$\bar{E}_j = (E_0 + j\varepsilon_v), \tag{10.70}$$

where ε_v is approximately equal to the energy required to create a single vacancy in a perfect crystal whose energy is E_0. Then

$$Z_j \approx g_j \, e^{-\bar{E}_j/(k_B T)}, \tag{10.71}$$

with g_j given in equation (10.67). Now suppose that the sum in equation (10.69) is dominated by the term $j = J$, so that

$$Z \approx g_J e^{-\bar{E}_J/(k_B T)}. \tag{10.72}$$

Let us now evaluate the Helmholtz free energy F, using equation (10.72),

$$F = -(k_B T) \ln Z$$
$$= -(k_B T)\{\ln g_J - \bar{E}_J/(k_B T)\}. \tag{10.73}$$

We use Stirling's formula, equation (10.52), to approximate $\ln g_J$. From equation (10.52), for large n,

$$\ln(n!) \approx \{(n + \tfrac{1}{2}) \ln(n) - [n - \tfrac{1}{2}\ln(2\pi)]\}$$
$$\approx n[\ln(n) - 1]. \tag{10.74}$$

Then, from equation (10.67) with equation (10.74),

$$\ln g_J = \{N(\ln N - 1) - J(\ln J - 1) - (N - J)[\ln(N - J) - 1)]\}. \tag{10.75}$$

In equation (10.75), all the terms with 1 in them cancel. We further approximate the last term in equation (10.75) by using

$$\ln(N - J) = \ln\left[N\left(1 - \frac{J}{N}\right)\right]$$
$$= \left\{\ln N + \ln\left(1 - \frac{J}{N}\right)\right\}$$
$$\approx \ln N, \tag{10.76}$$

since $J \ll N$ according to the assumption of low concentration: see equation (10.66). Thus, with equation (10.76), equation (10.75) reduces to

$$\ln g_J \approx (-J \ln J + J \ln N) = -J \ln\left(\frac{J}{N}\right). \tag{10.77}$$

From equation (10.73) with equations (10.77) and (10.70), we have

$$F = \left\{(k_B T)J \ln\left(\frac{J}{N}\right) + (E_0 + J\varepsilon_v)\right\}. \tag{10.78}$$

In terms of concentration C_v, equation (10.66), equation (10.78) becomes

$$F = \{E_0 + NC_v[(k_B T) \ln C_v + \varepsilon_v]\}. \tag{10.79}$$

Now consider experimental conditions of constant pressure p and temperature T. We have the combined first and second laws of thermodynamics in the form

$$dU = (T\,dS - p\,dV + N\mu_v\,dC_v), \tag{10.80}$$

where μ_v is the chemical potential for vacancies, U is internal energy, and S and V are entropy and volume respectively. The Gibbs free energy G is defined as

$$G = (U - TS + pV), \qquad (10.81)$$

whence the combined first and second laws give

$$dG = (-S\,dT + V\,dp + N\mu_v\,dC_v). \qquad (10.82)$$

It follows that the equilibrium condition at constant pressure and temperature is

$$dG(T, p, C_v) = 0 \qquad (10.83)$$

which requires, for $dC_v \neq 0$,

$$\left(\frac{\partial G}{\partial C_v}\right)_{T,p} = 0. \qquad (10.84)$$

Now the Gibbs and Helmholtz free energies are related as follows:

$$G = (F + pV). \qquad (10.85)$$

Thus equation (10.84), the equilibrium condition, is

$$\left(\frac{\partial G}{\partial C_v}\right)_{T,p} = \left\{\left(\frac{\partial F}{\partial C_v}\right)_{T,p} + p\left(\frac{\partial V}{\partial C_v}\right)_{T,p}\right\} = 0. \qquad (10.86)$$

For a wide range of solids, $F \approx G$, whence

$$p\left(\frac{\partial V}{\partial C_v}\right)_{T,p} \ll \left(\frac{\partial F}{\partial C_v}\right)_{T,p}, \qquad (10.87)$$

and the equilibrium condition, equation (10.86), becomes

$$\left(\frac{\partial F}{\partial C_v}\right)_{T,p} \approx 0. \qquad (10.88)$$

From equation (10.88) with equation (10.79) we can determine the equilibrium concentration of vacancies:

$$\left(\frac{\partial F}{\partial C_v}\right)_{T,p} = \{Nk_B T(\ln C_v + 1) + N\varepsilon_v\} = 0. \qquad (10.89)$$

For low concentration, suppose

$$C_v \leq 10^{-6}$$

whence

$$|\ln C_v| \approx 14 \gg 1.$$

The solution of equation (10.89), neglecting 1 relative to $\ln C_v$, is

$$C_v = e^{-\varepsilon_v/(k_B T)}. \qquad (10.90)$$

We note that ε_v must be calculated, for this purpose, at constant T and p, in simulation studies. How does equation (10.90) conform to our original assumption of low concentration? Consider $T \approx 300\,\text{K}$ and $\varepsilon_v \approx 1\,\text{eV}$. Then with $k_B = 8.62 \times 10^{-5}\,\text{eV/K}$ we find

$$C_v \approx 10^{-17}.$$

from equation (10.90). This certainly is consistent with the original assumption, and there is a wide class of materials and point defects for which equation (10.90) is a good approximation. The discussion of this section closely follows that of Agulo Lopez *et al.* (1988), sections 2.2–2.5.

10.5 Temperature dependence of diffusion: the Vineyard relation

Consider first a very simple example of diffusion: diffusion by the vacancy mechanism in a planar crystal. The system is illustrated in figure 10.1, where there is one vacancy in a square two-dimensional crystal, where one atom, the one at site A, will undergo a diffusion jump into the vacancy, which is shown by a small square at B. All atomic sites shown are equilibrium sites in the crystal when the diffusing atom is at A. Consider first the case that only the atom at A moves during the diffusion jump, and all the other atoms remain fixed during the jump. Sites A and B are energy minima for the atom at A. Equipotential lines for the motion of the atom at A are shown dashed. For any planar trajectory of the diffusing atom from A to B, there is a point of *maximum* energy which falls on the curve labelled S. The *lowest-energy point* on the curve S is at P. In moving along S in either direction from P, the energy rises, but in moving perpendicularly to S from P the energy drops in either direction. The point P is therefore called the *saddlepoint for the motion*: the potential energy surface has a saddle-like shape near P. In the present example, when only one atom moves, and that in a plane, its coordinates can be denoted (y_1, y_2), as in figure 10.1.

In a more realistic picture of the planar process, nearby atoms will be displaced somewhat from their original positions as the diffusing atom moves away from site A. In practice, this has a major effect in reducing the maximum energy rise (the *activation energy*) experience by the atom in a diffusion jump. We therefore speak of the configuration of the crystal at any point during the diffusion jump process, and denote this configuration in terms of the complete set of generalized coordinates $(y_1, y_2, \ldots, y_n)$. For a crystal consisting of N point-mass ions in a three-dimensional crystal, we should have $n = 3N$, and for shell-model ions, $n = 6N$. We denote a given configuration by an n-dimensional vector $\underline{y}$:

$$\underline{y} \equiv (y_1, y_2, \ldots, y_n). \tag{10.91}$$

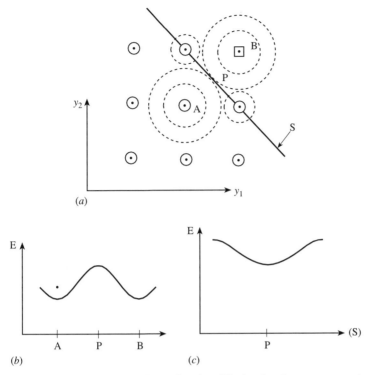

Figure 10.1. (*a*) Two-dimensional illustration for diffusion by the vacancy mechanism. Dashed curves are equipotential lines. All diffusion paths from A to B (the vacancy site) achieve an energy maximum on the curve S. The lowest such maximum is at P. (*b*) Total energy E (schematic) along the straight-line diffusion path A–P–B in part (*a*). (*c*) Total energy E (schematic) along the line S in part (*a*). Parts (*b*) and (*c*) together, with the graphs placed perpendicular to each other while points P coincide, illustrate the nature of the saddle point for energy E versus position in the plane of part (*a*).

In such a generalized vector terminology, $\underline{y}$ is in an n-dimensional hyperspace, and the line S in figure 10.1 represents an $(n-1)$-dimensional hypersurface.

We now approach the diffusion process from the viewpoint of classical statistical mechanics [see for example, Schrödinger (1952)]. We imagine an ensemble of systems, infinite in number, each in a thermal bath consisting of all the others. Each is in some mechanical state, specified by a configuration space position $\underline{y}$ and a corresponding momentum-space configuration $\underline{p}$,

$$\underline{p} = (p_1, p_2, \ldots, p_n), \qquad (10.92)$$

where p_j is the canonical momentum conjugate to y_j, equation (10.91). The union of configuration space and momentum space, called phase space, has points specified by $(\underline{y}, \underline{p})$, equations (10.91) and (10.92). At absolute temperature T, the systems of the statistical ensemble are distributed among the phase space points according to the distribution function

$$\rho(\underline{y},\underline{p}) = \rho_1 \exp \left\{ \frac{-\left[\varphi(\underline{y}) + \frac{1}{2m}p^2\right]}{k_B T} \right\} \qquad (10.93)$$

if all of the particles have the same mass m. In equation (10.93) $\varphi(\underline{y})$ is the potential energy of the system, whose equipotential hypersurfaces are represented schematically by dashed lines in figure 10.1. If $\rho(\underline{y},\underline{p})$, equation (10.93), is normalized, then it gives the probability density in phase space for finding the system to be in state $(\underline{y},\underline{p})$ at temperature T. The normalizing condition, from equation (10.93), is

$$1 = \int d^n y\, d^n p\, \rho(\underline{y},\underline{p})$$

$$= \int d^n y \exp[-\varphi(y)/(k_B T)] \int d^n p\, \rho_1 \exp[-p^2/(2mk_B T)]$$

$$= \int d^n y \exp[-\varphi(y)/(k_B T)]\rho_1 (2\pi m k_B T)^{n/2}. \qquad (10.94)$$

In equation (10.94) we have used the fact that

$$p^2 = \left(\sum_j p_j^2\right), \qquad (10.95)$$

and

$$\int_{-\infty}^{\infty} dp_j \exp[-p_j^2/(2mk_B T)] = (2mk_B T\pi)^{1/2}. \qquad (10.96)$$

If we denote the probability density of points in *configuration* space to be $\xi(\underline{y})$, where

$$\xi(\underline{y}) = \rho_0 \exp[-\varphi(\underline{y})/(k_B T)], \qquad (10.97)$$

then normalization gives

$$\frac{1}{\rho_0} = \int d^n y \exp[-\varphi(\underline{y})/(k_B T)]. \qquad (10.98)$$

From equation (10.94) with equation (10.98) we have

$$1 = \frac{\rho_1}{\rho_0}(2\pi m k_B T)^{n/2}, \qquad (10.99)$$

determining ρ_1 in terms of ρ_0, equation (10.98). In equations (10.94)–(10.99), all integrals are over the infinite range of phase space variables $(\underline{y},\underline{p})$.

Let us refer again to the simple example represented by figure 10.1. In that case configuration space and momentum space are both two dimensional. Consider the flux I of particles in the ensemble, from left to right across the line S. It is given by

$$I = \frac{N_A}{\tau}, \tag{10.100}$$

where N_A is the number of elements of the ensemble with the diffusing particle lying to the left of S, and τ is the mean time between particle jumps. Then from equation (10.100) the mean jump rate is

$$\Gamma \equiv \frac{1}{\tau} = \frac{I}{N_A}. \tag{10.101}$$

When we generalize to a multi-dimensional configuration space, N_A can be evaluated as follows:

$$N_A = \int_{V_A(\ell)} d^n y \int d^n p \, \rho(\underline{y}, \underline{p}), \tag{10.102}$$

since different elements of the ensemble are differentiated by different points in phase space. In equation (10.102), $V_A(\ell)$ is the hypervolume associated with the diffusion from site A to the left (ℓ) of the now hypersurface S. We can evaluate the integral over momenta in N_A, equation (10.102), as before, using equations (10.96) and (10.99) with equation (10.93),

$$N_A = \int_{V_A(\ell)} d^n y \exp[-\varphi(y)/(k_B T)](2\pi m k_B T)^{n/2} \rho_1$$

$$= \rho_0 \int_{V_A(\ell)} d^n y \exp[-\varphi(y)/(k_B T)]. \tag{10.103}$$

In a similar way, the generalized flux I can be evaluated. It is the number of phase space points per unit time passing from left to right through the hypersurface S. This is the density of phase space points at the hypersurface S, times the component of their velocity normal to S, provided it is outward from $V_A(\ell)$, integrated over S:

$$I = \int_{S(\underline{Y})} d^{n-1} y \, \hat{n} \cdot \int_> d^n p \left(\frac{1}{m} \underline{p} \right) \rho(\underline{y}, \underline{p}). \tag{10.104}$$

In equation (10.104), $\underline{Y}$ are points on S, to which the integral over $\underline{y}$ is limited; the n-dimensional dot product is $(\hat{n} \cdot \underline{p})$, where $\hat{n}$ is a unit normal vector on S, outward from $V_A(\ell)$; $>$ limits the integral over $\underline{p}$ to momenta such that $(\hat{n} \cdot \underline{p}) > 0$; and $(1/m) \cdot \underline{p}$ is the velocity of phase space points. Now for a given point $\underline{Y}$ on S, we can choose a coordinate system such that one of the coordinate axes, say y_1, is in the direction of $\hat{n}$ there, so that

$$\hat{n} = (1, 0, 0, \dots, 0). \tag{10.105}$$

In that case,

$$\hat{n} \cdot \underline{p} = p_1. \tag{10.106}$$

Then substituting for ρ in equation (10.104) from equation (10.93), using (10.95) and (10.106), we have

$$I = \int_{S(\underline{Y})} d^{n-1}\underline{y} \exp[-\varphi(\underline{y})/(k_B T)] \int d^n \underline{p} \frac{p_1}{m} \rho_1 \exp\left[-\frac{1}{2m} \sum_{j=1}^{n} p_j^2/(k_B T)\right]$$

$$= \rho_1 \int_{S(\underline{Y})} d^{n-1}\underline{y} \exp[-\varphi(\underline{y})/(k_B T)](2\pi m k_B T)^{(n-1)/2}$$

$$\times \int_0^\infty dp_1 \frac{p_1}{m} \exp\left[-\frac{p_1^2}{(2mk_B T)}\right]$$

$$= \frac{\rho_0}{(2\pi m k_B T)^{1/2}} \int_0^\infty du \frac{(2mk_B T)}{2m} e^{-u} \int_{S(\underline{Y})} d^{n-1}\underline{y} \exp[-\varphi(\underline{y})/(k_B T)]$$

$$= \rho_0 \left(\frac{k_B T}{2\pi m}\right)^{1/2} \int_{S(\underline{Y})} d^{n-1}\underline{y} \exp[-\varphi(\underline{y})/(k_B T)]. \tag{10.107}$$

We can now combine equations (10.103) and (10.107) to evaluate the mean jump rate Γ, equation (10.101):

$$\Gamma = \left\{ \frac{\left(\dfrac{k_B T}{2\pi m}\right)^{1/2} \displaystyle\int_{S(\underline{Y})} d^{n-1}\underline{y} \exp[-\varphi(\underline{y})/(k_B T)]}{\displaystyle\int_{V_A(\ell)} d^n \underline{y} \exp[-\varphi(\underline{y})/(k_B T)]} \right\}. \tag{10.108}$$

We now assume that the integrals in equation (10.108) are dominated by the region in which $\varphi(\underline{y})$ is smallest, because of the exponential functions. For the integral over $S(\underline{Y})$, $\varphi(\underline{y})$ is a minimum at the saddle point, which we denote $\underline{Y}_P$. For the integral over $V_A(\ell)$, $\varphi(\underline{y})$ is a minimum at $\underline{Y}_A$, indicated schematically at A in figure 10.1. We introduce Taylor expansions of $\varphi(\underline{y})$ for the respective integrals about $\underline{Y}_P$ and about $\underline{Y}_A$. Because these points are minima, we have

$$\left(\frac{\partial \varphi}{\partial y_i}\right)_{\underline{y}=\underline{Y}_P} = \left(\frac{\partial \varphi}{\partial y_i}\right)_{\underline{y}=\underline{Y}_A} = 0, \tag{10.109}$$

where the first partial derivative is restricted to the hypersurface S. If, on S and in $V_A(\ell)$ respectively, we denote

$$\underline{y} = (\underline{Y}_P + \underline{U}) \qquad \text{on S,} \tag{10.110a}$$

$$\underline{y} = (\underline{Y}_A + \underline{u}) \qquad \text{in } V_\ell(A), \tag{10.110b}$$

then we have

$$\varphi(\underline{y}) = \left\{ \varphi(\underline{Y}_P) + \frac{1}{2} \sum_{i,j} K_{ij} U_i U_j \right\} \qquad \text{on S,} \qquad (10.111a)$$

$$\varphi(\underline{y}) = \left\{ \varphi(\underline{Y}_A) + \frac{1}{2} \sum_{i,j} k_{ij} u_i u_j \right\} \qquad \text{in } V_A(\ell). \qquad (10.111b)$$

In equations (10.111),

$$K_{ij} = \left(\frac{\partial^2 \varphi}{\partial y_i \, \partial y_j} \right)_{\underline{y} = \underline{Y}_P} \qquad \text{on S,} \qquad (10.112a)$$

$$k_{ij} = \left(\frac{\partial^2 \varphi}{\partial y_i \, \partial y_j} \right)_{\underline{y} = \underline{Y}_A} \qquad \text{in } V_A(\ell). \qquad (10.112b)$$

In equations (10.111) and (10.112), K_{ij} and k_{ij} are elements of force constant matrices $\underline{\underline{K}}$ and $\underline{\underline{k}}$ respectively. In equations (10.111), the quadratic terms are of the form

$$\tfrac{1}{2} \underline{U}^T \cdot \underline{\underline{K}} \cdot \underline{U}. \qquad (10.113)$$

in matrix notation. The matrix $\underline{\underline{K}}$ can be diagonalized by a unitary transformation $\underline{\underline{M}}$:

$$\underline{\underline{K}}' = (\underline{\underline{M}} \cdot \underline{\underline{K}} \cdot \underline{\underline{M}}^{-1}), \qquad (10.114a)$$

$$\underline{U}' \equiv \underline{Q} = (\underline{\underline{M}} \cdot \underline{U}), \qquad (10.114b)$$

$$\underline{\underline{M}}^T = \underline{\underline{M}}^{-1}. \qquad (10.114c)$$

The transformed generalized coordinates $\underline{U}' \equiv \underline{Q}$ are normal modes; the diagonal elements K'_{ii} of $\underline{\underline{K}}'$ are the eigenvalues of $\underline{\underline{K}}$: they are related to the normal mode angular frequencies Ω_i as follows:

$$K'_{ij} = m\Omega_i^2 \delta_{ij}. \qquad (10.115)$$

Exactly similar results apply to the term $\underline{u}^T \cdot \underline{\underline{k}} \cdot \underline{u}$ in equation (10.111b), where

$$\underline{\underline{k}}' = (\underline{\underline{m}} \cdot \underline{\underline{k}} \cdot \underline{\underline{m}}^{-1}), \qquad \underline{u}' \equiv \underline{q} = (\underline{\underline{m}} \cdot \underline{u}), \qquad \underline{\underline{m}}^T = \underline{\underline{m}}^{-1}, \qquad (10.116)$$

where

$$k'_{ij} = m\omega_i^2 \delta_{ij}. \qquad (10.117)$$

Note in the above that the scalar m is the particle mass, and has nothing to do with the unitary transformation matrix $\underline{\underline{m}}$. With equations (10.113)–(10.117), the potential energies of equations (10.111) become

$$\varphi(\underline{y}) = \left\{ \varphi(\underline{Y}_P) + \frac{1}{2} \sum_i m\Omega_i^2 Q_i^2 \right\} \qquad \text{on S,} \qquad (10.118a)$$

$$\varphi(\underline{y}) = \left\{ \varphi(\underline{Y}_A) + \frac{1}{2} \sum_i m\omega_i^2 q_i^2 \right\} \qquad \text{in } V_A(\ell). \qquad (10.118b)$$

We now return to the mean jump rate Γ, equation (10.108). The hypersurface and hypervolume elements $d^{n-1}y$ and $d^n y$ respectively now become $(dQ_1 \ldots dQ_{n-1})$ and $(dq_1 \ldots dq_n)$ respectively. When the sums in equations (10.118) are substituted into equation (10.108), the integrals separate, with the following result:

$$\Gamma = \left(\frac{k_B T}{2\pi m} \right)^{1/2} \left\{ \frac{\exp\left[-\frac{\varphi(\underline{Y}_P)}{k_B T} \right] \prod_{j=1}^{(n-1)} \int_{-\infty}^{\infty} dQ_j \exp\left[-\frac{m\Omega_j^2 Q_j^2}{2k_B T} \right]}{\exp\left[-\frac{\varphi(\underline{Y}_A)}{k_B T} \right] \prod_{j=1}^{n} \int_{-\infty}^{\infty} dq_j \exp\left[-\frac{m\omega_j^2 q_j^2}{2k_B T} \right]} \right\}. \qquad (10.119)$$

The integrals are of the form

$$\int_{-\infty}^{\infty} dx \exp\left[-\frac{mw^2 x^2}{2k_B T} \right] = \left(\frac{2\pi k_B T}{m} \right)^{1/2} \frac{1}{w}. \qquad (10.120)$$

Thus, from equations (10.119) and (10.120),

$$\Gamma = \frac{1}{(2\pi)} \frac{\left(\prod_{j=1}^{n} \omega_j \right)}{\left(\prod_{j=1}^{(n-1)} \Omega_j \right)} \exp\left\{ -\frac{[\varphi(\underline{Y}_P) - \varphi(\underline{Y}_A)]}{k_B T} \right\}. \qquad (10.121)$$

In terms of natural frequencies ν_j about $\underline{Y}_A$ and ν'_j about $\underline{Y}_P$,

$$\omega_j = 2\pi\nu_j \qquad (10.122a)$$

$$\Omega_j = 2\pi\nu'_j, \qquad (10.122b)$$

we have, from equation (10.121),

$$\Gamma = \Gamma_0 \exp\left\{ -\frac{[\varphi(\underline{Y}_P) - \varphi(\underline{Y}_A)]}{k_B T} \right\}, \qquad (10.123)$$

where

$$\Gamma_0 = \left(\frac{\prod\limits_{j=1}^{n} \nu_j}{\prod\limits_{j=1}^{(n-1)} \nu'_j} \right). \tag{10.124}$$

Equation (10.123) with equation (10.124) constitutes the Vineyard relation [Vineyard (1957)]. In equation (10.123) we have the *migration energy*, or the *activation energy* for the given diffusion process, which we denote ε_a, where

$$\varepsilon_a = [\varphi(\underline{Y}_P) - \varphi(\underline{Y}_A)]. \tag{10.125}$$

This is simply the difference between the potential energy of the system in the activated configuration $\underline{Y}_P$, and in the initial equilibrium configuration $\underline{Y}_A$. Furthermore, the amplitude Γ_0 of the mean jump rate is determinable from the normal mode frequencies of oscillation for the system, ν'_j and ν_j in the corresponding configurations. Contemporary modelling methods at the level, for example, of the shell model of the previous chapter, are capable of determining these normal modes in good approximation. Note especially that neither configuration $\underline{Y}_P$ nor $\underline{Y}_A$ is the perfect crystal configuration: they are different defect configurations, and indeed, ν'_j are normal modes of vibration restricted to the hypersurface of activated configurations.

   The result given by equation (10.123) with equation (10.125) is the temperature dependence of diffusion, the so-called Arrhenius relation:

$$\Gamma = \Gamma_0 \, e^{-\varepsilon_a/(k_B T)}. \tag{10.126}$$

In cases where the theory of this chapter is valid, the standard semi-log plot of experimental values of $\ln \Gamma$ versus $(k_B T)^{-1}$ gives a straight line from whose negative slope ε_a can be read off.

Appendix to Chapter 10: Stirling's formula

We here sketch a derivation, based on Courant (1937), of Stirling's formula, which is an approximation to the factorial function $n!$, n an integer, for large n.
   Consider the definite integral I:

$$I = \int_1^n dx \ln x = [n \ln(n) - (n-1)]. \tag{A10.1}$$

The integral can be approximated as a sum of trapezoidal areas A_j between $n = j$ and $n = (j+1)$:

$$A_j = \tfrac{1}{2} [\ln(j) + \ln(j+1)]. \tag{A10.2}$$

Then

$$I \gtrsim \sum_{j=1}^{(n-1)} A_j = \left\{ \sum_{j=1}^{(n-1)} \ln(j) + \tfrac{1}{2}\ln(n) \right\}, \tag{A10.3}$$

where we have used $\tfrac{1}{2}\ln 1 = 0 = \ln 1$. Thus from equation (A10.3) we have

$$I \gtrsim \left\{ \sum_{j=1}^{n} \ln(j) - \tfrac{1}{2}\ln(n) \right\} = \ln(n!) - \tfrac{1}{2}\ln(n). \tag{A10.4}$$

Combining equations (A10.1) and (A10.4),

$$[n\ln(n) - (n-1)] \gtrsim \ln(n!) - \tfrac{1}{2}\ln(n) \tag{A10.5}$$

or

$$\ln(n!) \lesssim \{(n+\tfrac{1}{2})\ln(n) - n + 1\}. \tag{A10.6}$$

For large n, replace the inequality as follows:

$$\ln(n!) = \{(n+\tfrac{1}{2})\ln(n) - n + (1 - a_n)\}. \tag{A10.7}$$

Courant shows that $a_n < 1$. From equation (A10.7) we have

$$n! = e^{(1-a_n)}\sqrt{n}\, n^n\, e^{-n}. \tag{A10.8}$$

Courant further proves that

$$\lim_{n \to \infty} (e^{(1-a_n)}) = \sqrt{2\pi}. \tag{A10.9}$$

The reader is encouraged to look up and work through this derivation, which is well within the scope of undergraduate calculus. Combining equations (A10.8) and (A10.9), we obtain the result given in equation (10.50):

$$n! = \sqrt{2\pi n}\, n^n\, e^{-n}. \tag{A10.10}$$

Already at the level of equation (A10.5), however, by neglecting 1 and $\tfrac{1}{2}$ relative to n, we had the main qualitative result:

$$n! \sim n^n\, e^{-n}. \tag{A10.11}$$

Chapter 11

Point defects in crystals

11.1 Introduction

Among solid materials there are several categories, including crystalline, amorphous and nanostructured solids. Amorphous solids are completely lacking in the translational symmetries that characterize crystals, but do have short-range order arising from the fact that the material consists of atoms that are tightly bound to each other and that have definite spatial size, characterized by angstroms (10^{-10} m). Thus one atom is more likely to have nearest-neighbor atoms at a distance of approximately one atomic diameter, rather than at slightly more or less distance. Nanostructured solids consist of tightly bound crystallites, within each of which translational crystalline order is qualitatively evident, but limited to nanometer (10^{-9} m) size. Paradoxically, the mismatch of different crystallite orientations at their mutual boundaries produces a higher level of disorder than occurs in amorphous solids. The situation is clearly illustrated and discussed by Birringer (1989).

11.1.1 Crystals and defects

In this chapter we discuss an aspect of disorder in crystals that are nearly perfect. We do not discuss crystalline symmetry in general. An excellent reference for such a discussion is by Ashcroft and Mermin (1976, Chapter 7). We shall, however, give a brief, qualitative description. There are two essential elements in the definition of a crystal: the *basis* and the *Bravais lattice*. The Bravais lattice expresses the translational invariance of the crystal: the fact that for an infinite crystal, certain spatial translations leave the crystal invariant. The Bravais lattice is then a set of points, infinite in all three spatial dimensions, generated by the set of *all* translations that leave the infinite crystal invariant. The basis of the crystal is a collection of atoms, in a specific spatial relation to each other, such that applying to it all the translations of the Bravais lattice generates the whole infinite crystal. The basis determines the chemical composition of the crystal, as well as additional, non-translational, symmetries.

We refer to such an infinite crystal as an *ideal crystal*. Real crystals are finite, bounded by crystalline surfaces. A *perfect crystal* sample will be a finite segment of an ideal crystal. Relative to an ideal crystal, the surfaces of a perfect crystal may be thought of as a kind of defect: they break the translational symmetry. We have seen in Chapter 4 that, in the continuum model, surfaces have dynamical properties distinct from the bulk properties that are essentially characteristic of the infinite solid. In the discrete-atom picture of a real solid, the atomic ordering at a surface will always deviate from that in a perfect crystal. At the least, atomic positions at the surface will relax to equilibrium sites that differ slightly from perfect-crystal sites. In some crystals, atomic ordering on a surface crystal plane is different from that on the corresponding plane in an infinite crystal. A famous example is silicon [see, for example, Burns (1985), section 17-4c], where surface reconstruction is quite spectacular.

Apart from surfaces, which are one type of two-dimensional defect, there is a wide variety of other crystalline defects. In Chapter 5 we discussed dislocations, which are linear, one-dimensional defects. In this chapter we concentrate on point defects which on a macroscopic scale are zero-dimensional. Point defects involve one site or a small number of spatially concentrated atomic sites. A simple example is a *vacancy*: one atomic site from which the atom is missing: see figure 11.1. Another is an *interstitial*: an atom at a position that does not correspond to an atomic site in the perfect crystal: see figure 11.3. An interstitial may be an atom of one of the host species (i.e. one of the chemical components of the perfect crystal), or it may be an impurity. An impurity atom may also be a *substitutional*, substituting at a perfect crystal site for one of the perfect crystal atoms.

In this chapter we not only limit ourselves to point defects, but to a selection of point-defect types in *ionic crystals*. The prototypical ionic crystal is an insulator consisting of well-defined ions, each with net charge equal to a non-zero, usually integral multiple of e, the charge of the proton. Consider, for example, sodium chloride, NaCl. The electrically neutral sodium atom in free space has an electronic configuration $1s^2 2s^2 2p^6 3s^1$. [The meaning of such notation is of central importance in understanding the electronic structure of atoms, molecules and solids: see for example Goswami (1992), section 20.2.] Similarly, neutral free chlorine is $1s^2 2s^2 2p^6 3s^2 3p^5$. At interionic distances as close as those in a NaCl crystal, the total energy is much lower in the ionic states Na^+: $1s^2 2s^2 2p^6$ and Cl^-: $1s^2 2s^2 2p^6 3s^2 3p^6$, due to the Coulomb attraction between them. We can say that the crystal is bound primarily by an electron transfer from sodium to chlorine, relative to the electrically neutral atomic configurations. The sodium chloride crystal, called rocksalt, has a particularly simple structure: its ions are located on a simple cubic array of sites with alternating charge. The basis is a NaCl 'molecule', and the Bravais lattice is face-centered cubic (f.c.c.), in which the lattice points occupy not only simple cubic

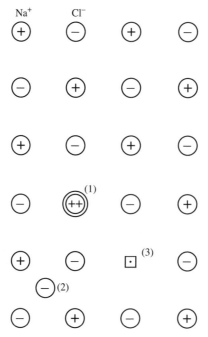

Figure 11.1. The rocksalt crystal structure exemplified by NaCl, and point defects: (1) substitutional cation impurity Mg^{2+}; (2) Cl^- anion self-interstitial half an interplanar distance in front of the other ions: see also figure 11.3; (3) Na^+ vacancy.

structure sites, but the centers of all cube faces in that structure as well. The edges of the primitive cubes of the Bravais lattice are twice the length of the nearest-neighbor interionic distances in the crystal. Many crystals have this rocksalt structures: all the alkali halides formed by combining alkali cations (Li^+, Na^+, K^+, Rb^+, Cs^+) with halide anions (F^-, Cl^-, Br^-, I^-), with the exception of body-centered cubic CsCl, CsBr and CsI; divalent MgO (Mg^{2+}, O^{2-}) and NiO (Ni^{2+}, O^{2-}) and related compounds, and many others. Figure 11.1 illustrates the rocksalt structure with NaCl. In figure 11.1 also are represented several point defects, namely, impurity substitutional (Mg^{2+}), self-interstitial (Cl^-), and sodium vacancy. A standard reference for crystal structures is Wyckoff (1963).

11.1.2 Modelling of point defects in ionic crystals

In Chapter 9, we have introduced the shell model as a classical atomistic model of an ionic crystal, and have discussed it at some length. The reader is referred to section 9.2 at this time. There is an important class of point defect problems for which the shell model by itself is satisfactory. For example, in section 11.2.1 we discuss impurity diffusion in alkali halides, based on such an approach. It does require, of course, that shell-model parameters for impurity–host short-range interactions be available, as well as those for the host crystal itself.

There are many point defect properties, however, where classical modelling alone is intrinsically inappropriate, as illustrated in sections 11.4 to 11.10. For example, classical modelling of an optical excitation process is inappropriate: indeed, the need to explain the Balmer series for the excitation spectrum of hydrogen was a major impetus for the development of quantum theory. [See for example Born (1951), section V.1.] In other problems also, however, it may seem more satisfactory, and be more accurate, to use a quantum-mechanical method. Thus, for example, in section 11.2.2 where the dissociation of a defect complex consisting of an O^- ion and a fluoride vacancy in BaF_2 is discussed, the diffusion of F^- by the vacancy mechanism is involved, and is analysed quantum-mechanically, although one might have used the shell model exclusively for this process, as in section 11.2.1.

Let us now discuss more fully the quantum-mechanical aspect of modelling point defects and their properties. In general, apart from the defect sites, a point defect produces a significant perturbation of the otherwise perfect crystal, involving only ions close to the defect sites. Thus, apart from ions close to the defect, the positions and electronic structures of ions are the same as those for the perfect crystal. The exception is when net electrostatic moments of the defect produce long or medium range deformation and polarization in the crystal such that the effect on the total energy is not negligible. The strongest such effect comes from defects with non-zero charge relative to the perfect crystal: this is the case of the monopole moment. The 'charge relative to the perfect crystal' is the deviation of the defect's charge from that of the perfect crystal region that it occupies. For example, referring to figure 11.1, a Mg^{2+} impurity ion substituting for a Na^+ ion in NaCl has a charge of $+1$ (units of e) relative to the perfect crystal. Henceforth, this will be referred to simply as the defect charge. There are cases where the dipole or quadrupole moment of the defect also needs to be taken into account.

We now return to the issue of perturbation of the crystal's electronic configuration by the defect. If this perturbation is of short range, apart from dipole polarization of more distant ions, then we note that only a small number of ions is involved. These ions will be regarded as a molecular cluster that includes the defect sites. Accordingly, well developed methods of quantum chemistry can be used to determine the detailed electronic configuration of the defect and its nearby region in the crystal. These methods are mentioned in Chapter 12, and one of them, the Hartree–Fock approximation with many-body perturbation theory (MBPT) correlation correction, is discussed in detail.

For point defects in ionic crystals, the molecular cluster analysis must include two crystal-related effects: both the short-range quantum-mechanical effects and the long-range Coulomb effects of the rest of the crystal upon the cluster. The long-range Coulomb effects are derived simply from the core and shell charges of shell-model ions; i.e. the region outside the defect cluster is represented by an embedding shell-model crystal. The short-range inter-action between the cluster and the embedding shell-model crystal may be represented at various levels. Most simply, it may be the shell-model Buck-ingham or other classical potential. This is how all the computations to be described in sections 11.3 to 11.10 have been done. Such an approach requires special treatment of the atomic orbital basis set for the cluster. A more realistic model would have a region surrounding the cluster represented by ionic pseudopotentials. This would subject electrons in the quantum cluster to the non-point charge nature of the ions, and depending on the pseudopotential, to Pauli and other interelectronic correlation effects. Still more realistic would be the inclusion in the Fock operator for the cluster calculation of terms that are more directly derived from the quantum mechanical treatment of the infinite defect crystal. This 'embedding problem' is discussed in some detail in section 12.5. The above methodology, up to embedding pseudopotentials combined with more distant shell-model embedding, is called the ICECAP method [Harding *et al.* (1985)]. We elaborate on this method in the Appendix to this chapter.

In the remainder of this chapter we discuss nine experimental properties and processes, and related computations, for point defects in ionic crystals. In the process, we introduce a wide variety of point defect types. In all cases, results are calculated for specific materials by the ICECAP program, and are applicable to the case of highly dilute defect concentration in the crystal. In many cases, computed results can be compared with experiment, and in all but one case the comparison is favorable. In that exception, the reliability of the calculation points to an essential aspect of the theory (quantum diffusion, section 11.9) that requires additional attention. Those computations for which no experimental results are available are similar enough to others for which experimental results do exist that the predictions are credible. Among these nine types of defect properties, one can discern a web of interconnections that will be commented on in section 11.11. All of the computed results are due to the author and his collaborators, as is the ICECAP methodology. Through their variety, I believe they give the reader a fair glimpse into the much wider general topic of defects in crystalline solids, not limited to the ionic type. Briefly, the properties to be discussed are: classical diffusion, charge-state stability, defect-complex stability, optical excitation, spin densities, local band-edge modification, electronic localization, quantum diffusion and local modes. The style of this chapter deviates from that of the rest of the book by being descriptive and computationally and experimentally oriented, rather than being oriented toward the theoretical structure of the subject. As such, it exemplifies the objective to whose achievement the other chapters of the book are directed, namely reliable computational simulation of solid materials.

At this point we direct the reader to other important sources for point defects in insulators. The book *Physics of Color Centers* edited by Fowler (1968) is still a mine of valuable information, as is Stoneham's *Theory of Defects in Solids* (1975). The periodical *Radiation Effects and Defects in Solids* (Gordon and Breach) is largely devoted to this topic, although important papers are also published in many other journals. For point defects in semiconductors and in metals, one may refer to Lannoo and Bourgouin (1981) and Bourgouin and Lannoo (1983), and to Leibfried and Breuer (1978) and Dederichs *et al.* (1980), respectively.

11.2   Classical diffusion

By classical diffusion we mean thermally activated hopping of ions from one site in the crystal to another. The process has been discussed in some detail in Chapter 10. Classical treatment of the ions is justified, although in section 11.2.2 the diffusing ion and its nearest neighbors are in fact treated quantum mechanically. Two examples are given: impurity diffusion in alkali halides, and dissociation of a vacancy-impurity defect complex.

11.2.1   Copper and silver diffusion in alkali halides

We have calculated the activation energies for diffusion of copper and silver in selected alkali halides [Meng *et al.* (1989)]. The results are given in table 11.1. Two mechanisms are considered: vacancy and interstitial, illustrated in figures 11.2 and 11.3 respectively. In each figure, the activated and initial configurations are illustrated, and it is the difference in the energies of these two configurations that determines the activation energy. [See also Chapter 10, equations (10.123)–(10.126).] The straight line diffusion path is indicated by a dashed line in each case.

Experimental results are available for three processes in KCl: for Cu^+ by vacancy and interstitial mechanisms, and for Ag^+ by the interstitial mechanism. We note that not only are the corresponding calculated values reasonably good, to ~ 0.1 eV, but the ordering of these three values is given correctly, a quite sensitive test of the computational method.

There is reason to question whether the straight-line diffusion path is the actual one for vacancy diffusion. The diffusing ion will follow the path of least resistance; the real activated configuration will be the one of lowest energy. Might not the ion pass somewhere below the edge of the diagonal plane, keeping a greater distance from the straddling negative ions whose electronic structures repel it? We have tested this conjecture in all Ag^+

Table 11.1. Calculated and experimental activation energies ΔE (eV) for vacancy and interstitial mechanisms of Cu^+ and Ag^+ impurities in some alkali halides.

Material	Mechanism	ΔE Calculated	ΔE Experiment
$KCl:Cu^+$	Vacancy	1.19	1.1 [a]
	Interstitial	0.78	0.83 [b]
$KCl:Ag^+$	Interstitial	0.83	0.95 [b]
$RbCl:Ag^+$	Vacancy [c]	0.44	–
$KCl:Ag^+$	Vacancy [c]	0.51	–
$NaCl:Ag^+$	Vacancy	0.93	–
$NaF:Ag^+$	Vacancy	0.72	–

[a] Henke *et al.* (1986).
[b] Pershitz and Kallenikova (1981).
[c] Non-collinear mechanism: see figure 11.4.

vacancy diffusion cases. As indicated in the table, we found that in two cases the activation energy was lower along a path deviating from the straight-line, the so-called non-collinear mechanism illustrated in figure 11.4. In these two cases, Ag^+ in RbCl and KCl, the activation energy is markedly lower than for the other, straight line cases.

The computed energies for table 11.1 were based on shell-model calculations in which all ions (cores and shells) were taken to be in equilibrium. This implies that the ionic positions of the host crystal follow the motion of the impurity ion instantaneously. All calculations were so-called lattice static, i.e. ionic vibration effects were completely ignored. In order to do such shell-model calculations, we need to have impurity-halide short-range

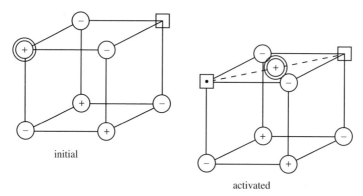

Figure 11.2. Diffusion by the vacancy mechanism of a monovalent substitutional cation impurity in an alkali halide: initial and activated configurations.

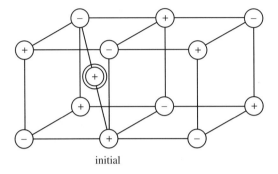

initial

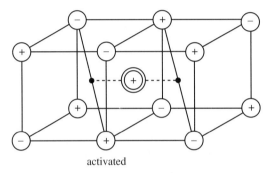

activated

Figure 11.3. Diffusion by the interstitial mechanism of a monovalent cation impurity in an alkali halide: initial and activated configurations.

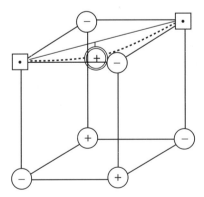

Figure 11.4. Activated configuration for diffusion of a monovalent cation substitutional impurity by the non-collinear vacancy mechanism in an alkali halide: compare with figure 11.2.

Buckingham potentials. While we considered the possibility of transferring potentials from pure silver halide and copper halide crystals, we concluded that such an approach is not, in general, reliable. Instead, we derived our potentials from embedded quantum clusters. The approach was as follows. Evaluate the total energy of a crystal, say KCl, containing a nearest-neighbor $AgCl_6$ quantum molecular cluster. Vary the nearest-neighbor distance, inward and outward from the perfect KCl crystal value. Use the resultant total energy versus distance curve to fit Buckingham potential parameters (B, ρ, C): see equation (9.4); i.e. determine (B, ρ, C) such that the curve from an all shell-model calculation fits that from the quantum cluster.

Although the results given in table 11.1, particularly the comparisons with experiment, are very encouraging, the method of calculation is such that many improvements could be contemplated. The reader is referred to the original paper [Meng *et al.* (1989)] for deeper insight into modelling details for such problems.

11.2.2   Dissociation of the oxygen-vacancy defect complex in BaF_2

Barium fluoride BaF_2 is one of a large class of high-density luminescent materials used for γ-ray detection in environmental, medical and high-energy particle accelerator applications. The subject has been reviewed in Weber *et al.* (1994). In practice, the crystal's high luminescent efficiency is degraded by radiation damage. At one time, oxygen was suspected to be responsible for this effect. Although this is no longer the case, it is worthwhile to understand the defect processes that were postulated, since they are undoubtedly representatives of processes that affect material properties in some situations.

The basis of the BaF_2 crystal is a molecular cluster $(Ba^{2+})_1 \cdot (F^-)_2$. The Bravais lattice is f.c.c. The crystal type is called the fluorite structure (for CaF_2). It is most easily visualized as a simple cubic array of F^- ions with Ba^{2+} ions at the center of every *second* cube of the fluoride sublattice. Thus every F^- ion has four nearest-neighbor Ba^{2+} ions in tetrahedral coordination, while every Ba^{2+} ion has eight nearest-neighbor F^- ions.

It is conjectured that in the crystal growth process, at relatively high temperature, oxygen impurity is unavoidable, and it occurs substitutionally for F^- in the filled-shell O^{2-} configuration $1s^2 2s^2 2p^6$. This impurity's net charge (-1) relative to the perfect crystal is compensated by a fluoride vacancy (net charge $+1$). Furthermore, the two defects occur at nearest-neighbor sites on the fluoride sublattice (second-neighbor sites in the crystal), forming a $(O^{2-} \cdot v_{F^-})$ defect complex, where v_{F^-} stand for the fluoride vacancy (see figure 11.5). This defect we refer to as dipolar: it consists of a net charge of $(+1)$ on one site and (-1) on the other, which at a large distance has predominantly an electric dipole moment. It is conjectured that this dipolar defect complex is stable in the presence of the crystal's

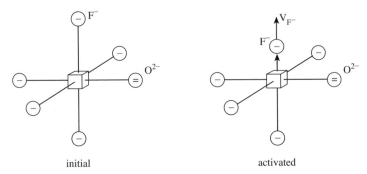

initial   activated

Figure 11.5. Dissociation process for a vacancy-impurity dipole defect complex $(O^{2-}) \cdot (v_{F^-})$ in BaF_2 (with Ba^{2+} ions omitted from the diagram). The process involves diffusion of the fluoride vacancy along the path shown in the activated configuration. The dipole complex is shown in the initial configuration.

intrinsic luminescent radiation and that it does not have excitation levels corresponding to the luminescence.

It is further conjectured that (a) γ-ray absorption (radiation damage) dissociates the $(O^{2-} \cdot v_{F^-})$ defect complex; (b) the free O^{2-} ion is unstable in the presence of a dissociated vacancy v_{F^-}, transferring an electron to the vacancy; (c) the resulting O^- ions have excitation energies that do approximately coincide with the intrinsic luminescence, absorbing it and thereby degrading the efficiency of the detector. Conjecture (c) appears not to be correct, as discussed in section 11.5.3, casting attention toward other defect complexes, possibly involving rare-earth impurities. Conjecture (a) can be represented by the process

$$[(O^{2-}) \cdot (v_{F^-}) + \hbar\omega] \rightarrow [O^{2-} + v_{F^-}]. \tag{11.1}$$

In equation (11.1), $\hbar\omega$ is the energy from radiation damage, possibly the γ-ray photon itself, that initiates the activation energy for diffusion of the vacancy away from the O^{2-} impurity. The process is illustrated in figure 11.5, where diffusion of the vacancy in one direction is equivalent to diffusion of an F^- ion in the opposite direction. We have calculated the activation energy as the difference in energy between activated and original configurations [Vail *et al.* (1998a)]. Our modelling is based on the embedded quantum clusters $(v_{F^-}) \cdot (O^{2-})_1 \cdot (F^-)_5 \cdot (Ba^{2+})_4$ as shown in figure 11.5, where oxygen and fluoride ions are given all-electron treatment (bare nuclei) and Ba^{2+} ions are represented by pseudopotentials. The calculated activation energy is 0.93 eV. This is similar to values obtained for impurities in the alkali halides (see table 11.1). The difference between the present calculation and those of section 11.2.1 is that there the process was modelled entirely in terms of the shell model, whereas here it is entirely in terms of quantum clusters (embedded, of course, in an infinite shell-model crystal).

We shall return to the problem of oxygen in BaF_2 in future sections: in section 11.4.2 the charge state stability of free O^{2-} ions in the presence of v_{F-} is discussed; in section 11.5.3 the optical excitation of O^- is discussed; and in section 11.7.2 the local modification by O^- of the conduction band edge is discussed. We also discuss the F center in BaF_2, in section 11.8.

11.3   Defect complex stability

It can happen that the atomistic ordering of a defect complex is in question, or the relative stability of alternative configurations may be of interest. An example of this situation is the $(F_2^+)^*$ center in NaF:Mg [Hofmann *et al.* (1985)]. This is one of many examples of an F-type center: electrons bound in vacancies. This particular defect is created by combining two defect complexes. One is an F_2^+ center: one electron bound in two nearest F^- vacancies, whence the superscript notation (+), indicating a net charge of (+1) relative to the perfect crystal. It would be electrically neutral if there were two electrons: it would then be a two-F center complex, whence the subscript 2. The second defect complex is a substitutional Mg^{2+} ion on an Na^+ site, compensated for its net charge (+1) by a nearest Na^+ vacancy: an impurity–vacancy dipole complex, similar to that seen with O^{2-} in BaF_2, section 11.2.2. These two defects are illustrated in figure 11.6.

How the two defects combine is a matter of interest. The simplest, most symmetrical combination is shown in figure 11.7. In that case, the excess electron of the F-type center would be symmetrically shared between the two F^- vacancies. The experimental work of Hofmann *et al.* (1985) shows that this is not the case: the ground-state character of the $(F_2^+)^*$ center is more like a weakly perturbed one-electron F center involving a single F^- vacancy. There is a large number of distinct possible unsymmetrical

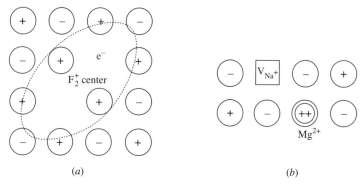

(a)   (b)

Figure 11.6. Two point defect complexes in NaF: (a) the F_2^+ center; (b) the impurity-vacancy dipole defect $(Mg^{2+}) \cdot (v_{Na^+})$.

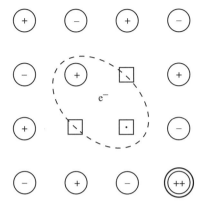

Figure 11.7. The symmetrical configuration of the $(F_2^+)^*$ center in NaF: see also figures 11.6(a) and (b).

configurations, particularly if one considers impurity–vacancy dipole configurations involving second-neighboring sites on the cation sublattice, as was done by Hofmann *et al.* (1985). We have studied 39 configurations in all at a relatively simple level of modelling [Vail *et al.* (1998b)]. From those we selected six that were lowest in total energy, and therefore most promising. All six were unsymmetrical, forcing the excess electron into a single F^- vacancy. These were analysed in terms of a nearest-neighbor embedded quantum cluster containing the six nearest-neighboring Na^+ ions of the F-type center. The calculated total energies of these six configurations are given in table 11.2. Two of these configurations were strongly preferred over the others on the basis of lower total energies. They are illustrated in figures 11.8 and 11.9. One, denoted configuration no. 6, figure 11.8, is a planar four-site defect. The other, configuration no. 24, figure 11.9, is non-coplanar. Within the limitations of the calculations, these are our candidates for the $(F_2^+)^*$ center's configuration. It may be that the defect occurs at ordinary temperatures in more than one configuration, all of approximately

Table 11.2. Computed total energies E (eV) of six configurations for the $(F_2^+)^*$ center in NaF : Mg.

Configuration no.	E
6	−26 647.35
24	−26 645.40
18	−26 619.52
31	−26 618.63
7	−26 598.58
15	−26 568.95

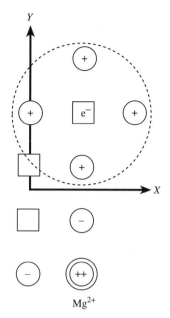

Figure 11.8. Low-energy planar configuration [no. 6: see Vail *et al.* (1998b)] of the $(F_2^+)^*$ center in NaF.

the same energy. Further calculations, which are extremely difficult, and particularly more detailed experimental characterization, are needed to determine these issues definitively. One possible contribution to such progress is discussed in section 11.6.3.

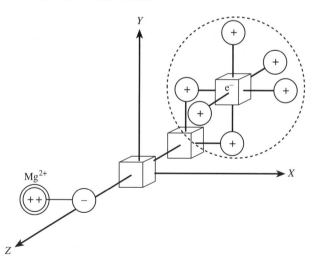

Figure 11.9. Low-energy non-planar configuration [no. 24: see Vail *et al.* (1998b)] of the $(F_2^+)^*$ center in NaF.

It is important to understand this defect at the level discussed above for practical reasons. There is a type of laser based on F-type centers that has unique value for studies of small organic molecules [Baldacchini (1989)]. To date, these lasers have the particular disadvantage of not being usable above liquid nitrogen temperature (77 K). At higher temperatures the defect complexes upon which they are based dissociate or become otherwise irreversibly modified. The $(F_2^+)^*$ center in NaF:Mg does not suffer from this disadvantage: its laser properties are room-temperature stable. Although $(F_2^+)^*$ center lasers are not marketable, it is hoped that by understanding this defect's stability it may be possible to develop other room-temperature F-center lasers that will be marketable.

11.4   Impurity charge-state stability

Since different charge states of an impurity species have distinct optical and chemical properties, it is important to be able to determine the charge state, either by experimental or computational means. In this section we discuss two such systems: nickel in MgO and oxygen in BaF_2.

11.4.1   Nickel in MgO

Simply on the basis of electrostatics, we expect nickel to be stable in MgO in charge state (+2), as a Ni^{2+} substitutional for a Mg^{2+} host ion. In terms of overall electrical neutrality for the crystal, one might ask whether diffuse distributions of Ni^+ and N^{3+} ions would be energetically favored, or not, over a diffuse distribution of Ni^{2+} ions. At high enough temperature, thermal activation would tend to bring Ni^+ ions (net charge -1) to form impurity dipole defect complexes with Ni^{3+} ions (net charge $+1$). Therefore, we consider only low enough temperature so that the diffuse distributions will be stable.

We have calculated the total energies of all three charge states of Ni^{n+} ($n = 1, 2, 3$), substitutional in MgO [Meng *et al.* (1990a)]. We conclude that:

$$(Ni^+ + Ni^{3+}) \rightarrow 2(Ni^{2+}) + 10.1 \, \text{eV}. \tag{11.2}$$

It is therefore clear that so-called disproportionation, in a homogeneous dilute mixture of Ni^+ and Ni^{3+}, is counterindicated quite strongly. The detailed electronic structure of the calculations for Ni^+ and Ni^{3+} show interesting features. For Ni^+, the net charge (-1) is not located primarily on the impurity, but on the second-neighbor Mg^{2+} ions, leaving the impurity ion essentially in charge state (+2). Similarly for Ni^{3+}, the net charge $(+1)$ resides primarily in the form of a hole in the electronic structure of nearest-neighbor O^{2-} ions, again leaving the impurity in charge state (+2). In the real crystal, where electronic and hole states of more distant ions are available, the electron trapping (hole loss) of Ni^+ and electron loss (hole capture) of Ni^{3+} might be more diffuse, possibly even fully spread out through the valence and conduction bands, respectively, of the host crystal. In our calculations, second neighbor embedded quantum clusters $(Ni^{n+})_1 \cdot (O^{2-})_6 \cdot (Mg^{2+})_{12}$, $n = 1$, 2, 3, were used, with the Mg^{2+} ions represented by pseudopotentials.

Experimental studies of MgO:Ni have failed to show evidence either of free, stable Ni^+ or Ni^{3+} ions. The Ni^{3+} ion has been stabilized experimentally by noncovalent impurity cations, notably Li^+ [Verwey *et al.* (1950)]. The conditions for Ni^+ to exist in defect complexes or metastably have not been clearly determined [Low and Offenbacher (1965)].

11.4.2 Oxygen in BaF_2

In section 11.2.2 we introduced the subject of oxygen in BaF_2, and discussed the defect–complex dissociation process represented by equation (11.1). After the complex dissociates into an O^{2-} ion and a fluoride vacancy v_{F^-}, one charged negatively and the other positively relative to the host crystal, the question of charge-state stability must be considered, just as for Ni^+ and Ni^{3+} in the previous section, equation (11.2). Electron transfer from O^{2-} to the vacancy would leave O^- and an F center F_c, each electrically neutral. The process is represented as follows:

$$(O^{2-} + v_{F^-}) \rightarrow (O^- + e^- + v_{F^-}) \rightarrow (O^- + F_c). \qquad (11.3)$$

Again, to determine the stable charge states, we need only evaluate the total energy of isolated O^{2-} ion and v_{F^-} on the one hand, and of isolated O^- and F_c on the other. We have done this in terms of embedded quantum clusters [Vail *et al.* (1998a)], with the following result:

$$(O^{2-} + v_{F^-}) \rightarrow (O^- + F_c + 1.4 \, eV). \qquad (11.4)$$

It is therefore clear that, in isolation, the stable charge state of oxygen in BaF_2 is O^-, when associated with a charge-compensating vacancy-type defect. We shall take up these defects again in section 11.5.3, where their optical excitations are discussed.

11.5 Optical excitation

In section 11.1.2, in the introduction to this chapter, we identified optical excitation processes as representative of defect problems that are intrinsically quantum mechanical. In this section we describe such processes in MgO (section 11.5.1), in NaF (section 11.5.2), and in BaF_2 (section 11.5.3). In all three sections, optical excitation of impurities is discussed. In sections 11.5.1 and 11.5.3, intrinsic processes are also discussed. In MgO, the excitation of O^{2-} (the Frenkel exciton problem) is addressed, and in BaF_2 the F center problem is commented on, both specifically and in the wider context of other host materials. Optical absorption by the F center in BaF_2 is discussed further in section 11.8.

Some general comments on modelling optical excitations in ionic crystals are in order here. All our calculations are for small, embedded quantum clusters. Thus, systems whose optical transitions are not localized within first, or at most second, neighbor distances require special treatment. This is particularly emphasized in the F-center discussion of section 11.5.3 and in section 11.8. Our calculations have two particular strengths. One is that optical transition energies are calculated as differences between many-body states, not single-particle states. The other is that, at least in sections 11.5.2 and 11.5.3, care is given to the inclusion of correlation effects.

The study of optical processes is relevant to a variety of materials properties. They cast light (no pun intended) on the intrinsic nature of the crystalline solid and determine technologically significant properties for optical applications of approximately perfect bulk samples. They determine properties of point defects, again for the purpose of characterizing the defects, and for technological applications. Finally, optical excitations in perfect and defected crystals are increasingly being used to tailor electronic and atomistic structures to specific properties and purposes. A recent work on this aspect of defect excitation is Itoh and Stoneham (2000).

In what follows, the choice of examples is quite narrow in a field where the experimental results are extremely far reaching. In particular, we do not give any examples of optical *de-excitation*, or emission, which is at least half of the subject. While our methods are capable of dealing with the emission process, we have not yet done any studies of that sort.

11.5.1   Frenkel exciton and impurity absorption in MgO

The optical excitation of a crystal may be viewed as the creation of an electron–hole pair. Such a combination is referred to as an *exciton*. An exciton localized in atomic dimensions is referred to as a *Frenkel exciton*. Pandey *et al.* (1989) have analysed the Frenkel exciton in MgO using the identical embedded quantum cluster (ICECAP) method whose results are presented throughout this chapter. Their results, obtained at the Hartree–Fock level for both singlet and triplet excited states, agree remarkably well with experiment. The computed singlet and triplet excitations are 7.77 eV and 7.73 eV respectively compared with experimental values of 7.76 eV and 7.69 eV [Roessler and Walker (1967) and (1966)]. These results indicate that, in this material, the exciton is highly localized, corresponding to a $2p^6 \rightarrow 2p^5 3s$ transition of O^{2-}. The Frenkel exciton may be regarded as a defect, involving only the local electronic structure, but not affecting the atomistic structure of the crystal.

Pandey and co-workers have also studied optical excitation of impurities in MgO. In the previously cited work [Pandey *et al.* (1989)] they considered the isovalent anion impurities S^{2-} and Se^{2-}, substitutional for O^{2-} in MgO. While experimental values are not available, singlet and triplet values calculated for S^{2-} were 7.12 eV and 7.09 eV respectively, and for Se^{2-}, 6.71 eV and 6.68 eV respectively. The systematic downward trend in excitation energy with increasing atomic number is to be noted. Pandey and Kunz (1990) have also calculated the $2p^6 \rightarrow 2p^5\,3s$ transition of F^- substitutional impurity in MgO. They obtain an excitation energy of 15.2 eV, about double that of the host O^{2-} Frenkel exciton. Since O^{2-} and F^- have the same number of electrons, ten, this result might seem surprising, until one recognizes that F^- is a positively charged defect in MgO. The positive charge comes from incompletely charge-compensated nearest-neighbor Mg^{2+} ions, whose contribution to the Madelung field deepens the effective potential well seen by the F^- ion's electrons. This has the effect of increasing the separation of the energy levels.

11.5.2 Cu^+ in NaF

In the previous section we discussed electric dipole excitations in MgO. In this section we shall deal with the dipole forbidden, two-photon excitations of Cu^+ ($d^{10} \rightarrow d^9\,s$), substitutional for Na^+ in NaF. We refer here to the ICECAP computations of Meng *et al.* (1988), augmented as they were by careful treatment of correlation correction. Meng's calculated average value over the excited states is 4.02 eV, compared with 4.01 eV from the experimental work of Berg and McClure (1989). Overall, this excellent agreement with experiment, while significant, is moderated by the fact that the computed ordering of excited states of different symmetries is not entirely in agreement with experiment. Meng's calculated value of crystal field splitting, 0.31 eV, also compares very well with the experimental value of 0.35 eV [Payne *et al.* (1984)]. We shall return to Cu^+ in NaF in section 11.10.

11.5.3 O^- in BaF_2

Substitutional oxygen impurity in BaF_2 has been introduced and discussed earlier, with respect to its origin in a defect complex, section 11.2.2, equation (11.1), and with respect to its charge state stability, section 11.4.2. The picture has been developed of O^- being created in two stages, first by radiation damage dissociating the $(O^{2-} \cdot v_{F^-})$ defect complex, equation (11.1), and second by electron transfer from O^{2-} to v_{F^-}, creating O^- and F center defects, equation (11.3). The issue has been raised whether O^- can then absorb the intrinsic luminescence of the crystal, thereby degrading its luminescent efficiency in γ-ray detectors.

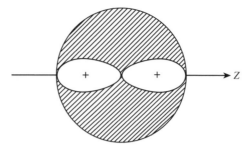

Figure 11.10. Electronic distribution of O$^-$ substitutional impurity ion in BaF$_2$, showing p$_z$-like hole, producing a prolate quadrupole moment.

The optical absorption energies of O$^-$ in BaF$_2$ are not known experimentally. Using the ICECAP method of embedded quantum clusters, with pseudopotential Ba^{2+} ions, and including correlation correction, we have calculated these energies for dipole-allowed, single photon absorption. We predict excitations of 8.95 eV and 9.74 eV [Vail *et al.* (1998a)]. This is to be compared with the intrinsic luminescence of BaF$_2$, which is 5.64 eV and 6.36 eV [Zhu (1994)]. We conclude that absorption by O$^-$ is not responsible for degrading the luminescence.

The calculated splitting of the O$^-$ absorption line in BaF$_2$ has an interesting physical origin. As we have mentioned, O^{2-} is a filled-shell ion 1s^2 2s^2 2p^6. It follows that O$^-$ will have one electron less than the filled shell: it will have one electron whose spin (say spin up) is not paired with another electron in the same spatial orbital. There is therefore a *hole* in the filled shell configuration, and it turns out that the missing spin-down electron is in a 2p-like state oriented along one of the crystal's cubic axes, say the z axis. We therefore have a p$_z$-like hole, with geometrical character (cos θ_z). The situation is illustrated schematically in figure 11.10. This configuration 1s^2 2s^2 2p^5 has a quadrupole moment that is *prolate* in the z direction. This in turn generates a local quadrupolar strain that is *oblate* in the z direction, tending to screen the quadrupole field of the O$^-$ ion. Excitations occur from O$^-$ p-like states that are symmetrical linear combinations of 2p$_x$ and 2p$_y$, denoted 2p$_{x,y}$. The unequal electronic configurations in the occupied spin up ↑ and spin down ↓ manifolds produce spin polarization, so that 2p$_{x,y}$↑ and 2p$_{x,y}$↓ have different single-particle energies. This arises from applying the so-called *unrestricted* Hartree–Fock approximation: see Chapter 12. The fact that the final states of these two transitions are correspondingly spin polarized accounts for the absorption-line splitting. It turns out that the final states are not oxygen-like localized states, but are locally perturbed parts of the conduction band. This feature of O$^-$ in BaF$_2$ will be discussed further in section 11.7.2.

In the picture of the radiation damage process presented above and in sections. 11.2.2 and 11.4.2, O$^-$ is not responsible for absorption of the luminescence. Experimental radiation damage studies [Woody *et al.* (1989)] show a broad optical absorption band centered at 2.18 eV, probably due to a variety of point defects, undoubtedly including the F center, whose absorption energy is accurately known to be 2.03 eV [Cavenett *et al.* (1967)]. Thus it is clear that the F center is also not responsible for luminescent absorption.

While our calculations for the F center optical absorption in BaF_2 could be included in this section, we defer it to section 11.8, where we emphasize the spatial diffuseness of its excited state.

11.6   Spin densities

The hyperfine interactions of a defect electron's magnetic moment with the magnetic moments of nuclei in a solid provide a particularly sensitive test of the electron's wave function throughout a localized region. More correctly, what is involved is a many-electron molecular cluster containing the defect. If it has an unpaired electron, as in the previous section (O^- in BaF_2), then the isotropic part of the hyperfine interaction is proportional to the electronic spin density at the nuclear position, represented by

$$\langle \Psi | \sum_j \delta(\vec{R} - \vec{r}_j)\vec{S}_j | \Psi \rangle \tag{11.5}$$

where $|\Psi\rangle$ is the many-electron wave function, r_j is the position vector of electron j, $\vec{S}_j$ is the electron's spin, and $\vec{R}$ is the nuclear position. The subject of spin resonance is discussed by Seidel and Wolf (1968) and by Stoneham (1975). For nuclei that are inequivalent with respect to the electronic wave function, the spin resonance can be resolved by a combination of nuclear and electron spin resonance effects: ENDOR, standing for electron-nuclear double resonance [Feher (1959), Seidel and Wolf (1968)]. We now discuss three related defects for which spin densities have been computed, for one of which experimental results are also available.

11.6.1   F center in NaF

We have determined the F-center ground state in NaF, using a second-neighbor embedded cluster [Vail and Yang (1993)]. The spin densities at the six equivalent nearest-neighbor Na^+ ions and twelve equivalent second-neighbor F^- ions are converted to MHz units for the isotropic hyperfine constant. These can be compared with the experimental values of Seidel and Wolf (1968). The results for Na^+ are 80:107 and for F^- 32:97 (computed:experimental), all in units of MHz. The 25% discrepancy between computation and experiment for nearest neighbors may seem large, unless one realizes the sensitivity of the result to the detailed shape of the electronic spin density and to the positions of the nuclei in the locally deformed crystal. The factor of three discrepancy for second neighbors suggests that the computations represent an F center that is significantly more localized than the experimentally observed one.

11.6.2 F_2^+ center in NaF

The F_2^+ center in NaF has three inequivalent sets of nearest-neighbor Na^+ ions, as can be seen from figure 11.6. Their coordinates are represented, in units of nearest-neighbor spacing, by (0.5, −0.5, 0), (1.5, 0.5, 0) and (0.5, 0.5, 1.0) in the absence of local strain. Our calculated results [Vail and Yang (1993)] show these coordinates to be modified by the defect to (0.55, −0.55, 0), (1.55, 0.49, 0) and (0.49, 0.49, 1.05). This represents the expansion of the surrounding crystal by the Coulomb forces that act on the Na^+ ions, arising from the net positive charge of the defect. The isotropic hyperfine constants for these sets of Na^+ ions are, respectively, 189 MHz, 12 MHz and 31 MHz. This result indicates that the F_2^+ center is most strongly localized about its center of symmetry rather than about the centers of its two F^- vacancies. This qualitative result is borne out by the experimental spin resonance results of Hofmann *et al.* (1985), although they were unable to resolve the individual components. Thus no quantitative comparison with experiment is available. Our calculated results were based on an embedded quantum cluster containing the ten nearest-neighbor Na^+ ions of the F_2^+ center.

11.6.3 $(F_2^+)^*$ center in NaF

This four-site point defect complex has been introduced in section 11.3. In figures 11.8 and 11.9 we presented our two best candidates for its stable ground-state configuration. The computed energies for those two configurations, table 11.2, are too close to choose between them. If the spin densities could be determined experimentally, it might be possible to choose between them on the basis of symmetry and if our modelling were accurate enough, on the basis of quantitative comparison with experiment. Our calculated spin densities are shown in table 11.3 [Vail *et al.* (1998b)]. As a result of reflection symmetry in the plane of configuration no. 6, figure 11.8, two of its spin densities are degenerate, unlike configuration 24, which has no symmetry. Unfortunately, it is unlikely that the ENDOR results could give sufficient accuracy to allow us to distinguish between two such configurations, and it is also unlikely that our modelling is accurate enough to be reliable for the distinctions that appear in table 11.3. This modelling is based on the embedded quantum cluster of six nearest neighbors of the F^- vacancy that contains the excess electron.

Table 11.3. Spin densities of the six nearest neighbors of the $(F_2^+)^*$ center in NaF:Mg in the two lowest-energy configurations, nos. 6 and 24. Units: $\hbar/(2a_0^3)$. Atomic sites specified relative to the F-center position.

Configuration no. 6		Configuration no. 24	
Atomic site	Spin density	Atomic site	Spin density
$(1,0,0)$	0.0914	$(1,0,0)$	0.0942
$(-1,0,0)$	0.1029	$(-1,0,0)$	0.1051
$(0,1,0)$	0.0874	$(0,1,0)$	0.0945
$(0,-1,0)$	0.1090	$(0,-1,0)$	0.1021
$(0,0,1)$	0.0930	$(0,0,1)$	0.0958
$(0,0,-1)$	0.0930	$(0,0,-1)$	0.0957

11.7   Local band-edge modification

Point defects can locally alter the chemical nature of the valence and/or conduction band edges in insulators. In other words, excitation from the valence band may come predominantly from ions of one chemical species in the perfect crystal, but from another species in excitations that involve the defect. Similarly, excitations to the conduction band may go predominantly to ions of one species in the perfect crystal, but to another species when the defect is involved. When the defects come to have large concentrations, the optical and other properties of the material may be fundamentally changed. Such behavior is apparently crucial to the high-temperature superconducting property of some cuprate materials [Müller and Bednorz (1987) and Chu (1987)].

In this section we discuss both possibilities: in section 11.7.1 the local modification of the top of the valence band in Li-doped NiO, and in section 11.7.2 the spin-splitting of the bottom of the valence band in BaF_2 into barium and fluoride sub-bands by O^- impurity.

11.7.1   Valence band edge in NiO:Li

In NiO, the top of the valence band is known experimentally to be dominated by nickel 3d orbitals [Eastman and Freeouf (1975)]. When lithium impurity is introduced substitutionally in increasing concentration, experimental study [Kuiper *et al.* (1989)] shows that the hole associated with Li^+ resides primarily in oxygen rather than in nickel. We have found indications of these characteristics by using small embedded quantum clusters [Meng *et al.* (1990b)]. First, from an Ni^{2+}-centered nearest-neighbor cluster in a NiO crystal (rocksalt structure, divalent ions) we observed some heavy admixture of Ni^{2+} 3d orbitals with O^{2-} 2p orbitals at the top of the valence band, qualitatively in agreement with Eastman and Freeouf (1975). Second, we considered a similar nickel-centered cluster with a lithium impurity added at a second-neighbor position. This cluster will be discussed further as to detailed electronic distribution in section 11.8.2. The point to be noted here, however, is that the role of the Ni^{2+} 3d orbitals was now completely suppressed at the top of the valence band, leaving pure oxygen 2p character. While this is a local effect in our modelling, which only applies to a highly dilute (extremely low concentration) lithium content, it is reasonable to expect it to show up as a bulk property at the concentrations (1:20 to 1:2) studied by Kuiper *et al.* (1989).

11.7.2   Conduction band edge in $BaF_2:O^-$

The optical excitation process for substitutional oxygen in charge state (-1), O^- in BaF_2 has been discussed in section 11.5.3, and the role of O^- in radiation damage of BaF_2 has been discussed there and in sections 11.2.2 and 11.4.2. The second-neighbor oxygen-centered embedded quantum cluster, containing four Ba^{2+} pseudopotential ions and six all-electron F^- ions has been analysed in detail, computationally [Vail *et al.* (1998a)]. We have already mentioned some features, such as the quadrupole strain and spin polarization effects. These led us to consider two excitations, both from $2p_{x,y}$-like levels, in spin-up and spin-down manifolds respectively. The triplet excited state, dipole forbidden, must be projected out of each of

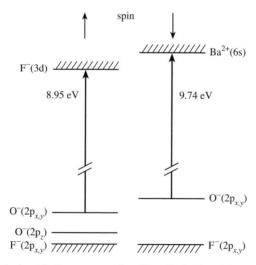

Figure 11.11. Optical excitation process for O^- substitutional impurity ion in BaF_2, showing spin polarization and local chemical differentiation by spin of the bottom of the conduction band.

these, after correlation correction. The results show that the spin-down excitation, in the manifold containing the hole (see figure 11.11), has a Ba^{2+} 6s-like excited state (excitation energy 9.74 eV). The spin-up excitation, by contrast, has fluoride F^- 3d-like character in its final state (excitation energy 8.95 eV). We therefore conclude that O^- substitutional impurity in BaF_2 has the effect of splitting the bottom of the conduction band, at least locally, into spin-polarized, chemically distinct sub-band edges.

11.8   Electronic localization

In section 11.3 we introduced the $(F_2^+)^*$ four-site defect in NaF:Mg, and returned to it in section 11.6.3. In section 11.3 we presented the qualitative computational result, in agreement with experimental spin resonance measurements [Hofmann *et al.* (1985)], that the defect was unsymmetrical with respect to the two nearest anion vacancies (see figures 11.8 and 11.9) [Vail *et al.* (1998b)]. This has the effect of reducing the otherwise symmetrical F_2^+ center, with its electron shared equally by the two vacancies, figure 11.5, to a perturbed F center, with the excess electron localized in one vacancy only. In this case then, electronic localization is associated with symmetry breaking in the atomistic configuration.

We remind the reader that in section 11.6.2, we deduced from spin density calculations the result, also determined experimentally [Hofmann *et al.* (1985)], that the F_2^+ center in NaF, while symmetrical (figure 11.6), is strongly localized about the center of symmetry, rather than spread out into two lobes localized near the vacancy centers. This then represents another kind of localization from that which occurs with the $(F_2^+)^*$ center.

In section 11.7.1, we introduced the subject of lithium substitutional impurity in NiO, discussing its effect on the valence band edge of the crystal. Other calculations on this system [Meng *et al.* (1990b)] cast further light on the electronic configuration associated with lithium in this case. First, since lithium is to substitute for Ni^{2+} with no charge-compensating defect, it must go into NiO with two electrons missing. In charge state (+2), Li^{2+} has one 1s-like electron. We have found, in our calculations, that Li^{2+} in NiO is unstable with respect to electron capture from its O^{2-} neighbors, leaving it in the $1s^2$ charge state Li^+, associated with an oxygen-like hole. The lithium ion's tendency to form the filled $1s^2$ shell Li^+ configuration overwhelms the tendency of the crystal's Madelung potential to hold it in the 1s configuration Li^{2+}. We have further considered two possibilities for this oxygen-like hole: Is it shared equally among the Li^+ ion's six nearest-neighbor oxygen ions, or is it trapped by a single oxygen ion, which would then be in charge state (−1), as O^-? The answer is the latter: the lithium impurity actually exists as a dipolar defect consisting of an Li^+ ion substituting for an Ni^{2+} ion, and a nearest neighbor positively charged defect consisting of an O^{2-} binding a hole, amounting to an O^- ion. In all these calculations, relaxation of the surrounding crystal to equilibrium with the particular defect was included.

The issue addressed here is electronic localization, actually *hole* localization on a single oxygen ion rather than the more diffuse sharing of it among more oxygen neighbors of the impurity lithium. The stabilization of the $(Li^+) \cdot (O^-)$ defect dipole of C_{4v} symmetry relative to the O_h symmetrical configuration is calculated to be 3.1 eV. The system also illustrates the charge-state stabilization of lithium as Li^+ in NiO, and as such might have been discussed in section 11.4.

In section 11.2.2 we discussed how the F center in BaF_2 is related to oxygen impurity through the radiation damage process of equations (11.1) and (11.3). Its optical excitation energy is known experimentally to be 2.03 eV [Cavenett *et al.* (1967)]. We have studied this process, using the same kind of embedded cluster as for the O^- impurity, in this case $F_c \cdot (Ba^{2+})_4 \cdot (F^-)_6$ [Vail *et al.* (2002)]. The computed results indicate an excitation energy of 3.33 eV, comparing badly with the experimental value. We are therefore led to suspect that the true excited state of this F center, although inevitably bound to the vacancy, is more spatially diffuse than our quantum cluster, with the excess electron overlapping far more ions. With the ICECAP modelling method, ions outside the quantum cluster are represented by the classical shell model. Thus, while they have appropriate charges and positions, they lack quantum-mechanical features, namely spatial extent, exchange and correlation. Furthermore, it is known that electrons in diffuse quantum states interact quantum-mechanically with the vibrations of the crystal to an extent that, for strongly ionic materials, significantly affects their properties, as discussed in Chapter 6. We have therefore investigated the possibility of a diffuse excited state for the F center in BaF_2. It has been known experimentally for some time that the F-center excited states in KI [Mollenauer and Baldacchini (1972)] and in KBr [Baldacchini and Mollenauer (1973)] are diffuse, as documented by ENDOR analysis (see section 11.6). Indeed, it is suspected that many F centers have diffuse excited states.

We have found that, with static shell-model embedding, a diffuse excited state gives an estimate of the excitation energy of 2.56 eV, which is much lower, and closer to experiment, than found from localization within the cluster. We have therefore systematically corrected the ICECAP calculation, replacing the excited state's interaction with the shell model crystal by a polaronic description [Fröhlich (1963)]. The periodic potential of the crystal is replaced by an effective band mass, and the interaction of the electron with the vacancy is replaced by a point charge and an effective dielectric constant [Fröhlich (1963)]. The interaction with phonons is represented by a polaronic factor [see equation (6.73)] which, due to the relatively strong electron–phonon coupling in BaF_2, must be evaluated by Feynman's method [Feynman (1955)]. The result of all this is an estimated excitation energy of 2.04 eV, in almost exact agreement with experiment. While such close agreement must be in part fortuitous, given the many limitations of the model applied in these calculations, the fact that the polaronic treatment is of the correct order of magnitude, and in the right direction, is encouraging.

11.9   Quantum diffusion

The classical description of the diffusion process has been discussed in Chapter 10, and in section 11.2.2 we showed how this process can be analysed in the static-crystal approximation based entirely on embedded quantum clusters. While this picture of an atom, described classically or quantum-mechanically, hopping from site to site within the crystal is undoubtedly valid, given the wealth of experimental data with which it conforms, another view of mass transport in a crystal needs to be considered also. The alternative view is based on the quantum-mechanical Bloch states of a particle in the periodic potential of the crystal. Bloch states are not associated with single sites, but rather are extended throughout the crystal. The semi-classical theory of conduction is based on wave packets constructed from Bloch states [see Ashcroft and Mermin (1976), Chapter 13].

In descriptive terms, the site-to-site hopping mechanism needs to be activated by vibrational energy from surrounding atoms, and is also facilitated by the way that such vibrations open up gaps in the periodic potential. Thus it is favored at higher temperatures. The Bloch-like motion relies on the periodic potential not being too much disrupted by crystal vibrations, or by phonon scattering. It is therefore favored at lower temperatures. Quantum-mechanical behavior of this latter type is favored for particles whose masses are small compared with those of the crystal's ions: electrons, obviously, and decreasingly, muons and muonium, protons and hydrogen, helium and lithium.

Considerable work has been done to develop a unified theoretical treatment of diffusion that spans both mechanisms through a broad enough temperature range. Work up to 1995 is summarized by McMullen *et al.* (1995). The following results in this section are based on the theoretical framework established by Flynn and Stoneham (1970), Kagan and Klinger (1974) and McMullen and Bergersen (1978). Two physical parameters are essential: the intersite transfer matrix element $t^{(0)}$ [see for example Ashcroft and Mermin (1976), Chapter 10] and the particle–phonon coupling constant $g^{(0)}$. If these are evaluated for a static crystal, then the effects on them of crystal vibration, referred to as phonon renormalization, need to be taken into account subsequently.

Very well characterized experimental results for diffusion of muonium (Mu) as a function of temperature have been published for KCl [Kiefl *et al.*

(1989)] and for NaCl [Kadono *et al.* (1990)]. Muonium is a hydrogenic atom consisting of a nucleus that is a positive muon μ^+ binding one electron. The small mass of μ^+ compared with that of the proton in hydrogen, ~ 0.1, renders Mu an excellent test of the theory, and provides a striking illustration of the transition from the coherent (Bloch-like) to incoherent (site-to-site) diffusion mechanisms with rising temperature.

Based on embedded quantum cluster calculations for interstitial Mu in NaF [McMullen *et al.* (1995)] we have estimated the parameters $t^{(0)}$ and $g^{(0)}$. These values have then been used to compute the Mu hop rate as a function of temperature. The intrinsic similarity between KCl and NaF, and the observed experimental similarity for this process between KCl and NaCl, encourage us to believe that muonium diffusion will be qualitatively similar in all three crystals. The calculated results for NaF are plotted along with the experimental results for KCl in figure 11.12. While the temperatures for minima of the two data sets are comparable, the hop-rate magnitudes as computed are too low by a factor $\sim 10^{-8}$. This teaches a powerful lesson, or two. First, to the extent that the discrepancy is due to neglect of phonon renormalization, the latter effect must be very large, and therefore cannot be included as a perturbation, contrary to some suggestions in the literature [see for example Kagan and Prokofev (1990)]. We believe

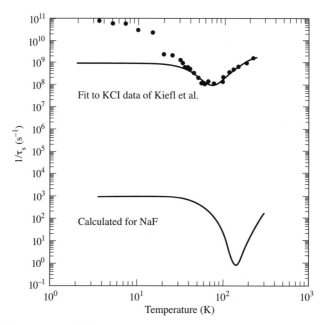

Figure 11.12. Hop rates $1/\tau_s$ for interstitial muonium diffusion as a function of temperature: calculated for NaF; experimental data points for KCl [see Kiefl *et al.* (1989)], with a parametric fit of the theory.

that the discrepancy arises largely from the use of $t^{(0)}$, the unrenormalized tunnelling matrix element, which is underestimated by a factor ~300, compared with what is required for a fit of theory to experiment near the curve's minimum, as shown in figure 11.12. The limitations of our embedded cluster method do not seem to admit any discrepancy of this magnitude, considering the method's wide-ranging success in comparisons with other experimental properties, as reported in the rest of this chapter. Another lesson, therefore, is that if phonon renormalization turns out to be as important as our results suggest, namely zeroth order, then the general formulation should not be carried too far without it. The present results emphasize the importance of quantitative modelling of real materials to provide guidance in the development of theoretical formalism. Furthermore, the deviations at very low temperature between experimental data and a parametric fit of the theory (figure 11.12) suggest a further fundamental weakness of the theory.

11.10   Effective force constants for local modes

We have discussed local-mode vibrations in a simple, one-dimensional example in Chapter 8. It is possible to estimate the effective force constant and local phonon mode frequency from some of our defect calculations.

In section 11.5.2 we introduced the Cu^+ substitutional impurity in NaF. By varying the size of the nearest-neighbor embedded quantum cluster, Meng *et al.* (1988) deduced the force constant K and frequency ν of this 'breathing' mode to be $10.0\,eV\ a_0^{-2}$ and $(35.2/h)\,meV$. The latter result agrees quite well with the experimental value $(24.8/h)\,meV$ [Payne *et al.* (1984)]. In the above units, a_0 is the Bohr radius and h is Planck's constant. The force constant K is found by fitting a parabola to the curve of energy E versus displacement x for nearest-neighbor F^- ions,

$$E = \tfrac{1}{2}Kx^2, \qquad \text{or} \qquad K = \frac{2E}{x^2}, \qquad (11.6)$$

and the frequency ν is given by the harmonic oscillator expression,

$$\nu = \frac{\omega}{2\pi} = \frac{1}{2\pi}(K/M)^{1/2}, \qquad (11.7)$$

where M is the mass of all six fluoride ions.

We have reported the results of similar calculations for Cu^{2+} and Ni^{2+} substitutional impurities in MgO [Meng *et al.* (1990a)], a system first discussed in section 11.4.1. The effective force constants for metal–oxygen interaction are respectively, $K = 31.3\,eV\ a_0^{-2}$ for Cu^{2+} and $35.3\,eV\ a_0^{-2}$ for Ni^{2+}. These values lead to local phonon mode energy predictions of $69.8\,meV$ for Cu^{2+} and $71.4\,meV$ for Ni^{2+}.

11.11   Summary

We have presented in some detail a selection of properties of point defects in ionic crystals. For each defect discussed, computations have been reported, based on quantum-molecular clusters embedded with some physical consistency in an infinite shell-model crystal. In a good smattering of cases, comparison with experimental results has been possible. In all calculations, a common physical model has been used, applying a single computer program bearing the acronym ICECAP [Harding *et al.* (1985)]. This fact enables us to assess the reliability of the model and method across a considerable range of materials, defect types and defect properties. We believe that this gives the reader a good introduction to the science of defects, even though the topics covered necessarily represent only a small fraction of the subject. We now review these topics to indicate some of the relationships among them.

We establish four categories: *quantitative* defect properties for which comparable experimental data are available, and similar properties for which they are not, leaving us with computationally based predictions; and *qualitative* defect properties that are correspondingly experimentally known or predictive. We begin by summarizing quantitative results.

In section 11.2.1, activation energies for vacancy and interstitial diffusion of Cu^+ and interstitial diffusion for Ag^+, all in KCl, are computed, table 11.1. Good quantitative agreement is obtained with experiment, for the activation energies and their ordering. Vacancy activation energies are quantitatively predicted for Ag^+ in four other alkali halides, in two of which the qualitative feature of a non-collinear mechanism, figure 11.4, is predicted. In table 11.1, the calculations are all based on shell-model representation of the impurity, derived from subsidiary embedded quantum cluster calculations. The diffusing ion is a noble metal impurity cation in all cases. In section 11.2.2, by contrast, vacancy diffusion of the host anion F^- in BaF_2 is analysed with full embedded quantum cluster modelling rather than by the shell model exclusively. The context is that of defect complex dissociation, namely for the dipolar O^{2-}–fluoride vacancy complex. Here the resultant activation energy of 0.93 eV is predictive, with no experimental value available.

Three kinds of optical excitation process are discussed in section 11.5, all in quantitative terms. One, in section 11.5.1, is an intrinsic excitation in MgO, the Frenkel exciton involving the O^{2-} anion. Excellent agreement is obtained with experiment for both singlet and triplet excitations. Another, in section 1.5.3, is the O^- substitutional impurity ion in BaF_2, whose singlet line is split by spin polarization. This result is predictive. Also predictive are the calculated excitations of substitutional impurity S^{2-} and Se^{2-} anions in MgO. Third, in section 11.5.2, the calculated excitation of the substitutional impurity Cu^+ cation in NaF agrees well with experiment, as does its crystal field splitting. The optical absorption of the F center in BaF_2 is discussed in section 11.8, where good quantitative agreement with experiment is obtained when a polaronic correction is introduced into the calculations to account for the diffuseness of the excited state.

Quantitative spin densities calculated for three related F-type centers in NaF are presented in section 11.6. In section 11.6.1, fair agreement is obtained between calculated and experimental spin density for nearest neighbors of the F center, an intrinsic defect. Experimental values are not available for the F_2^+ center, another intrinsic defect, but the quantitative calculated values of section 11.6.2 agree with a qualitative experimental feature mentioned later, namely electronic localization. For the $(F_2^+)^*$ center, a four-site impurity-related defect, basically a perturbed F center, quantitative spin densities for inequivalent nearest neighbors are calculated in section 11.6.3. Experimental values are unavailable and are likely to remain so. Furthermore, the calculations are so much more difficult than and different from the others mentioned here that they cannot be assumed to be reliable. They do illustrate, however, how different atomistic configurations can be distinguished through different spin density patterns (table 11.3).

Local phonon mode frequencies have been calculated for three substitutional impurity cations. For Cu^+ in NaF, fair quantitative agreement with experiment has been obtained. For Cu^{2+} and Ni^{2+} in MgO, experimental values are not available, so the calculated values are predictive. For Cu^+ in NaF, the local mode energy is 35.2 meV (calculated) and 24.8 meV (experimental), while for Cu^{2+} and Ni^{2+} in MgO the calculated values are 69.8 meV and 71.4 meV respectively.

We now turn to qualitative results, which can be just as revealing as quantitative results are about the nature of the defect solid and about the computational model's reliability.

The stability of a defect complex has been exemplified by a study of the $(F_2^+)^*$ center in NaF:Mg in section 11.3. Quantitative analysis of a large number of configurations led to two qualitative conclusions. First, the symmetric configuration, figure 11.7, is not energetically favored. This conclusion conforms with experiment. Second, two particular configurations, quite close in energy, figures 11.8 and 11.9, have significantly lower energies than other configurations. One of these is planar, invariant under a single symmetry operation; the other is non-coplanar, with no symmetry.

Charge-state stability of substitutional impurity ions has been studied in two systems. In section 11.4.1, we determined that nickel in MgO is stable in charge state (+2) as Ni^{2+}, compared with an equal mixture of Ni^{3+} and Ni^+. This qualitative result conforms with the experimental fact that neither Ni^{3+} or Ni^+ are found free-standing in MgO. In fact, our calculations show that Ni^{3+} transforms to Ni^{2+} with an associated hole in neighboring oxygen ions, and Ni^+ transforms to Ni^{2+} with an excess electron associated with neighboring magnesium ions. The disproportionation energy required to transfer an electron from one Ni^{2+} ion to another, creating the polarization fields in the crystal associated with Ni^+ and Ni^{3+}, is estimated to be 10.1 eV. In section 11.4.2, oxygen in BaF_2 associated with a vacancy is found to be stable in charge state (-1) as O^-, associated with the one-electron F center. Again, to transfer the F-center electron to oxygen, making it the filled shell ion O^{2-}, the energy required is calculated to be 1.4 eV. While the first of these examples involves cations (nickel) and the second involves anions (oxygen), in both cases the stable charge states are uncharged point defects, Ni^{2+} in MgO and O^- and F center in BaF_2.

Local modification of electronic band edges have been described in two systems in section 11.7. The effect is described as chemical, since the atomic species that dominates the band-edge states is affected by the point defect. In section 11.7.1 we showed how the impurity lithium anion, substitutional in NiO, suppresses the nickel 3d character of the valence-band edge, exposing pure oxygen 2p character. This conforms with experimental results at significant concentration. The substitutional lithium must have charge $(+2)$. However it is found to exist as filled-shell Li^+ with an associated hole in neighboring oxygen ions. In section 11.7.2 the substitutional impurity cation O^- in BaF_2 is again discussed. The ion's unpaired electron creates a symmetry-breaking electric quadrupole moment and a local screening quadrupole strain. The spin polarization, particularly, splits the bottom of the conduction band into barium and fluoride parts, the splitting showing up energetically in the optical absorption process. Thus in section 11.7, a cation lithium impurity in NiO modifies the top of the host crystal's valence band, as found experimentally, and an anion oxygen impurity in BaF_2 locally modifies the bottom of the conduction band. For the latter effect, experimental results are unavailable.

Electronic localization is discussed for four defects, in section 11.8. Lithium substitutional impurity in NiO consists of an Li^+ cation plus a hole. In fact, the hold localization is on a single oxygen ion, and is energetically favored over distribution on six nearest neighbors by 3.1 eV. This results in an impurity dipole complex consisting of the negative defect Li^+ and the positive defect O^- in NiO. We might therefore characterize this defect as another example of charge-state stability and also of defect-complex stability, both of which have already been discussed for other systems. In the F_2^+ center in NaF, the excess electron is found to be localized about the defect's center of symmetry, while in the $(F_2^+)^*$ center it is forced into one vacancy by symmetry breaking. For both of these F-type centers, experimental results support the qualitative conclusions derived from computational modelling. The spatial diffuseness of the F-center excited state in BaF_2 predicted by our calculations leads to quantitative agreement with experimental optical absorption.

Quantum diffusion in alkali halides, as described in section 11.9, has been well characterized experimentally. Application of the standard theory without phonon renormalization is made possible by physically reliable embedded quantum cluster calculations. In this case, spectacular *disagreement* with experiment teaches us something useful, namely that the renormalization effects must be large, ruling out perturbative treatment. Qualitative agreement between experiment and computation is obtained for the crossover temperature between coherent and incoherent diffusion mechanisms.

The network of results summarized in this section involves different materials, defect types, and processes; some qualitative results, some quantitative; some predictive results, some supported by experiment. As such it supports the view that the methodology (ICECAP) based on shell-model embedding for quantum molecular clusters can be widely reliable for point defects in insulators when properly applied.

Appendix to Chapter 11: the ICECAP method

The ICECAP method referred to in section 11.1.2 has been reviewed by Vail *et al.* (1991). We give a brief outline in this appendix.

In order to model a point defect in an ionic crystal, we must calculate the total energy of the crystal containing the defect, in a particular physical state. We represent the point defect and its immediate surroundings in the crystal as a molecular cluster, treated quantum mechanically. The rest of the crystal is represented by the shell model (section 9.2). We begin with an infinite, perfect shell-model crystal. We define a *cluster region* in this crystal: a set of shell-model ions that will be replaced by the defect quantum molecular cluster. In general, the defect quantum cluster will have a charge distribution differing from that of the perfect-crystal shell-model ions in the cluster region. This cluster charge distribution must be in equilibrium with shell-model ion displacements and polarizations in the embedding classical crystal.

In order to accomplish this, we first replace the cluster region with a classical representation of the defect, in terms of shell-model ions and additional point charges (when needed), at specific sites. Consider, as an example, the O^- ion in BaF_2, figure 11.10. The ion has a prolate quadrupole moment, as discussed in section 11.5.3. Thus, in addition to charge (-1) at the oxygen ion site, we would add two charges, each of charge $q > 0$ at positions $\pm u$ on the z axis, and a further charge $(-2q)$ at the ionic site to maintain total charge (-1). These *quadrupole simulators* induce a polarization field in the surrounding shell-model crystal when the total energy of the classical defected crystal is minimized with respect to all core and shell positions. The ICECAP program contains a sub-program [HADES: see Norgett (1974)] that performs this operation.

With initial values of point charge simulators (q in the BaF_2:O^- example) and their positions (u in the example) having determined the embedding polarization field, we now replace the cluster region with a quantum-molecular cluster. The quantum cluster, in general, consists of nuclei, electrons and atomic pseudopotentials. In the example, suppose we take a second-neighbor cluster consisting of the O^- ion, its four nearest-neighbor Ba^{2+} ions, and its six second-neighbor F^- ions. We must specify the number of electrons. For all-electron treatment of this cluster, from the periodic table of the elements we conclude that we need 9 electrons for O^-, 54 electrons for each Ba^{2+} ion, and 10 electrons for each F^- ions, for a total of 285 electrons. If this leads to calculations that are prohibitively large, as in our case, then we can replace the Ba^{2+} ions by pseudopotentials. A pseudopotential is a relatively simple effective potential that approximates the effect of the ion on electrons outside the ion. Many forms of pseudo-potential have been derived and tabulated. The ICECAP program accepts three types: KKLP [Kunz and Klein (1978)], BHS [Bachelet *et al.* (1982)], and TOP [Topiol *et al.* (1978)]. In this case, the quantum cluster consists of four Ba^{2+} pseudopotentials, one oxygen nucleus (charge $+8$), six fluorine nuclei (charges $+9$), and 69 electrons.

The total energy and many-body wave function of the electrons in the presence of the fixed nuclei, pseudopotentials, and the shell-model point charges of the embedding crystal, are now determined by a Hartree–Fock program [Kunz (1982)] incorporated in ICECAP. The Hartree–Fock approximation and its application are discussed in chapter 12, especially in sections 12.2 and 12.3. This requires us to specify a basis set within which the Fock equation is solved exactly. The solution is, of course, by no means exact, because the chosen basis set is finite, rather than complete in the mathematical sense. Many useful tabulations of atomic orbital basis sets exist. One that we have found useful for point-defect calculations is by Huzinaga (1984). If one uses relatively small basis sets, as we do, it will generally be necessary to optimize the basis functions by minimizing the total energy of the defected crystal with respect to exponential coefficients and contraction coefficients: see equation (12.110) and discussion following it.

The total energy of the defected crystal, from the ICECAP calculation described above, is the sum of three terms. These are: (1) the total energy of the polarized embedding shell-model crystal, relative to the energy of the unpolarized (perfect) embedding; (2) the Coulomb interaction energy among nuclei and pseudopotentials of the quantum cluster, and between them and the embedding shell-model ions; and (3) the Hartree–Fock energy of the cluster electrons, including their interactions with nuclei and pseudopotentials of the cluster and with the shell-model embedding ions. A fourth contribution to the total energy can be added: correlation (see section 12.6).

For a given state of the defected crystal, two more computational processes need to be carried out. One is to achieve consistency between point-charge simulators and the Hartree–Fock solution. This is achieved by minimizing the total energy, as defined in the previous paragraph, with respect to variation of charges and positions of the simulators, denoted q and u respectively in the example of $BaF_2:O^-$. The other process is to minimize the total energy with respect to nuclear and pseudopotential positions within the quantum cluster. The various optimizations mentioned (basis set, point-charge simulators, and molecular cluster configuration) need to be carried through to mutual consistency ('global' total energy minimization) by iteration.

Optical excitations are often thought of in terms of the Condon approximation, in which ionic positions do not relax to equilibrium with the electronic distribution of the excited state during the transition. The ICECAP program enables us to take this approach, by keeping nuclei and pseudopotentials fixed in the positions they have in the ground state, and also maintaining the polarization of the embedding crystal in the ground state configuration.

The embedding problem is discussed in section 12.5. Basically, it consists of ensuring that the cluster electrons experience, in some way, the quantum-mechanical effects of nearby ions in the embedding region, even though these ions are in fact classical shell-model ions, with no actual quantum-mechanical properties. The preferred method of embedding is to have the outer ions of the molecular cluster represented exclusively by pseudopotentials.

Chapter 12

Theoretical foundations of molecular cluster computations

12.1 Introduction

In the previous chapter we described a computational model for point defects in insulators. In it, the local region containing the defect is to be analysed quantum-mechanically, and the rest of the crystal is to be described in terms of a classical atomistic model. In the applications described in Chapter 11, all atomic positions and configurations were assumed to be static. The quantum-mechanical part of the model consists of the electrons associated with a small molecular cluster containing the point defect, including the interaction of the electrons with nuclei in the cluster and with the classical atoms of the embedding region. This chapter is devoted to the theoretical background of such a quantum-mechanical calculation.

In sections 12.2 to 12.6 we describe a particular approach to embedded molecular cluster calculation, based on the Hartree–Fock approximation (sections 12.2 and 12.3), including correlation correction (section 12.6). In a crystal, quantum-mechanical aspects of cluster-embedding interaction must be considered (sections 12.4 and 12.5). In section 12.7 some general considerations are presented regarding the N-body problem that may in future lead to improved computational methods, namely density functional methods. Currently, and in fact since the 1950s, a single-particle density functional method has provided a useful alternative to the Hartree–Fock based approach, as discussed in section 12.7.2.

While molecular cluster methods are very important in the theory of solid materials for the calculation of point-defect and other localized electronic properties, and indeed for electronic band structure, their significance is in fact far broader, and fast increasing. With the development of quantum-mechanical embedding schemes, the method will become increasingly applicable to semiconductor and metallic materials. Already, molecular cluster methods are of great value in the analysis of local properties on surfaces [see for example Sushko *et al.* (2000)]. Of course, such methods originated in the field of quantum chemistry, where the analysis of isolated molecules is of major importance. But there is another area of science where these and other methods of theoretical and experimental atomic-scale physics are important and proliferating. This is the field of molecular biology. The methods of this chapter apply to local electronic and related properties of large biomolecules, and to smaller free biomolecules. Beyond this, however, is the fast-emerging field in which biomolecules, either alone or combined with inorganic atomistic elements, have properties that allow radical new device components to be fabricated. As a single example we cite Bhyrappa *et al.* (1999), who created a molecular complex with a metallic ion at its center such that, depending on the complex's detailed configuration, differently shaped natural biomolecules would be selected to bind to the metallic ion. At the American Physical Society March 2000 meeting there was a series of six focused sessions on the subject of molecular scale electronics [see *APS Bulletin*, 2000, **45**(1), 20]. The key to the explosive growth in sub-nanoscale technology now taking place in the physical and life sciences is the ability to control, by chemical and physical means, the detailed atom-by-atom configurations of molecular clusters. This is where the material of this chapter comes in, for complementary computational modelling of known or postulated clusters.

12.2 Hartree–Fock approximation

12.2.1 The approximation

The Hartree–Fock approximation, and the Fock equation of the next section, are standard subjects in quantum chemistry and solid state physics. As such, they are introduced in many books. Some of the best are Reitz (1955, sections I–III), Slater (1963, especially Chapter 5 and Appendices 4–9), and Szabo and Ostlund (1982, Chapters 2 and 3).

The theoretical description of a solid, in terms of its electrons, nuclei and crystal structure, is formulated in sections 7.1 and 7.2, and the first paragraph of section 7.3. We refer the reader to these sections now. We begin from the last paragraph of section 7.5, in which the nuclei are viewed as classical point charges in equilibrium with the electrons: equations (7.13) and (7.75). These equations require that the many-electron wave function $\psi_\lambda(\underline{r})$, with $\underline{r} = (\vec{r}_1, s_1; \vec{r}_2, s_2; \ldots; \vec{r}_N, s_N)$, should be an eigenstate of the static crystal hamiltonian H_s, equation (7.77) or equivalently, of the last two terms of H_s which in this chapter we simply denote H:

$$H = \sum_{j=1}^{N} \left\{ -\frac{\hbar^2}{2m}\nabla_j^2 - \frac{e^2}{4\pi\varepsilon_0}\left[\sum_J Z_J|\vec{r}_j - \vec{R}_J|^{-1} - \frac{1}{2}\sum_{j'}{}' |\vec{r}_j - \vec{r}_{j'}|^{-1} \right] \right\}.$$

$$(12.1)$$

It is common to re-express this in terms of Bohr–Hartree atomic units, where energy is measured in units of hartree [$1\,\mathrm{Hy} = 2\,\mathrm{Ry} = (me^4/\hbar^2) = 27.2\,\mathrm{eV}$] and length is measured in bohr [$1\,\mathrm{bohr} \equiv a_0 = (4\pi\varepsilon_0\hbar^2)/(me^2) = 0.529\,\mathrm{A}$]. In these units, equation (12.1) reduces to

$$H = \sum_{j=1}^{N}\left\{-\frac{1}{2}\nabla_j^2 - \sum_J Z_J|\vec{r}_j - \vec{R}_J|^{-1} + \frac{1}{2}\sum_{j'}{}'|\vec{r}_j - \vec{r}_{j'}|^{-1}\right\}. \tag{12.2}$$

In equations (12.1) and (12.2), R_J is the nuclear position, and the double sum over j and j' omits the term $j = j'$, indicated by the prime on $\sum$. This hamiltonian is appropriate for the electrons in a molecular cluster, but does not contain the Coulomb interaction energy among the nuclei.

Now consider the many-electron wave function $\psi_\lambda(\underline{r})$, corresponding to the energy eigenvalue E_λ:

$$H|\psi_\lambda\rangle = E_\lambda|\psi_\lambda\rangle. \tag{12.3a}$$

Let us first think of Fourier analysing an *arbitrary* function $f(\underline{r})$ of one-particle variables in terms of a complete orthonormal set of single-particle basis functions $\{\varphi_k(\underline{r}_j)\}$, where $\underline{r}_j = (\vec{r}_j, s_j)$ are single-particle position and spin coordinates. Then orthonormality is expressed as

$$\int dr_j\, \varphi_k^*(\underline{r}_j)\varphi_{k'}(\underline{r}_j) = \delta_{k,k'}, \tag{12.3b}$$

where the symbol $\int dr_j$ forms the inner product of spin orbitals φ_k and $\varphi_{k'}$ in both configuration and spin subspaces $\vec{r}_j$ and s_j respectively. The Fourier series for $f(\underline{r})$ then becomes

$$f(\underline{r}_1, \underline{r}_2, \ldots, \underline{r}_N) = \sum_{k_1, k_2, \ldots, k_N} c_{k_1, k_2, \ldots, k_N}\varphi_{k_1}(\underline{r}_1)\varphi_{k_2}(\underline{r}_2), \ldots, \varphi_{k_N}(\underline{r}_N), \tag{12.4}$$

where in the summation each index k_j ranges over the whole infinite orthonormal set $\{\varphi_k(\underline{r}_j)\}$. There are two popular approximations to the many-electron wavefunction arising from equation (12.4). One is the Hartree approximation, which begins by taking a single term from the sum, with the condition that $(k_1, k_2, \ldots, k_N)$ are all different. This satisfies the Pauli exclusion principle intuitively by having each electron's coordinates restricted to a single function, distinct from all the others. Unfortunately, this renders the electrons distinguishable. The other approximation is the Hartree–Fock approximation.

The Hartree–Fock approximation arises upon applying the Pauli principle in its general form to the many-particle function $f(\underline{r})$, equation (12.4). If $f(\underline{r})$ is to be a many-electron wave function $\psi_\lambda(\underline{r})$, then it must be odd with respect to pairwise interchange of any pair of particles,

$$P_{ij}\psi_\lambda(\underline{r}) = -\psi_\lambda(\underline{r}), \tag{12.5}$$

for all (i, j) with $i \neq j$, where P_{ij} is the operator that interchanges particle coordinates i and j:

$$P_{ij}\psi_\lambda(\underline{r}_1, \ldots, \underline{r}_i, \ldots, \underline{r}_j, \ldots, \underline{r}_N) = \psi_\lambda(\underline{r}_1, \ldots, \underline{r}_j, \ldots, \underline{r}_i, \ldots, \underline{r}_N). \quad (12.6)$$

In equation (12.4), now consider all those terms for which the sets of indices $(k_1, k_2, \ldots, k_N)$ are the same, with the k_j only permuted amongst themselves. If any two, say k_i and k_j, are equal, then the action of P_{ij}, equation (12.6), in interchanging $\underline{r}_i$ and $\underline{r}_j$ produces no change, and in particular does not introduce the negative sign of equation (12.5), regardless of the values of the coefficients $c_{k_1, \ldots, k_N}$ of these terms. Such coefficients must therefore be zero, and the only terms in equation (12.4) that may be non-zero are those for which the indices $(k_1, k_2, \ldots, k_N)$ are all different. Denote such a set of indices by $\underline{k}$ without regard for the ordering of $k_1, k_2, \ldots, k_N$.

Because the sums in equation (12.4) all range over the infinite complete set $\{\varphi_k(\underline{r}_j)\}$, all permutations of $(k_1, k_2, \ldots, k_N)$ occur. Let P_N be a particular permutation of $(k_1, k_2, \ldots, k_N)$. Then P_N consists of a non-unique set of pairwise interchanges. Two different sets of pairwise interchanges giving P_N differ by an even number of pairwise interchanges. Thus, for given $\underline{k}$, the non-zero terms in equation (12.4) are

$$\sum_{P_N} P_N(c_{k_1, \ldots, k_N} \varphi_{k_1}, \ldots, \varphi_{k_N}). \quad (12.7)$$

The sets of terms for different sets of indices $\underline{k}$ in equation (12.4) are linearly independent. Thus every pairwise interchange applied to equation (12.7) must give a negative sign. Such a pairwise interchange of coordinates r_i and r_j is indistinguishable from the corresponding interchange of indices k_i and k_j. Thus, from equation (12.7) and equation (12.5) applied to a particular set of terms defined by $\underline{k}$,

$$P_{ij} \sum_{P_N} P_N(c_{k_1, \ldots, k_N} \varphi_{k_1}, \ldots, \varphi_{k_N}) = - \sum_{P_N} P_N(c_{k_1, \ldots, k_N} \varphi_{k_1}, \ldots, \varphi_{k_N}). \quad (12.8)$$

Consider two specific permutations $\tilde{P}_N$ and $P_{ij}\tilde{P}_N$ from the sum in the right-hand side of equation (12.8). They are:

$$(\tilde{P}_N c_{k_1, \ldots, k_N})(\tilde{P}_N \varphi_{k_1}, \ldots, \varphi_{k_N}) + (P_{ij}\tilde{P}_N c_{k_1, \ldots, k_N})(P_{ij}\tilde{P}_N \varphi_{k_1}, \ldots, \varphi_{k_N}). \quad (12.9)$$

Under the operation of P_{ij} as in equation (12.8), the two products $\tilde{P}_N \varphi_{k_1}, \ldots, \varphi_{k_N}$ and $(P_{ij}\tilde{P}_N \varphi_{k_1}, \ldots, \varphi_{k_N})$ interchange, and this interchange must produce a sign change, because all different specific permutations $\tilde{P}_N$ are linearly independent. It follows from equation (12.9) that

$$(P_{ij}\tilde{P}_N c_{k_1, \ldots, k_N}) = -(\tilde{P}_N c_{k_1, \ldots, k_N}). \quad (12.10)$$

This condition is satisfied in general if and only if

$$(\tilde{P}_N c_{k_1, \ldots, k_N}) \propto (-1)^{\delta_{\tilde{P}_N}} \quad (12.11)$$

where $\delta_{\tilde{P}_N}$ is the number of pairwise interchanges in $\tilde{P}_N$. While this number is not unique, $(-1)^{\delta_{\tilde{P}_N}}$ is unique. Thus, apart from normalization, the terms for a given set of N indices $\underline{k}$ in equation (12.4) are, from equations (12.7) and (12.11),

$$\sum_{P_N} P_N(-1)^{\delta_{P_N}} (\varphi_{k_1}, \ldots, \varphi_{k_N}). \tag{12.12}$$

It follows that the most general many-fermion wave function, satisfying equation (12.5) is, from equations (12.4) and (12.12),

$$\psi(\underline{r}) = \sum_{\underline{k}} c_{\underline{k}} \sum_{P_N} P_N(-1)^{\delta_{P_N}} \prod_{\underline{k}} \varphi_{k'_j}(\underline{r}_j), \tag{12.13}$$

where $\underline{k}$ is a distinct set of N indices $(k'_1, \ldots, k'_N)$, all different, chosen from the infinite set of indices $(k_1, \ldots, k_N, k_{N+1}, \ldots)$, $c_{\underline{k}}$ are arbitrary constants, subject to normalization, $\prod_{\underline{k}} \varphi_{k'_j}(\underline{r}_j)$ is the product $(\varphi_{k'_1}, \ldots, \varphi_{k'_N})$, and P_N is a permutation of $(k'_1, \ldots, k'_N)$.

Equation (12.13) is commonly written as

$$\psi(\underline{r}) = \sum_{\underline{k}} c_{\underline{k}} A \prod_{\underline{k}} \varphi_{k'_j}(\underline{r}_j), \tag{12.14}$$

where the antisymmetrizing operator A is defined by equations (12.13) and (12.14) as

$$A = \sum_{P} P(-1)^{\delta_P}, \tag{12.15}$$

where in equation (12.15) we have suppressed the subscript N for the N-body system. While equation (12.14) represents the most general many-fermion wave function, the most general many-boson wave function (symmetric under pairwise interchange), has the form

$$\tilde{\psi}(\underline{r}) = \sum_{\underline{k}} c_{\underline{k}} S \prod_{\underline{k}} \varphi_{k'_j}(\underline{r}_j), \tag{12.16}$$

where the symmetrizing operator is

$$S = \sum_{P} P. \tag{12.17}$$

We can now specify the Hartree–Fock approximation. Using the language of Hilbert space (single-particle function space in the present case), each term in equation (12.14), corresponding to a different set $\underline{k}$ of N single particle functions, comes from a different N-dimensional manifold of Hilbert space. If a single term from equation (12.14), i.e. a single N-dimensional manifold, dominates the wave function $\psi(\underline{r})$, then we have the Hartree–Fock approximation:

$$\psi(\underline{r}) \approx nA \prod_{\underline{k}} \varphi_{k'_j}(\underline{r}_j) \tag{12.18}$$

where n is the normalizing factor. Referring to equation (12.15) with equation (12.18), we see that an alternative form for $\psi(\underline{r})$ is in terms of a determinant, the so-called Slater determinant:

$$\psi(\underline{r}) = n \cdot \det \begin{pmatrix} \varphi_1(\underline{r}_1) & \varphi_2(\underline{r}_1) & \cdots & \varphi_N(\underline{r}_1) \\ \varphi_1(\underline{r}_2) & \varphi_2(\underline{r}_2) & \cdots & \varphi_N(\underline{r}_2) \\ \vdots & & & \\ \varphi_1(\underline{r}_N) & \varphi_2(\underline{r}_N) & \cdots & \varphi_N(\underline{r}_N) \end{pmatrix}. \qquad (12.19)$$

In equation (12.19) we have simplified the notation $(k_1, k_2, \ldots, k_N)$ to $(1, 2, \ldots, N)$, labelling the N single-particle functions, which we reiterate are an orthonormal set.

12.2.2 Normalization

We must now learn how to calculate with Hartree–Fock wave functions, or Slater determinants. First, a general result. Consider a symmetrical N-body operator $O(\underline{r})$ such that $PO(\underline{r}) = O(\underline{r})P$. Its matrix element between two Slater determinants is, from equations (12.18) and (12.15),

$$\int d\tau_{\underline{r}}\, \psi_1^*(\underline{r}) O(\underline{r}) \psi_2(\underline{r}) = n^2 \int d\tau_{\underline{r}} \left\{ \sum_P P(-1)^{\delta_p} \prod_k \varphi_{k_j}^*(\underline{r}_j) \right\}$$

$$\times O(\underline{r}) \left\{ \sum_{P'} P'(-1)^{\delta_{P'}} \prod_l \varphi_{l_j}(\underline{r}_j) \right\}, \qquad (12.20)$$

where the sets $\underline{k}$ and $\underline{l}$ may differ. In each term in the sum over P, we now relabel the variables $(\underline{r}_1, \underline{r}_2, \ldots, \underline{r}_N)$ as $(\underline{r}'_1, \underline{r}'_2, \ldots, \underline{r}'_N)$ where

$$(\underline{r}'_1, \underline{r}'_2, \ldots, \underline{r}'_N) = P^{-1}(\underline{r}_1, \underline{r}_2, \ldots, \underline{r}_N). \qquad (12.21)$$

Now

$$P^{-1} \cdot d\tau_{\underline{r}} = d\tau_{\underline{r}}, \qquad \text{unchanged};$$

$$P^{-1} \cdot P \prod_k \varphi_{k_j}^*(\underline{r}_j) = \prod_k \varphi_{k_j}^*(\underline{r}_j);$$

and P^{-1} applied to $O(\underline{r})$ leaves $O(\underline{r})$ unchanged. We thus have, for equation (12.20),

$$\int d\tau_{\underline{r}}\, \psi_1^*(\underline{r}) O(\underline{r}) \psi_2^*(\underline{r}) = n^2 \sum_P \int d\tau_{\underline{r}} \left\{ \prod_k \varphi_{k_j}^*(\underline{r}_j) \right\}$$

$$\times O(\underline{r}) \left\{ \sum_{P'} P^{-1} \cdot P' \cdot (-1)^{\delta_{P'} + \delta_P} \prod_l \varphi_{l_j}(\underline{r}_j) \right\}, \qquad (12.22)$$

since $(-1)^{\delta_P} = (-1)^{\delta_{P^{-1}}}$, and if we let $P^{-1} . P' = P''$, then

$$\sum_{P'} P^{-1} P' (-1)^{\delta_{P'} + \delta_{P^{-1}}} = \sum_{P''} P'' (-1)^{\delta_{P''}} = A. \qquad (12.23)$$

It follows that

$$\int d\tau_{\underline{r}} \, \psi_1^*(\underline{r}) O(\underline{r}) \psi_2(\underline{r}) = n^2 \sum_P \int d\tau_{\underline{r}} \left\{ \prod_k \varphi_{k_j}^*(\underline{r}_j) \right\} O(\underline{r}) \left\{ A \prod_l \varphi_{l_j}(\underline{r}_j) \right\}$$

$$= n^2 . N! \int d\tau_{\underline{r}} \, \psi_1^{(H)^*}(\underline{r}) O(\underline{r}) \psi_2^{(HF)}(\underline{r}). \qquad (12.24)$$

since $\sum_P$ is $N!$ identical terms, and where Hartree (H) and Hartree–Fock (HF) wave functions are

$$\psi_1^{(H)}(\underline{r}) = \prod_k \varphi_{k_j}(\underline{r}), \qquad (12.25)$$

$$\psi_2^{(HF)}(\underline{r}) = A \prod_l \varphi_{l_j}(\underline{r}_j). \qquad (12.26)$$

We use equation (12.24) first for normalization, by considering $O(\underline{r})$ to be the identity operator, independent of $\underline{r}$, and $\psi_1 = \psi_2$:

$$\int d\tau_{\underline{r}} \, \psi^*(\underline{r}) \psi(\underline{r}) = 1$$

$$= n^2 . N! \int dr_2 \, dr_2, \ldots, dr_N \{ \varphi_1^*(\underline{r}_1), \ldots, \varphi_N^*(\underline{r}_N) \}$$

$$\times A \{ \varphi_1(\underline{r})_1, \ldots, \varphi_N(\underline{r}_N) \} \qquad (12.27)$$

where we have replaced $(k_1, \ldots, k_N)$ by $(1, \ldots, N)$. But only the identity permutation in A, equation (12.15), gives a non-zero result for equation (12.27), due to orthonormality of the set $\{ \varphi_k(\underline{r}_j) \}$. It follows that

$$1 = n^2 . N!, \qquad (12.28)$$

whence the normalized N-electron Slater determinant is, from equations (12.18) and (12.28):

$$\psi(\underline{r}) = (N!)^{-1/2} A \prod_k \varphi_j(\underline{r}_j). \qquad (12.29)$$

12.2.3 Total energy

Next we consider the total energy, as estimated in Hartree–Fock approximation,

$$E \approx \langle \psi | H | \psi \rangle, \qquad (12.30)$$

where $|\psi\rangle$ is a Slater determinant, equation (12.29), with H given by equation (12.2). We note first that H is a symmetrical operator:

$$PH = HP. \qquad (12.31)$$

Thus we apply equation (12.24), with $\psi_1 = \psi_2$, $O = H$, and $n^2 N! = 1$ [equation (12.28)]:

$$E \approx \langle \psi | H | \psi \rangle$$

$$= \int dr_1, \ldots, dr_N \{\varphi_1^*(\underline{r}_1), \ldots, \varphi_N^*(\underline{r}_N)\} H . A\{\varphi_1(\underline{r}_1), \ldots, \varphi_N(\underline{r}_N)\}. \quad (12.32)$$

First consider the single-particle terms H_1 in H, equation (12.2):

$$H_1 = \sum_j h_1(\underline{r}_j); \qquad h_1(\underline{r}_j) = \left\{ -\frac{1}{2}\nabla_j^2 - \sum_J Z_J |\underline{r}_j - \underline{R}_{J_0}|^{-1} \right\}. \quad (12.33)$$

Without loss of generality, consider the term $j = 1$ from equation (12.33) in equation (12.32). It is

$$\int dr_1, \ldots, dr_N \{\varphi_1^*(\underline{r}_1), \ldots, \varphi_N^*(\underline{r}_N)\} h_1(\underline{r}_1) A\{\varphi_1(\underline{r}_1), \ldots, \varphi_N(\underline{r}_N)\}. \quad (12.34)$$

Now, again from orthonormality, only the identity permutation in A gives a non-zero result. With this identity term we have, from equation (12.34),

$$\int dr_1 \, \varphi_1^*(\underline{r}_1) h_1(\underline{r}_1) \varphi_1(\underline{r}_1) \int dr_2 \, \varphi_2^*(\underline{r}_2) \varphi_2(\underline{r}_2) \cdots \int dr_N \, \varphi_N^*(\underline{r}_N) \varphi_N(\underline{r}_N)$$

$$= \int dr_1 \, \varphi_1^* h_1 \varphi_1 = \langle \varphi_1 | h_1 | \varphi_1 \rangle. \quad (12.35)$$

In equation (12.35) we have introduced Dirac notation for single-particle functions. From equations (12.33) and (12.35) we have the single-particle energy E_1:

$$E_1 = \langle \psi | H_1 | \psi \rangle = \sum_{j=1}^{N} \langle \varphi_j | h_1 | \varphi_j \rangle. \quad (12.36)$$

Next consider the two-particle terms H_2 in the hamiltonian, equation (12.2):

$$H_2 = \frac{1}{2} \sum_{j,j'}' h_2(\underline{r}_j, \underline{r}_{j'}), \qquad h_2(\underline{r}_j, \underline{r}_{j'}) = |\vec{r}_j - \vec{r}_{j'}|^{-1}. \quad (12.37)$$

Similar to equation (12.34) we now have a term in the energy:

$$\int dr_1, \ldots, dr_N \{\varphi_1^*(\underline{r}_1)\varphi_2^*(\underline{r}_2), \ldots, \varphi_N^*(\underline{r}_N)\} h_2(\underline{r}_1, \underline{r}_2) A \varphi_1(\underline{r}_1)\varphi_2(\underline{r}_2), \ldots, \varphi_N(\underline{r}_N)\}.$$

$$(12.38)$$

Now from the antisymmetrizer A, only two terms give non-zero results. They are

$$\{\varphi_1(\underline{r}_1)\varphi_2(\underline{r}_2) - \varphi_1(\underline{r}_2)\varphi_2(\underline{r}_1)\}\varphi_3(\underline{r}_3), \ldots, \varphi_N(\underline{r}_N),$$

that is, from the two permutations that differ only in the ordering of $\underline{r}_1$ and $\underline{r}_2$; $\underline{r}_3$ must be in φ_3, etc. for orthonormality to give non-zero. We thus have, for the two-particle energy E_2,

$$E_2 = \langle\psi|H_2|\psi\rangle$$

$$= \frac{1}{2}\sum_{j,j'}{}'\left\{\int dr_1\,dr_2\,\varphi_j^*(\underline{r}_1)\varphi_{j'}^*(\underline{r}_2)h_2(\underline{r}_1,\underline{r}_2)\right.$$

$$\left. \times\, [\varphi_j(\underline{r}_1)\varphi_{j'}(\underline{r}_2) - \varphi_j(\underline{r}_2)\varphi_{j'}(\underline{r}_1)]\right\}$$

$$= \frac{1}{2}\sum_{j,j'}{}'\langle\varphi_j\varphi_{j'}|h_2(1 - P_{jj'})|\varphi_j\varphi_{j'}\rangle, \tag{12.39}$$

where we have introduced Dirac-like notation for products of two single-particle function, and the pairwise interchange operator $P_{jj'}$ from equation (12.6).

We now collect equations (12.36) and (12.39) to get the total energy:

$$E = (E_1 + E_2)$$

$$= \sum_j\left\{\langle\varphi_j|h_1|\varphi_j\rangle + \frac{1}{2}\sum_{j'}{}'\langle\varphi_j\varphi_{j'}|h_2(1 - P_{jj'})|\varphi_j\varphi_{j'}\rangle\right\}. \tag{12.40}$$

12.2.4 Charge density and exchange charge

It is tempting to give a simplified interpretation of equation (12.40). The single-particle term is the sum of one-electron energies for particles in states, $|\varphi_j\rangle$, $j = 1, 2, \ldots, N$. The two-particle term has two parts. The first is the Coulomb repulsive energy between pairs of electrons (the factor $\frac{1}{2}$ eliminating double counting) in states $|\varphi_j\rangle$ and $|\varphi_{j'}\rangle$,

$$\langle\varphi_j\varphi_{j'}|h_2|\varphi_j\varphi_{j'}\rangle = \int dr\,dr'\,\varphi_j^*(\underline{r})\varphi_{j'}^*(\underline{r}')|\vec{r} - \vec{r}'|^{-1}\varphi_j(\underline{r})\varphi_{j'}(\underline{r}')$$

$$= \int dr\,dr'\,\frac{\rho_j(\underline{r})\rho_{j'}(\underline{r}')}{|\vec{r} - \vec{r}'|}, \tag{12.41}$$

where

$$\rho_j(\underline{r}) = |\varphi_j(\underline{r})|^2 \tag{12.42}$$

is the charge density (in the given units) of a particle in state $|\varphi_j\rangle$. The other two-particle term is called the *exchange* term. It consists of terms like

$$- \int dr\, dr' \frac{\varphi_j^*(\underline{r})\varphi_{j'}^*(\underline{r}')\varphi_j(\underline{r}')\varphi_{j'}(\underline{r})}{|\vec{r}-\vec{r}'|}. \qquad (12.43)$$

If $\varphi_j(\underline{r})$ are eigenstates of a component of spin, say the z-component, then we see that the generalized integral in equation (12.43) gives zero from spin orthogonality for orbitals φ_j and $\varphi_{j'}$ of opposite spin, since h_2 is independent of spin. Only pairs of orbitals φ_j and $\varphi_{j'}$ with *parallel* spins contribute to the exchange term. In equation (12.40) for the total energy, the two-particle terms in $j = j'$ are omitted, because the classical Coulomb interaction among particles is evaluated without including the infinite self energy of a point charge interacting with itself: see also equation (12.37). However, from equations (12.41)–(12.43), we see that the quantum-mechanical Coulomb and exchange terms for $j = j'$ are not necessarily zero, but they are equal and opposite. Henceforth we therefore include these terms $j = j'$ in the two-particle interaction energy in equation (12.40), knowing that they cancel.

It is natural to think of the single-particle part of the total energy, E_1, equation (12.36), as the sum of N single-particle energies, with one particle in each of the orbitals $\{\varphi_j\}$ that make up the Slater determinant, equation (12.19). This picture can be extended to include the two particle part, E_2, equation (12.39), by considering the particle density, $\rho(\vec{r})$. Its operator $\rho_{op}(\vec{r})$ is:

$$\rho_{op}(\vec{r}) = \sum_{j=1}^{N} \delta(\vec{r}-\vec{r}_j). \qquad (12.44)$$

By analogy with the derivation of equation (12.36), which applies to any sum over identical single-particle operators,

$$\rho(\vec{r}) = \langle \psi|\rho_{op}|\psi\rangle = \sum_{j=1}^{N} \langle \varphi_j|\delta(\vec{r}-\vec{r}_j)|\varphi_j\rangle$$

$$= \sum_{j=1}^{N} \int dr_j |\varphi_j(\underline{r}_j)|^2 \delta(\vec{r}-\vec{r}_j) = \sum_{j=1}^{N} |\varphi_j(\underline{r})|^2 = \sum_{j=1}^{N} \rho_j(\underline{r}). \qquad (12.45)$$

The last equation in (12.45) introduces the density $\rho_j(\underline{r})$ that was defined in equation (12.42). Thus from equations (12.40) and (12.41), the Coulomb contribution to the total energy (including the term $j = j'$) is

$$\frac{1}{2}\sum_{j,j'} \int dr\, dr' \frac{\rho_j(\underline{r})\rho_{j'}(\underline{r}')}{|\vec{r}-\vec{r}'|} = \frac{1}{2}\int dr\, dr' \frac{\rho(\underline{r})\rho(\underline{r}')}{|\vec{r}-\vec{r}'|}, \qquad (12.46)$$

where we have also used equation (12.45). The exchange contribution, from equation (12.43), summed over j and j', can be written in a similar form, as follows:

$$-\frac{1}{2} \sum_{\substack{j,j' \\ (\parallel \text{spins})}} \int dr \, dr' \frac{\varphi_j^*(\underline{r}) \varphi_{j'}^*(\underline{r}') \varphi_j(\underline{r}') \varphi_{j'}(\underline{r})}{|\vec{r} - \vec{r}'|}$$

$$= -\frac{1}{2} \sum_j \int dr \, dr' \frac{\rho_j(\underline{r}) \rho_{j,\text{ex}}(\underline{r}';\underline{r})}{|\vec{r} - \vec{r}'|}, \qquad (12.47)$$

where

$$\rho_{j,\text{ex}}(\underline{r}';\underline{r}) = \sum_{\substack{j' \\ (\parallel \text{spins})}} \frac{\varphi_{j'}^*(\underline{r}') \varphi_{j'}(\underline{r}) \varphi_j^*(\underline{r}) \varphi_j(\underline{r}')}{\varphi_j^*(\underline{r}) \varphi_j(\underline{r})}. \qquad (12.48)$$

In words, the exchange charge density $\rho_{j,\text{ex}}(\underline{r}';\underline{r})$ can be thought of as a charge density at $\underline{r}'$ due to the Pauli principle, which is seen by an electron in state φ_j; more precisely, by that part of the wave function φ_j in the vicinity of $\underline{r}$.

We have presented the exchange contribution in the form of equation (12.48) because it is sometimes encountered in the literature, and is helpful in visualizing the exchange effect, as follows. Let us write the two-electron energy, from equations (12.46) and (12.47) as follows:

$$\frac{1}{2} \sum_j \int dr \, dr' \frac{\rho_j(\underline{r})\{\rho(\underline{r}') - \rho_{j,\text{ex}}(\underline{r}';\underline{r})\}}{|\vec{r} - \vec{r}'|}. \qquad (12.49)$$

Let us see what total charge is involved in the charge density seen by an electron in state φ_j. It is, from equation (12.49),

$$\int dr' \{\rho(r') - \rho_{j,\text{ex}}(\underline{r}',\underline{r})\}. \qquad (12.50)$$

Now, from equation (12.45),

$$\int dr' \, \rho(\underline{r}') = \sum_{j=1}^N \int dr' \, |\varphi_j(\underline{r}')|^2 = N, \qquad (12.51)$$

from the normalization of the φ_j. From equation (12.48),

$$\int dr' \, \rho_{j,\text{ex}}(\underline{r}';\underline{r}) = \int dr' \sum_{j'} \varphi_{j'}^*(\underline{r}') \varphi_j(\underline{r}') \frac{\varphi_{j'}(\underline{r})}{\varphi_j(\underline{r})}$$

$$= \sum_{j'} \delta_{j'j} \frac{\varphi_{j'}(\underline{r})}{\varphi_j(\underline{r})} = 1. \qquad (12.52)$$

Thus from equations (12.50)–(12.52) we find the total effective charge for the interaction of a single elecron with all the others to be

$$\int dr'\{\rho(\underline{r}') - \rho_{j,\text{ex}}(\underline{r}';\underline{r})\} = (N-1). \tag{12.53}$$

In our formulation, the Coulomb effect, including self-interaction, is contributed by all N electrons, while the exchange interaction subtracts a total charge equal to that of one electron, contributed to by all electrons with spins parallel to that of the single electron in question. While the Kronecker delta in equation (12.52) might suggest that exchange simply subtracts off the single-electron self-energy, the explicit form of $\rho_{j,\text{ex}}(\underline{r}';\underline{r})$, equation (12.48), shows that the effective exchange charge distribution is not simply that of φ_j, but involves all electrons of a given spin.

12.2.5 The single-particle density functional

We now proceed to an important result, namely that the Hartree–Fock approximation to the total energy of the system is a unique functional of the so-called single-particle density. We introduce the single-particle density matrix $\rho_1(\underline{r},\underline{r}')$ as a generalization of equation (12.45):

$$\rho_1(\underline{r},\underline{r}') = \sum_{j=1}^{N} \varphi_j^*(\underline{r})\varphi_j(\underline{r}'). \tag{12.54}$$

We note that in terms of this notation

$$\rho_1(\underline{r},\underline{r}) = \rho(\underline{r}). \tag{12.55}$$

Equation (12.54) can also be expressed in Dirac notation as

$$\rho_1(\underline{r},\underline{r}') = \sum_{j=1}^{N} \langle \underline{r}|j\rangle\langle j|\underline{r}'\rangle, \tag{12.56}$$

where $|j\rangle$ stands for $|\varphi_j\rangle$, as it appears in equation (12.45). We then arrive at the concept of the single-particle density operator ρ_1, expressed in matrix form, from equation (12.56), as

$$\rho_1 = \sum_{j=1}^{N} |j\rangle\langle j|. \tag{12.57}$$

This is referred to as the Fock–Dirac density. Along with two-body and n-body density matrices, it will be discussed further in section 12.7. In terms of ρ_1 we see from equations (12.46), (12.47) and (12.54) that the two-body part of the total energy is

$$E_2 = \frac{1}{2}\int dr\,dr'\,\frac{\{\rho_1(\underline{r},\underline{r})\rho_1(\underline{r}',\underline{r}') - \rho_1(\underline{r},\underline{r}')\rho_1(\underline{r}',\underline{r})\}}{|\vec{r}-\vec{r}'|}. \tag{12.58}$$

Equation (12.58) is important in terms of general theory because it gives the exchange energy, as well as the Coulomb energy, as an explicit functional of the single-particle density, ρ_1. The same can be done trivially for the single-particle part of the energy, E_1. From equations (12.33) and (12.36), it is

$$
\begin{aligned}
E_1 &= \sum_{j=1}^{N} \langle \varphi_j | h_1 | \varphi_1 \rangle \\
&= \sum_{j=1}^{N} \langle \varphi_j | \left(-\frac{1}{2}\nabla^2 - \sum_J Z_J | \vec{r} - \vec{R}_{J_0} |^{-1} \right) | \varphi_j \rangle \\
&= \sum_j \int dr\, \varphi_j^*(\vec{r}) \left\{ -\frac{1}{2}\nabla^2 - \sum_J Z_J | \vec{r} - \vec{R}_{J_0} |^{-1} \right\} \varphi_j(\underline{r}) \\
&= \sum_j \int dr\, \varphi_j^* \left(-\frac{1}{2}\nabla^2 \right) \varphi_j - \sum_J \int dr\, \frac{Z_J \rho_1(\underline{r},\underline{r})}{|\vec{r} - \vec{R}_{J_0}|} \\
&= -\int dr\, dr'\, \delta(\underline{r} - \underline{r}') \frac{1}{2}\nabla'^2 \rho_1(\underline{r},\underline{r}') - \sum_J \int dr\, Z_J \frac{\rho_1(\underline{r},\underline{r})}{|\vec{r} - \vec{R}_J|}. \quad (12.59)
\end{aligned}
$$

In equation (12.59), the operators ∇^2 and ∇'^2 involve differentiation with respect to the components of $\vec{r}$ and $\vec{r}'$, respectively. Combining equations (12.58) and (12.59) we see that the total N-electron energy $E = (E_1 + E_2)$ is an explicit functional of the Fock–Dirac density ρ_1. Indeed, the single-particle energy E_1, equation (12.59), is a linear functional of ρ_1. We leave it to the reader to show that, using a trick similar to that applied to the ∇^2 term in equation (12.59), the two-particle energy E_2 can be expressed as the integral of an explicit operator acting on a single product of two ρ_1.

12.3 The Fock equation

12.3.1 The variational derivation

In section 12.2.1 we described the Hartree–Fock approximation in terms of an N-dimensional manifold in the Hilbert space of single-particle functions. We said that the approximation would be useful if a single such manifold dominates the many-electron wave function. We now consider how to specify this manifold.

According to equation (12.40), the total energy E in Hartree–Fock approximation is

$$
E = \left\{ \sum_{j=1}^{N} \langle j | h_1 | j \rangle + \frac{1}{2} \sum_{j,j'} \langle jj' | h_2 (1 - P_{jj'}) | jj' \rangle \right\}. \quad (12.60)
$$

In equation (12.60) we have replaced the ket $|\varphi_j\rangle$ simply by $|j\rangle$, and we have included the terms $j = j'$ in the two-body part, as discussed following equation (12.43). The total energy E in equation (12.60) is clearly determined by the choice of N-dimensional manifold spanned by the set $\{|1\rangle, |2\rangle, \ldots, |N\rangle\} \equiv \{|j\rangle\}_N$. The variational principle would then say that the optimal set $\{|j\rangle\}_N$ satisfies the condition that E is a minimum with respect to variation of the N-dimensional manifold. The following approach, so far as the author knows, is due to Pryce (1961).

Consider first the single-particle terms in the total energy E, equation (12.60). They constitute a part of the trace of h_1 in the Hilbert space of single-particle functions. Thus, if p is a projection operator onto an N-dimensional manifold,

$$\sum_{j=1}^{N} \langle j|h_1|j\rangle = \mathrm{Tr}(h_1 \cdot p). \tag{12.61}$$

The two-particle part of E, equation (12.60), can be expressed in similar terms:

$$\frac{1}{2}\sum_{j,j'} \langle jj'|h_2(1 - P_{jj'})|jj'\rangle = \frac{1}{2}\mathrm{Tr}\{\bar{h}_2 p \cdot p'\} \tag{12.62}$$

where

$$\bar{h}_2(\underline{r},\underline{r}') = h_2(\underline{r},\underline{r}')[1 - P(\underline{r},\underline{r}')] \tag{12.63}$$

and where p acts in a Hilbert space of functions of $\underline{r}$, and p' acts in a Hilbert space of functions of $\underline{r}'$.

We note some properties of these projection operators. First, *commutativity*:

$$[p, p'] = 0. \tag{12.64}$$

Second, *idempotency*:

$$p^2 = p. \tag{12.65}$$

Proof: If $|u\rangle$ lies in the manifold, and $|v\rangle$ lies outside, orthogonal to it, then

$$p|u\rangle = |u\rangle, \qquad p|v\rangle = 0,$$

and so if, for arbitrary $|\varphi\rangle$,

$$|\varphi\rangle = (|u\rangle + |v\rangle),$$

then

$$p|\varphi\rangle = |u\rangle$$

and

$$p^2|\varphi\rangle = p|u\rangle = |u\rangle = p|\varphi\rangle.$$

Third, *hermiticity*. Consider

$$|\varphi_1\rangle = (|u_1\rangle + |v_1\rangle)$$
$$|\varphi_2\rangle = (|u_2\rangle + |v_2\rangle)$$

with $|u_i\rangle$ in the manifold and $|v_i\rangle$ outside, so that

$$\langle\varphi_1|p|\varphi_2\rangle = \langle\varphi_1|u_2\rangle = \langle u_1|u_2\rangle \tag{12.66}$$

$$= \langle u_2|u_1\rangle^* = \{(\langle u_2| + \langle v_2|)|u_1\rangle\}^*$$

$$= \langle\varphi_2|u_1\rangle^* = \langle\varphi_2|p|\varphi_1\rangle^*. \tag{12.67}$$

In these equations, $*$ means complex conjugate. The equality of the first and last expressions in equations (12.66) and (12.67) respectively, for arbitrary $|\varphi_1\rangle$ and $|\varphi_2\rangle$, is the definition of hermiticity for an operator p. Fourth:

$$\mathrm{Tr}(p) = N. \tag{12.68}$$

Proof: If $\{|j\rangle\}$, $j = 1, 2, \ldots, \infty$, is an orthonormal basis, and if $\{|j\rangle\}_N$, $j = 1, 2, \ldots, N$, is an N-dimensional subset, then if p projects onto $\{|j\rangle\}_N$:

$$\mathrm{Tr}(p) = \sum_{j=1}^{\infty}\langle j|p|j\rangle = \sum_{j=1}^{N}\langle j|p|j\rangle = \sum_{j=1}^{N}\langle j|j\rangle = N.$$

We now return to equations (12.61) and (12.62). We may use them to write the total energy E as follows:

$$E = \mathrm{Tr}\{[h_1 + \tfrac{1}{2}\mathrm{Tr}'\,\bar{h}_2 \cdot p']p\}, \tag{12.69}$$

where Tr and Tr$'$ are traces over the whole infinite-dimensional Hilbert spaces of functions of $\underline{r}$ and $\underline{r}'$ respectively. However, because of the operators p and p' in equation (12.69) the traces are effectively reduced to the finite N-dimensional subspace.

The variational principle identifies the stationary states of the system as satisfying the condition that, for small variation δp of the N-dimensional projection p, the variation δE of E will be approximately zero. Thus consider

$$(E + \delta E) = \{\mathrm{Tr}[h_1(p + \delta p)] + \tfrac{1}{2}\,\mathrm{Tr}\,\mathrm{Tr}'[\bar{h}_2(p + \delta p)(p' + \delta p')]\}$$

or

$$\delta E = 0 = \{\mathrm{Tr}(h_1 \cdot \delta p) + \tfrac{1}{2}\,\mathrm{Tr}\,\mathrm{Tr}'[\bar{h}_2(p\delta p' + p'\delta p)]\}. \tag{12.70}$$

If we think of the trace as a sum of matrix elements, all of which are expressible as integrals, then we recognize that interchanging $\underline{r}$ and $\underline{r}'$ in the two-particle integrals changes nothing. Thus,

$$p'\delta p = p\delta p'. \tag{12.71}$$

Thus, equation (12.70) reduces to

$$\mathrm{Tr}\{[h_1 + \mathrm{Tr}'\,\bar{h}_2 \cdot p'] \cdot \delta p\} = 0, \tag{12.72}$$

where Tr$'$ is the trace in the Hilbert space of functions of $\underline{r}'$. Here, in equation (12.72) and later, in equation (12.94) for the Fock operator, we note that the two-particle term in $\bar{h}_2$ is not preceded by a factor $\frac{1}{2}$, as occurs in the expression for the total energy, equation (12.40), and later in equations (12.99) and (12.100).

Now imagine that the N-dimensional manifold determined by equation (12.72) is spanned by the orthonormal basis $\{|k\rangle\}_N$, and that these N basis functions are the first N elements of the basis for the infinite-dimensional Hilbert space. In such a basis, the projection operator p has a matrix representation of the form

$$p = \begin{pmatrix} I_N & 0 \\ 0 & 0 \end{pmatrix} \tag{12.73}$$

where I_N is the identity $N \times N$ matrix. We note that this satisfies our fourth condition, $\text{Tr}(p) = N$, equation (12.68). Now let us write δp in the same representation, subject to hermiticity, equation (12.66) and (12.67),

$$\delta p = \begin{pmatrix} a & b \\ b^\dagger & c \end{pmatrix}, \tag{12.74}$$

where $b^\dagger$ is the complex conjugate transpose of b, and a and c are hermitian. We now apply the idempotency requirement, (equation (12.65), to the variation

$$(p + \delta p)^2 \approx (p^2 + p \cdot \delta p + \delta p \cdot p) = (p + \delta p), \tag{12.75}$$

valid to first order, which, with $p^2 = p$ reduces to

$$(p \cdot \delta p + \delta p \cdot p) = \delta p. \tag{12.76}$$

In the matrix notation of equations (12.73) and (12.74) this becomes

$$\left\{ \begin{pmatrix} a & b \\ 0 & 0 \end{pmatrix} + \begin{pmatrix} a & 0 \\ b^\dagger & 0 \end{pmatrix} \right\} = \begin{pmatrix} a & b \\ b^\dagger & c \end{pmatrix} \tag{12.77}$$

or

$$\begin{pmatrix} 2a & b \\ b^\dagger & 0 \end{pmatrix} = \begin{pmatrix} a & b \\ b^\dagger & c \end{pmatrix}. \tag{12.78}$$

Equation (12.78) requires $a = c = 0$, whence

$$\delta p = \begin{pmatrix} 0 & b \\ b^\dagger & 0 \end{pmatrix}. \tag{12.79}$$

Let us now write equation (12.72) in the form

$$\text{Tr}(h \cdot \delta p) = 0 \tag{12.80}$$

where

$$h \equiv [h_1 + \text{Tr}'(\bar{h}_2 \cdot p')]. \tag{12.81}$$

Note that h in equation (12.81) is a single particle operator. Writing h in a matrix form corresponding to that for δp, equation (12.79), we find that equation (12.80) becomes

$$\mathrm{Tr}\left\{ \begin{pmatrix} h_{11} & h_{12} \\ h_{12}^{\dagger} & h_{22} \end{pmatrix} \begin{pmatrix} 0 & b \\ b^{\dagger} & 0 \end{pmatrix} \right\} = \mathrm{Tr}\begin{pmatrix} h_{12}b^{\dagger} & h_{11}b \\ h_{22}b^{\dagger} & h_{12}^{\dagger}b \end{pmatrix} = 0. \qquad (12.82)$$

For arbitrary submatrix b, i.e. arbitrary variation δp of the projection, this requires that $h_{12} = 0$. Thus,

$$h = \begin{pmatrix} h_{11} & 0 \\ 0 & h_{22} \end{pmatrix}. \qquad (12.83)$$

Now what, exactly, has this energy extremalization process taught us about the optimal N-dimensional manifold from which we shall construct the Slater determinant, equation (12.19)? First, we note that the operator h, equation (12.81), is the physical entity involved, equation (12.80). To see how h is involved, we consider its eigenvectors, using the matrix representation, equation (12.83), where the first N orthonormal basis vectors are those that extremalize the Hartree–Fock estimate of the total energy. The form of equation (12.83) comes from the fact that the projection p is idempotent, equations (12.65) and (12.73). Let $\underline{a}_j$ be an eigenvector of h, equation (12.83), consisting of subvectors $\underline{a}_j^{(1)}$ and $\underline{a}_j^{(2)}$ in and outside the N-dimensional manifold, respectively. Then, from equation (12.83),

$$\underline{\underline{h}} \cdot \underline{a}_j = \varepsilon_j \underline{a}_j \qquad (12.84)$$

or

$$\begin{pmatrix} \underline{\underline{h}}_{11} & 0 \\ 0 & \underline{\underline{h}}_{22} \end{pmatrix} \begin{pmatrix} \underline{a}_j^{(1)} \\ \underline{a}_j^{(2)} \end{pmatrix} = \varepsilon_j \begin{pmatrix} \underline{a}_j^{(1)} \\ \underline{a}_j^{(2)} \end{pmatrix} \qquad (12.85)$$

whence

$$\underline{\underline{h}}_{11} \cdot \underline{a}_j^{(1)} = \varepsilon_j \underline{a}_j^{(1)}, \qquad (12.86)$$

$$\underline{\underline{h}}_{22} \cdot \underline{a}_j^{(2)} = \varepsilon_j \underline{a}_j^{(2)}. \qquad (12.87)$$

Equations (12.86) and (12.87) allow solutions with $\underline{a}_j^{(2)} = 0$, $j = 1, 2, \ldots, N$ and with $\underline{a}_j^{(1)} = 0$ for $j > N$. The eigenvectors therefore are split into two classes, those in and those outside of the N-dimensional manifold, respectively, with the optimal choice of this manifold. Without the variational constraint, we should have had

$$\underline{\underline{h}} = \begin{pmatrix} \underline{\underline{h}}_{11} & \underline{\underline{h}}_{12} \\ \underline{\underline{h}}_{12}^{\dagger} & \underline{\underline{h}}_{22} \end{pmatrix}, \qquad (12.88)$$

leading to the eigenvalue equations

$$(\underline{h}_{11} \cdot \underline{a}_j^{(1)} + \underline{h}_{12} \cdot \underline{a}_j^{(2)}) = \varepsilon_j \underline{a}_j^{(1)}, \tag{12.89}$$

$$(\underline{h}_{12}^\dagger \cdot \underline{a}_j^{(1)} + \underline{h}_{22} \cdot \underline{a}_j^{(2)}) = \varepsilon_j \underline{a}_j^{(2)}. \tag{12.90}$$

These eigenvectors cannot include those with $\underline{a}_j^{(2)} = 0$, except for the unique and uninteresting case $\underline{a}_j^{(1)} = 0$, i.e. $\underline{a}_j = 0$.

Returning to equation (12.86), we see that the optimal N-dimensional manifold is spanned by the eigenvectors $\underline{a}_j^{(1)}$, which are eigenvectors of h_{11}. This eigenvalue equation is called the Fock equation, and the operator h is called the Fock operator. Explicitly,

$$h = [h_1 + \text{Tr}'(\bar{h}_2 \cdot p')], \tag{12.81}$$

$$h_1 = \left\{ -\frac{1}{2}\nabla^2 - \sum_J Z_J |\vec{r} - \vec{R}_{J_0}|^{-1} \right\}, \tag{12.33}$$

$$\bar{h}_2 = \{ h_2(\underline{r},\underline{r}')[1 - P(\underline{r}.\underline{r}')] \}, \tag{12.63}$$

whence

$$\text{Tr}'(\bar{h}_2 \cdot p') = \left\{ \sum_{j=1}^N \langle j' | h_2(\underline{r},\underline{r}')[1 - P(\underline{r},\underline{r}')] | j' \rangle \right\}, \tag{12.91}$$

where in this case the prime in $|j'\rangle$ indicates integration is over $\underline{r}'$, and where

$$h_2(\underline{r},\underline{r}') = (|\vec{r} - \vec{r}'|^{-1}). \tag{12.37}$$

We now change the notation for h to

$$h \equiv F, \tag{12.92}$$

see equation (12.81), hereby defining the Fock operator F. The Fock equation is now

$$F|j\rangle = \varepsilon_j |j\rangle. \tag{12.93}$$

In equation (12.93) the Fock operator F is

$$F = \left\{ -\frac{1}{2}\nabla^2 - \sum_J Z_J |\vec{r} - \vec{R}_{J_0}|^{-1} + \sum_{j'=1}^N \langle j' | \bar{h}_2 | j' \rangle \right\}, \tag{12.94}$$

see equations (12.33, 12.37, 12.38, 12.63, 12.81, and 12.91). Note that F is a hermitian operator, with infinitely many eigenvalues, in general. Thus the Fock equation (12.93) does not give us the N-dimensional manifold for the Slater determinant explicitly. Any choice of N of its eigenstates $|j\rangle$ will extremalize the total energy; i.e. all excited states are stationary for the given hamiltonian H, equation (12.2).

The Fock operator is a hermitian single-particle operator and, as we have mentioned, it is useful, though not precisely correct, to think of the

Hartree–Fock approximation as consisting of N electrons, each in one of the Fock eigenstates $|j\rangle$, $j = 1, 2, \ldots, N$. The Fock operator consists of three parts: kinetic energy, potential energy due to nuclei $J = 1, 2, \ldots$, and a further potential energy due to electron–electron Coulomb and exchange interactions. This final term is called the *self-consistent field*, because it is not a given potential, but depends on the solution $\{|j\rangle\}_N$ itself.

12.3.2 Total energy algorithm

The total energy, equation (12.69) is

$$E = \mathrm{Tr}\{[F - \tfrac{1}{2}\mathrm{Tr}'(\bar{h}_2 \cdot p')]p\}, \tag{12.69}$$

where $F \equiv h$ is given in equation (12.94). Now, from equation (12.93),

$$\mathrm{Tr}(F \cdot p) = \sum_{j=1}^{N} \varepsilon_j. \tag{12.95}$$

Furthermore, from equation (12.81),

$$\mathrm{Tr}(F \cdot p) = \mathrm{Tr}\{[h_1 + \mathrm{Tr}'(\bar{h}_2 \cdot p')]p\}. \tag{12.96}$$

We thus arrive at two alternative expressions for E. From equations (12.69) and (12.95),

$$E = \left\{ \sum_{j=1}^{N} \varepsilon_j - \frac{1}{2}\mathrm{Tr}\,\mathrm{Tr}'(\bar{h}_2 \cdot p' \cdot p) \right\}. \tag{12.97}$$

From equations (12.69) and (12.96),

$$E = \mathrm{Tr}\{[h_1 + \tfrac{1}{2}\mathrm{Tr}'(\bar{h}_2 \cdot p')]\,p\} = \frac{1}{2}\left(\sum_{j=1}^{N} \varepsilon_j + \mathrm{Tr}\,h_1 \right). \tag{12.98}$$

More explicitly equations (12.97) and (12.98) are

$$E = \sum_{j=1}^{N} \left\{ \varepsilon_j - \frac{1}{2}\sum_{j'=1}^{N} \langle jj'|\bar{h}_2|jj'\rangle \right\}, \tag{12.99}$$

and

$$E = \frac{1}{2}\sum_{j=1}^{N}(\varepsilon_j + \langle j|h_1|j\rangle). \tag{12.100}$$

12.3.3 Solution of the Fock equation

Suppose we want to calculate the electronic distribution and total energy of a specific molecular cluster: that is, where the nuclei are specified both as to chemistry, Z_J, and as to atomic positions $\vec{R}_J$: see equation (12.2). We cannot solve the Fock equation exactly, even if we ignore the self-consistent field (scf), which we do not want to do: see equations (12.39) and (12.40). There are three elements in the most commonly applied strategy. First, the scf is handled iteratively: that is, it is first formed from a zeroth-order initial guess at what the solution $\{|j\rangle\}_N$ will be, and with this specific potential replacing the scf, the Fock equation is solved, giving a first approximation to $\{|j\rangle\}_N$, whence a first-order form for the scf, and iterated to consistency. Formally, given the zeroth-order guess $\{|j\rangle^{(0)}\}_N$, from equations (12.93), (12.94) and (12.33), solve

$$\left\{h_1 + \sum_{j=1}^{N} {}^{(0)}\langle j'|\bar{h}_2|j'\rangle^{(0)}\right\}|j\rangle^{(1)} = \varepsilon_j^{(1)}|j\rangle^{(1)}, \tag{12.101}$$

then solve

$$\left\{h_1 + \sum_{j=1}^{N} {}^{(1)}\langle j'|\bar{h}_2|j'\rangle^{(1)}\right\}|j\rangle^{(2)} = \varepsilon_j^{(2)}|j\rangle^{(2)}, \tag{12.102}$$

and continue to iterate until the sets of vectors $\{|j\rangle^{(n)}\}_N$ and $\{|j\rangle^{(n+1)}\}_N$ are equal, to within a predetermined limit, or until the total energy fails to change by more than a predetermined tolerance from nth order to $(n+1)$st order.

The second element of strategy is to think of the eigenvector as expanded in terms of a complete set $\{|k\rangle\}$. Then, formally,

$$|j\rangle = \sum_{k} a_j(k)|k\rangle, \tag{12.103}$$

and the Fock equation is of the form

$$F\left\{\sum_{k} a_j(k)|k\rangle\right\} = \varepsilon_j\left\{\sum_{k} a_j(k)|k\rangle\right\}. \tag{12.104}$$

In equation (12.104), the Fock operator F depends quadratically on the coefficient $a_j(k)$ through the scf, but since the scf is fixed at each level of iteration, solution of the Fock equation becomes a linear problem. This problem is then solved by standard matrix methods. Thus, take the inner product with $|k'\rangle$ from the left in equation (12.104) and obtain

$$\sum_{k}\langle k'|F|k\rangle a_j(k) = \varepsilon_j \sum_{k}\langle k'|k\rangle a_j(k). \tag{12.105}$$

This is an equation for matrix multiplication:

$$\underline{F} \cdot \underline{a}_j = \varepsilon_j \underline{S} \cdot \underline{a}_j \tag{12.106}$$

where $\langle k'|F|k\rangle$ is an element of the square matrix $\underline{F}$, $\langle k|k'\rangle$ is an element of the so-called overlap matrix $\underline{S}$, and $\underline{a}_j$ is the column matrix determining the eigenvector $|j\rangle$, equation (12.103). If the basis $\{|k\rangle\}$ is orthonormal, then $\underline{S}$ is the identity matrix, $\underline{I}$.

Equation (12.106) could be solved by standard matrix computational methods for the eigenvalue problem if the complete set $\{|k\rangle\}$ were not infinite, which it is. In this case, the dimensionality of the matrix representation is infinite. This leads us to the third element of the strategy, namely to choose a basis set $\{|k\rangle\}_{N'}$ such that, for the given physical problem, a finite set of $N' \geq N$ elements is, for practical purposes, approximately complete. The key to this step is that, for essentially all terrestrial material problems, the material consists of recognizable atoms. By this we mean that internuclear distances in the material are comparable with or greater than atomic diameters. Let us see how this works out in a quantitative example. Consider the NaF crystal. Each ion is in the electronic configuration $1s^2 2s^2 2p^6$, Na being in charge state $+1$ and F in charge state -1. We can estimate the ionic diameters in free space in terms of the radial part of the 2s hydrogenic-type orbital. Such an orbital contains an exponential factor,

$$\psi \approx e^{-\lambda r}, \qquad \lambda = \frac{Z}{2}, \tag{12.107}$$

in Bohr units, where Z is the nuclear charge in units of proton charge. In the case of Na, $Z = 11$, and for F, $Z = 9$. We define a range R' in terms of $|\psi|^2 \approx e^{-2\lambda r}$, namely,

$$R' = (2\lambda)^{-1}, \tag{12.108}$$

so that

$$|\psi|^2 \approx e^{-r/R'}. \tag{12.109}$$

Thus R' is the distance at which the exponential factor in $|\psi|^2$ drops to $e^{-1} \approx 0.37$ of its maximum value. Then $(2R')$ is a measure of the ion's diameter D, namely,

$$\text{for Na}^+, \qquad D \approx 2/11 = 0.18\,a_0,$$
$$\text{for F}^-, \qquad D \approx 2/9 = 0.22\,a_0,$$

where a_0 is one Bohr. The nearest-neighbor spacing of the NaF crystal is $4.37\,a_0$. We see that for this highly ionic material, the internuclear distant is $\sim 20\times$ the ionic diameters. The explicit atomicity of matter is clearly demonstrated by a variety of experimental techniques. None is more graphic than that of field ion microscopy [see Müller and Tsong (1969)], one of the finest experimental developments not to have won a Nobel prize, in this author's opinion. Since atoms are so clearly recognizable in condensed matter, we shall, as our third strategic element, give an atomic-like representation for the basis functions $|k\rangle$ of the expansion of the Fock eigenstates, equation (12.103).

Atomicity is represented mainly by using functions that are localized on nuclear sites, and possibly other sites also. Each localized function is associated with a spherical harmonic. A finite set encompassing a sufficient range of spherical harmonics is required. The radial localization is expressed in terms of gaussian functions, rather than the so-called Slater type exponentials of equation (12.107), for technical reasons of computational efficiency: see Szabo and Ostlund (1982, Section 3.5.1). However, close to the nuclei, the actual radial dependence is Slater-like, with a cusp, rather than bell-shaped like a gaussian. The radial dependences therefore require linear combinations of gaussians. We now express all this formally. The space–spin representation of the atomic-like basis functions is

$$\langle \underline{r}|k \rangle = n_k \left(\sum_i d_i \exp[-\alpha_i |\vec{r} - \vec{R}_J|^2] \right) Y_l^m(\Omega) . \eta_k(s). \qquad (12.110)$$

In equation (12.110), n_k is a normalizing factor, and in the restricted or unrestricted Hartree–Fock approximation, $\eta_k(s)$ is a spin eigenstate (an eigenstate of a component, say S_Z, of the spin). Also, the notation in equation (12.110) is simplified, because for a given function $|k\rangle$, the set of linear coefficients d_i and the set of exponential coefficients α_i are in general unique to $|k\rangle$, and the choice of nuclear site J and the order of the spherical harmonic (l, m) are specific to $|k\rangle$. Thus the notation might show d_i, α_i, J, l and m all as functions of k, for all i. The linear coefficients d_i are called *contraction coefficients*. The angular position Ω in equation (12.110) is in spherical polar coordinates centered on site J. The function $|k\rangle$ is called an *atomic orbital (AO) basis function*. A single exponential with its spherical harmonic and spin eigenstate, normalized, is called a *primitive atomic orbital*.

We have referred to restricted and unrestricted Hartree–Fock approximations (RHF and UHF respectively) in the preceding paragraph. In RHF, the molecular cluster contains an even number, $N = 2M$, of electrons, which are assigned in pairs to pairs of AOs that differ only as to spin. The spatial $\vec{r}$-dependent parts of the Fock eigenstates are then determined from matrices of dimension M only, as we shall see. The UHF approximation is used for calculations with unequal numbers of spin-up and spin-down electrons, necessarily including clusters with an odd number of electrons. The Fock equations for spin-up and spin-down manifolds are then solved separately, using a common scf, of course. If $\eta_k(s)$ in equation (12.110) is not to be a spin eigenstate, the method is referred to as generalized Hartree–Fock (GHF).

Let us now return to the Fock equation (12.106). We note first that our AO basis set, equation (12.110), is non-orthogonal, so the overlap matrix $\underline{\underline{S}} \neq \underline{\underline{I}}$. We also note that, since we need N Fock eigenstates $\underline{a}_j$, we must use at least N atomic orbital basis functions $|k\rangle$, equation (12.110). In practice, acceptable accuracy may require a larger number $N' > N$. In that case, solution of the Fock equation (12.106) will produce $N' > N$ eigenstates and eigenvalues. The decision as to which set of N to choose from N', in order to evaluate the total energy and N-electron wave function, is dictated by the nature of the state that we wish to study: the ground state or a particular excited state. The set of N states that we choose span the so-called *occupied manifold*. The remaining states define the *virtual manifold*. An eigenstate $|j\rangle$, equation (12.103), defined in terms of its linear coefficients $a_j(k)$, or the eigenvector $\underline{a}_j$ of equation (12.106), is from equation (12.103) a *linear combination of atomic orbitals*: LCAO. Since in general a single eigenstate $|j\rangle$ is in this way associated spatially with the whole molecule, it is called a *molecular orbital*: MO, or more fully an LCAO-MO.

Further details of how the Fock equation is solved in practice, along with additional theoretical considerations involving the nature of the AO basis set, are to be found in standard textbooks on quantum chemistry, for example Szabo and Ostlund (1982). Those aspects of the computations that can be reduced to a formal sequence of operations are incorporated in highly sophisticated, reliable, flexible, user-friendly programs that are readily available. For molecular clusters, the GAUSSIAN program is probably pre-eminent at present. GAUSSIAN is revised annually, each year adding new features of physical or computational value (see GAUSSIAN in the bibliography). Electronic band structure is similarly served by the CRYSTAL program (see CRYSTAL in the bibliography). Apart from formal processes that can be incorporated in a program, however, there are elements of craft in quantum cluster computation, having to do with the development of suitable AO basis sets for particular problems. In simplest terms, this often involves a subsidiary process of computationally determining contraction and exponential coefficients, d_i and α_i respectively, equation (12.110), that are in some sense optimal.

12.4   Localizing potentials

Consider the solutions of a problem in Hartree–Fock approximation. In general we seek occupied and unoccupied eigenstates of the Fock operator, $|j_{\text{occ}}\rangle$ and $|j_{\text{virt}}\rangle$, with $j_{\text{occ}} = 1, 2, \ldots, N$ and $j_{\text{virt}} = N + 1, \ldots, N' > N$, where 'occ' and 'virt' stand for occupied and virtual. There may be convergence problems associated with the AO basis set chosen, for a given Fock operator F. Also, we may want, or need, to have a subset of the eigenstates with a particular mathematical or physical property, at least approximately. For example, we might want, for purposes of computation or visualization, to have eigenstates that correspond (approximately) to a single atom or group of atoms in a molecular cluster, or to a single molecular cluster within an extended solid. The latter case will in fact be discussed in detail in the next section, section 12.5. The best that we can do in these cases is to modify the Fock operator so that the modified eigenstates still correspond to the occupied and virtual manifolds of the original problem with the chosen AO basis set, while satisfying our requirement of localization or other property. This can be accomplished by adding a so-called *localizing potential* to the Fock operator F. A good reference for this subject is Gilbert (1964).

Formally, consider

$$F' = (F + A_F),\qquad(12.111)$$

where A_F is the localizing potential, and F' the modified Fock operator. Let $\{|i\rangle\}_N$ be the occupied manifold of F. We require F' to be hermitian, and therefore so must A_F be

$$A_F^\dagger = A,\qquad(12.112)$$

where $\dagger$ indicates hermitian conjugate. Next consider the eigenstates of F':

$$F'|\tilde{j}\rangle = \tilde{\varepsilon}_j|\tilde{j}\rangle.\qquad(12.113)$$

For $|\tilde{j}\rangle, j = 1, 2, \ldots, N$, to span the occupied manifold of F, we required

$$|\tilde{j}\rangle = \sum_{i=1}^{N} c_{ij}|i\rangle,\qquad(12.114)$$

where the transformation between orthonormal sets spanning the same manifold, represented by the matrix c_{ij}, is unitary. Thus,

$$F|\tilde{j}\rangle = \sum_{i=1}^{N} c_{ij}\varepsilon_i|i\rangle\qquad(12.115)$$

and

$$A_F|\tilde{j}\rangle = \sum_{i=1}^{N} c_{ij}A_F|i\rangle.\qquad(12.116)$$

Combining equations (12.115), (12.116), (12.111) and (12.113):

$$F'|\tilde{j}\rangle = (F + A_F)|\tilde{j}\rangle = \sum_{i=1}^{N} c_{ij}(\varepsilon_i + A_F)|i\rangle = \tilde{\varepsilon}_j|\tilde{j}\rangle.\qquad(12.117)$$

From equation (12.117), the vector $(A_F|i\rangle)$ must lie in the occupied manifold, because $|\tilde{j}\rangle$ on the right-hand side does. The localizing potential must therefore project onto the occupied manifold.

We recall that the Fock operator F is completely and uniquely determined by the Fock–Dirac density ρ_1, equation (12.56), see equation (12.58). We further note that ρ_1 projects onto the occupied manifold. Consider an arbitrary state vector $|\psi\rangle$,

$$|\psi\rangle = \left\{\sum_{j_{occ}} c_j|j\rangle + \sum_{j'_{virt}} d_{j'}|j'\rangle\right\},\qquad(12.118)$$

where 'occ' and 'virt' are occupied and virtual manifolds respectively, which are mutually orthogonal:

$$\langle j_{occ}|j'_{virt}\rangle = 0.\qquad(12.119)$$

Then, from equation (12.57),

$$\rho_1 = \sum_{j'_{occ}} |j'\rangle\langle j'|. \tag{12.57}$$

Trivially then, applying equation (12.57) to equation (12.118),

$$\rho_1|\psi\rangle = \sum_{j_{occ}} c_j|j\rangle. \tag{12.120}$$

Because A_F must project onto the occupied manifold, and because of the projection property of ρ_1, we consider A_F in the form

$$A_F = A'_F \cdot \rho_1. \tag{12.121}$$

Hermiticity, equation (12.112), then requires

$$A_F^\dagger = \rho_1 \cdot A_F'^\dagger = A'_F \cdot \rho_1, \tag{12.122}$$

since ρ_1 is hermitian. The solution of equation (12.122) for A'_F is

$$A'_F = \rho_1 \cdot A \tag{12.123}$$

where A is hermitian:

$$A^\dagger = A. \tag{12.124}$$

From equations (12.121) and (12.123),

$$A_F = \rho_1 \cdot A \cdot \rho_1, \tag{12.125}$$

where A is an arbitrary hermitian operator, which may be chosen to be a single-particle operator. Equation (12.125) is the definition of a localizing potential. Thus A must have matrix elements that involve only a single electron coordinate in $\underline{r}$ space.

We must now examine the relationship between the N-electron system and the modified Fock equation. From equations (12.93), (12.111), (12.113) and (12.114), we have:

$$F|j\rangle = \varepsilon_j|j\rangle,$$

$$F'|\tilde{j}\rangle = (F + A_F)|\tilde{j}\rangle = \tilde{\varepsilon}_j|\tilde{j}\rangle,$$

$$|\tilde{j}\rangle = \sum_i c_{ij}|i\rangle.$$

Recall that, from equations (12.69) and (12.94), the total energy E is:

$$E = \text{Tr}\{[F - \tfrac{1}{2}\text{Tr}'(\bar{h}_2 \cdot p')] \cdot p\}, \tag{12.126}$$

where, with the projection operators p and p', the traces reduce to:

$$\text{Tr}_N(O) \equiv \sum_{j=1}^{N} \langle j|O|j\rangle,$$

in which O is a hermitian operator. But the trace is independent of basis set, so:

$$\mathrm{Tr}_N(O) = \sum_{j=1}^{N} \langle \tilde{j}|O|\tilde{j}\rangle.$$

Now, from equations (12.126) and (12.113) we have:

$$E = \{\mathrm{Tr}_N[F' - \tfrac{1}{2}\mathrm{Tr}'_N(\bar{h}_2)] - \mathrm{Tr}_N(A_F)\}, \tag{12.127}$$

where the traces are evaluated in terms of matrices represented in the basis $\{|\tilde{j}\rangle\}_N$. Thus, having solved the modified Fock equation, we can use the resulting matrices directly to evaluate the total energy in terms of the appropriate traces, equation (12.127). We note that, in equation (12.127),

$$\mathrm{Tr}_N(F') = \sum_{j=1}^{N} \tilde{\varepsilon}_j, \tag{12.128}$$

from equation (12.113). Equation (12.99) or (12.100) can be used also, rewritten in terms of $\tilde{\varepsilon}_j$ and $|\tilde{j}\rangle$, to evaluate the first term in equation (12.127), but then the second term must not be forgotten.

The Slater determinant constructed from the eigenstates of the modified Fock operator will be a different N-electron wave function from that derived from the original Fock operator. Thus we must obtain the original Fock eigenstates. This can be done most easily simply by solving

$$F|j\rangle = \varepsilon_j|j\rangle,$$

using the Fock matrix F as given in the modified basis $|\tilde{j}\rangle$, which is available from the solution of equation (12.113).

Much more detailed discussion of localizing potentials is to be found in Gilbert (1964).

12.5 Embedding in a crystal

12.5.1 Introduction

When we use a quantum-molecular cluster in a computation to represent local electronic properties in a crystal, we must take account of how the local electronic structure is affected by the rest of the crystal. For example, in a highly ionic crystal, the electronic charge density is highly localized about each ion. It is then plausible to think of there being a specific number of electrons on each ion, and therefore to think of a molecular cluster as being electronically isolated, approximately, from the rest of the crystal. In that case, only the Madelung field, that is the Coulomb field of the ions of the embedding crystal, affects the cluster. This assumes, of course, that the computational cluster is localized precisely as dictated by the surrounding crystal. This localization arises, of course, from effects from the electrons of embedding ions, including quantum-mechanical Coulomb, exchange and correlation effects, in combination with attraction by the nuclei of the cluster, plus the embedding Madelung field.

The quantum-mechanical embedding effects, in the context described above, can be accounted for quite well by a set of subsidiary embedded cluster calculations. Consider, for example, a binary ionic crystal, KCl. An atomic orbital basis set $\{|k\rangle\}$ [see equation (12.110)] can then be developed, accurately reflecting the quantum-mechanical effects surrounding each ionic species, K^+ and Cl^- in the example. The procedure is as follows. Consider two embedded quantum clusters, one centered on each of the ionic species. In an ionic crystal, the embedding might be represented by a shell-model crystal. For nearest-neighbor clusters, the two clusters would be (KCl_6) and (K_6Cl) respectively. With a plausible initial basis set, such as those for free ions, optimize the K^+ basis set in the (KCl_6) cluster. Optimization means, in this case, to minimize the total energy of the crystal, cluster plus embedding, with respect to variation of both exponential coefficients α_i and contraction coefficients d_i, equation (12.110), keeping the Cl^- basis set fixed. Then introduce the optimized K^+ basis set into the (K_6Cl) cluster and, keeping this basis set fixed, optimize the Cl^- basis set. Iterate this process between clusters to convergence, i.e. until the total energy cannot be further reduced, within a specified accuracy.

This process should give quite a good picture of perfect-crystal ions. However, it does not give us a good description of the perfect crystal's properties. The reason is that the clusters that determine the electronic distribution within the ions do not give wave functions with the periodicity of the crystal lattice, nor do they give an appropriate density of states, because so few electrons are involved in the cluster. The basis sets derived from perfect crystal embedded clusters may, however, be used with very good effect to analyse point-defect properties, provided certain conditions are satisfied. Basically, a point-defect embedded-cluster calculation must contain enough atoms so that the outer atoms of the cluster are essentially unperturbed by the defect relative to their perfect crystal configurations. In that case, the point defect sites and the perturbed near neighbors will be subject to realistic quantum-mechanical effects from the surrounding crystal. The problem is that satisfying these conditions will often be prohibitive, in terms of present-day computational capacity and speed. Then, if a cluster of practical size is such that the outer ions of the cluster are perturbed by the defect, this perturbation needs to be determined variationally, by total energy minimization. But when we minimize the total energy by varying the electronic configuration of the outer ions, two effects are in operation. One physically valid effect is the tendency of the defect to perturb the ion. The other effect, however, is spurious. It represents the fact that the outer ions can, in general, reduce the total energy by allowing their electrons to spread out, or polarize, into the surrounding embedding region, which does not interact quantum-mechanically with the cluster. The best that we can do in this case is to limit the variational flexibility, by keeping the host-ion basis sets fixed. This does not prevent electronic charge redistribution amongst the basis functions from being affected by both physically valid and physically spurious effects, however.

The question that follows from the preceding discussion is whether we can limit the calculation of local defect properties to a small molecular cluster, containing only host ions that are perturbed by the defect. If so, the Fock operator [equations (12.94) and (12.37)] for the cluster must be modified by adding a term that accounts for the quantum-mechanical effect of the embedding on the cluster. This question in turn leads us to a deeper analysis of the problem than we have contemplated so far. The point is that, if we consider the whole crystal in a rigorous quantum-mechanical way, we conclude that it has few, if any, eigenstates that are localized on the cluster region, as implied by our basis set, equation (12.110), where the index J labelling ionic nuclei is limited to a small cluster region. Thus, while we might consider, as we shall do, using the Fock operator for the whole crystal to determine a set of cluster-localized LCAO-MOs from which to construct the cluster Hartree–Fock wave function, this crystal Fock operator does not, in general, possess such cluster-localized eigenstates: see equations (12.110), (12.103) and (12.29), where $\varphi_j(\underline{r}) \equiv \langle \underline{r}|j \rangle$. The question which now arises is: If the crystal does not possess cluster-localized eigenstates, then in what way are cluster-localized states related to local properties of the crystal?

12.5.2 Approximate partitioning with a localizing potential

From our knowledge of localizing potentials for the Fock equation, section 12.4, we are led to consider looking for a localizing potential A_F such that the modified Fock equation for the whole crystal possesses a sufficient number of at least approximately cluster-localized eigenstates: recall

$$F'|\tilde{j}\rangle \equiv (F + A_F)|\tilde{j}\rangle = \tilde{\varepsilon}_j|\tilde{j}\rangle. \tag{12.113}$$

In equation (12.113) we emphasize that F is the Fock operator for the *whole crystal*. What we want, then, is for A_F to form eigenstates $\{|\tilde{j}\rangle\}$ which are linear combinations of the eigenstates $\{|j\rangle\}$ of F, such that $\{|\tilde{j}\rangle\}$ are as strongly localized about the cluster region as possible. To be specific, suppose that the crystal is macroscopic, so that the total number of electrons $N_{\text{tot}} \gtrsim 10^{23}$, whereas the cluster region will involve N electrons, with $N \lesssim 10^3$. Then for the eigenstates $\{|\tilde{j}\rangle\}$ to include N_{tot} occupied states, we would need an AO basis set of at least this many atomic orbitals, distributed over all the nuclei of the crystal. One would expect that linear combinations of such a large number of linearly independent basis functions could form some, perhaps many, orthonormal combinations that are approximately localized in the cluster region. We say only approximately localized because even the large crystal-wide AO basis set is not complete. In any case, we want to order the modified Fock eigenvalues $\tilde{\varepsilon}_j$ according to strength of cluster localization. We could then solve approximately the modified Fock equation for the first N' eigenstates $\{|\tilde{j}\rangle\}_{N'}$ and eigenvalues $\{|\tilde{\varepsilon}_j\rangle\}_{N'}$ using the cluster-centered basis $\{|k\rangle\}_{N'}$. This approximate solution would be more accurate relative to F', equation (12.113), than the corresponding cluster-localized approximate solutions in the same AO manifold, would be relative to F. The reason is simply that the eigenvectors being found from the molecular cluster computation in both cases, $(F', |\tilde{j}\rangle)$ and $(F, |j\rangle)$, are precisely localized in the cluster with amplitudes that decay gaussian exponentially outside the cluster, since they are based only on the N' AOs that are centered on sites within the cluster. In other words, only N'-dimensional matrices are used for F' and F. At the same time, while F' has many eigenvectors that are approximately localized in the cluster region, and others that are correspondingly depleted, the original Fock operator F has few if any such eigenvectors. Thus, while solutions with F and F' for the whole crystal will give identical total energies, based on a given AO basis set, the same will not be true for approximate, strictly cluster-localized solutions for a cluster energy based on traces Tr_N and Tr'_N over the cluster manifold only, using equation (12.127).

We can expand on these ideas analytically. Suppose that $F' = (F + A_F)$ for the whole crystal has eigenstates $\{|\tilde{j}\rangle\}$, a complete set with N_{tot} occupied, where $N_{tot} \gtrsim 10^{23}$ is the total number of electrons. Arbitrarily, let us divide $\{|\tilde{j}\rangle\}$ into two sets $\{|\tilde{j}\rangle\}_0$ and $\{|\tilde{j}\rangle\}_1$, where $\{|\tilde{j}\rangle\}_0$ consists of the N' most strongly cluster-localized elements of $\{|\tilde{j}\rangle\}$: N' must in fact be equal to the number of AOs centered in the cluster region. For the moment, assume that $\{|\tilde{j}\rangle\}_0$ are *perfectly localized* within the cluster region, where the cluster region has spatially sharp boundaries, so that $\{|\tilde{j}\rangle\}_0$ are *exactly zero* outside these boundaries. Similarly, assume that $\{|\tilde{j}\rangle\}_1$ are exactly zero inside these cluster boundaries. Then $\mathrm{Tr}_{tot}(O)$, the trace over the N_{tot}-dimensional occupied manifold of the crystal, for arbitrary operator O separates precisely into cluster and embedding parts,

$$\mathrm{Tr}_{tot}(O) = \sum_{j=1}^{N_{tot}} \langle \tilde{j}|O|\tilde{j}\rangle$$

$$= \left\{ \sum_{j=1}^{N'} \langle \tilde{j}|O|\tilde{j}\rangle + \sum_{j'>N'} \langle \tilde{j'}|O|\tilde{j'}\rangle \right\} \qquad (12.129)$$

or

$$\mathrm{Tr}_{tot}(O) = \{\mathrm{Tr}_0(O) + \mathrm{Tr}_1(O)\}. \qquad (12.130)$$

Then from equation (12.98), the total energy E_{tot} of the whole crystal is

$$E_{\text{tot}} = \text{Tr}_0\{[h_1 + \tfrac{1}{2}\text{Tr}_0'(\bar{h}_2 \cdot p')] \cdot p\}$$
$$+ \text{Tr}_1\{[h_1 + \tfrac{1}{2}\text{Tr}_1'(\bar{h}_2 \cdot p')] \cdot p\}$$
$$+ \text{Tr}_0\,\text{Tr}_1'(\bar{h}_2 \cdot p' \cdot p). \tag{12.131}$$

At this point, a further technical question may occur to the reader. In practice, the analysis of an embedded cluster will include determination of atomic positions within the cluster, usually variationally. It may also involve varying the AO basis set within the cluster. Both of these variations affect the cluster manifold, and with it the embedding manifold. Thus each of the traces Tr_0 and Tr_1 are different for each change of atomic positions or of basis set, in equation (12.131). However, if the embedding region is, to good approximation, unperturbed, then its manifold is essentially unaffected by variations of the cluster, and it can be approximately determined once, for use throughout the cluster analysis. In that case the second term in equation (12.131) will be fixed, and in the third term, $\text{Tr}_1'(\bar{h}_2 \cdot p')$ will be a given function of $\underline{r}$.

In equation (12.131), h_1 contains the electron–nucleus interaction [see equation (12.33)], ranging over all nuclei of the crystal. In order that the first term in equation (12.131) may represent the cluster energy, it is necessary to add electrons to the nuclei outside the cluster; a corresponding remark applies to the second term. We therefore rewrite equation (12.131) as

$$E_{\text{tot}} = \text{Tr}_0\{[h_1 + \tfrac{1}{2}\text{Tr}_0'(\bar{h}_2 \cdot p') + \tfrac{1}{2}\text{Tr}_1'(\bar{h}_2 \cdot p')] \cdot p\}$$
$$+ \text{Tr}_1\{[h_1 + \tfrac{1}{2}\text{Tr}_1'(\bar{h}_2 \cdot p') + \tfrac{1}{2}\text{Tr}_0'(\bar{h}_2 \cdot p')] \cdot p\}. \tag{12.132}$$

In thus separating the total crystal energy into cluster and embedding parts, we see that evaluation of the cluster energy requires knowledge of the embedding eigenstates as well, in the term $\text{Tr}_1'(\bar{h}_2 \cdot p')$.

Recall that the result of equation (12.132) is based on perfect cluster localization for N' eigenvectors of the whole-crystal modified Fock operator. Since any localizing potential that we are likely to use will not produce this result, the equation for the cluster energy E will only be approximate:

$$E \approx \text{Tr}_0\{h_1 + \tfrac{1}{2}\text{Tr}_0'(\bar{h}_2 \cdot p') + \tfrac{1}{2}\text{Tr}_1'(\bar{h}_2 \cdot p')] \cdot p\}. \tag{12.133}$$

We persist in writing $\text{Tr}'(\bar{h}_2 \cdot p')$ in terms of its cluster and embedding parts Tr_0' and Tr_1' because, of course, we shall not obtain the solution for the embedding part from the solution of the cluster modified Fock equation. The embedding eigenstates will have to be obtained from calculations that are independent of the detailed solution that we will generate for the cluster. Once an approximate solution has been obtained for these embedding eigenstates, they can be kept fixed in the modified cluster Fock equation; that is, they will not be updated during the iterative process for the self-consistent field. If these eigenstates are taken to be known, and fixed, then in equation (12.133) the term

$$\tfrac{1}{2}\mathrm{Tr}_1'(\bar{h}_2 \cdot p') \cdot p \qquad (12.134)$$

is a given single-particle operator. Returning to equation (12.33) for h_1, we note that

$$h_1 = \left\{ -\frac{1}{2}\nabla^2 - \sum_J Z_J |\vec{r} - \vec{R}_J|^{-1} \right\}. \qquad (12.135)$$

Here the sum over J includes all the nuclei in the crystal. Let us denote the combined single-particle terms $\bar{h}_1$ from equation (12.133), using equations (12.134) and (12.135), as follows:

$$\bar{h}_1 = \left\{ -\frac{1}{2}\nabla^2 - \sum_{J(\mathrm{cl})} Z_J |\vec{r} - R_J|^{-1} \right.$$

$$\left. + \left[-\sum_{J(\mathrm{emb})} Z_J |\vec{r} - R_J|^{-1} + \frac{1}{2}\mathrm{Tr}_1'(\bar{h}_2 \cdot p') \cdot p \right] \right\}. \qquad (12.136)$$

In equation (12.136), $J(\mathrm{cl})$ and $J(\mathrm{emb})$ refer to nuclear sites in the cluster and embedding regions. We see that the expression in [] brackets represents nuclei and their associated electrons, i.e. whole ions, in the embedding region. In fact, it is the potential $V_{\mathrm{emb}}(\underline{r})$ seen by cluster electrons due to the embedding region. Thus a point-charge shell-model representation of the embedding region may be adequate if quantum-mechanical effects are negligible. If not, some other, quantum-mechanically based approximate determination of the occupied embedding Fock eigenstates $\{|\tilde{j}\rangle\}_1$, defining Tr_1' in equation (12.136), can be used.

We note that the operators p and p' in equations (12.133) and (12.136) project onto the occupied manifold of the whole crystal. But in these equations, the traces Tr_0 and Tr_1 span only cluster and embedding sub-manifolds respectively. Thus p and p' become redundant, and will be omitted. Now, from combining equations (12.133) and (12.136) we have, for the cluster energy,

$$E \approx \mathrm{Tr}_0\{\bar{h}_1 + \tfrac{1}{2}\mathrm{Tr}_0'(\bar{h}_2)\}. \qquad (12.137)$$

We re-emphasize that this formula, equation (12.137), can only be a useful approximation, representing the contribution to the total energy from the cluster region, if it is based on whole-crystal eigenstates that separate into cluster and embedding localized sets respectively. These must be eigenstates of the modified Fock operator F', equation (12.113).

In terms of F', the expression for the total energy, from equation (12.69), becomes

$$E_{\mathrm{tot}} = \mathrm{Tr}\{[F - \tfrac{1}{2}\mathrm{Tr}'(\bar{h}_2 \cdot p')] \cdot p\} \qquad (12.69)$$

$$= \mathrm{Tr}\{[(F' - A_F) - \tfrac{1}{2}\mathrm{Tr}'(\bar{h}_2 \cdot p')] \cdot p\}. \qquad (12.138)$$

Now, from equations (12.125) and (12.57), we can write

$$A_F = \sum_{i,i'} |\tilde{i}\rangle\langle\tilde{i}|A|\tilde{i}'\rangle\langle\tilde{i}'|, \tag{12.139}$$

where we do not need to restrict the sums to occupied states, since that is accomplished by the operator p in equation (12.138). Since $\{|\tilde{j}\rangle\}$ are a complete orthonormal set, we have, from equation (12.139) that

$$\mathrm{Tr}A_F = \mathrm{Tr}\,A, \tag{12.140}$$

where A is a single-particle operator. The total energy algorithm is then, from equations (12.138) and (12.140),

$$E_{\mathrm{tot}} = \mathrm{Tr}\{[(F' - A) - \tfrac{1}{2}\mathrm{Tr}'(\bar{h}_2 \cdot p')] \cdot p\}, \tag{12.141}$$

in which

$$\mathrm{Tr}(F' \cdot p) = \left(\sum_{j=1}^{N_{\mathrm{tot}}} \tilde{\varepsilon}_j\right), \tag{12.142}$$

with $\tilde{\varepsilon}_j$ the occupied eigenvalues of F', equation (12.113). As in equation (12.132), we can write E_{tot} as the sum of cluster energy E, and embedding energy E_1. From equation (12.141),

$$E_{\mathrm{tot}} = (E + E_1), \tag{12.143}$$

$$E = \mathrm{Tr}_0\{F' - A - \tfrac{1}{2}\mathrm{Tr}_1'(\bar{h}_2) - \tfrac{1}{2}\mathrm{Tr}_0'(\bar{h}_2)\}, \tag{12.144}$$

$$E_1 = \mathrm{Tr}_1\{F' - A - \tfrac{1}{2}\mathrm{Tr}_0'(\bar{h}_2) - \tfrac{1}{2}\mathrm{Tr}_1'(\bar{h}_2)\}. \tag{12.145}$$

Concentrating on the cluster energy, equation (12.144), we reiterate that, in practice, $\mathrm{Tr}_1'(\bar{h}_2)$ must be determined from subsidiary computation, yielding a single-particle operator. This operator, combined with a similar term in F' (with factor $(+1)$ instead of $(-\tfrac{1}{2})$) represents the electronic effects from embedding ions (see equation 12.136)). As in equation (12.142), we now have in equation (12.144)

$$\mathrm{Tr}_0(F') = \sum_{j=1}^{N} \tilde{\varepsilon}_j, \tag{12.146}$$

the sum of the occupied eigenvalues.

12.5.3 Summary

In summary, local properties in a crystal can be determined computationally from a localized, embedded molecular cluster, if and only if the total energy of the crystal can be separated into cluster and embedding parts. Since the total energy is a trace over occupied states, the occupied states must divide, at least approximately, into cluster and embedding localized sets. In general this is not the case for the Fock eigenstates. The condition may be satisfied, however, by eigenstates of a modified Fock operator containing a localizing potential. The total energy algorithm for the cluster, as well as those for the embedding region and the whole crystal, are expressible in terms of the modified Fock eigenstates and eigenvalues. The cluster energy is given in equation (12.144). The modified Fock equation for the cluster is

$$F'|\tilde{j}\rangle = \tilde{\varepsilon}_j|\tilde{j}\rangle, \tag{12.113}$$

where

$$F' = \{\bar{h}_1 + \mathrm{Tr}'_0(\bar{h}_2) + A_F\}. \tag{12.147}$$

In equation (12.147), the single-particle term $\bar{h}_1$ is

$$\bar{h}_1 = \left\{-\frac{1}{2}\nabla^2 - \sum_{J(\mathrm{cl})} Z_J|\vec{r} - \vec{R}_{J_0}|^{-1} + V_{\mathrm{emb}}(\vec{r})\right\}, \tag{12.136}$$

from equation (12.136), and the embedding potential V_{emb} is

$$V_{\mathrm{emb}}(\vec{r}) = \left\{-\sum_{J(\mathrm{emb})} Z_J|\vec{r} - \vec{R}_J|^{-1} + \frac{1}{2}\mathrm{Tr}'_1(\bar{h}_2)\right\}. \tag{12.148}$$

From equations (12.130) and (12.63), the partial trace in equation (12.148) is

$$\mathrm{Tr}'_1(\bar{h}_2) = \sum_{j'(\mathrm{emb})} \langle\tilde{j}'|\vec{r} - \vec{r}'|^{-1} \cdot [1 - P(\underline{r}, \underline{r}')]|\tilde{j}'\rangle. \tag{12.149}$$

Similarly in equation (12.147), the trace is

$$\mathrm{Tr}'_0(\bar{h}_2) = \sum_{j'(\mathrm{cl})} \langle\tilde{j}'|\vec{r} - \vec{r}'|^{-1} \cdot [1 - P(\underline{r}, \underline{r}')]|\tilde{j}'\rangle. \tag{12.150}$$

Equations (12.149) and (12.150) represent the two-particle cluster-embedding and intra-cluster interactions, respectively. Also in equation (12.147), the localizing potential A_F is

$$A_F = \rho_1 \cdot A \cdot \rho_1 \tag{12.125}$$

from equation (12.125) where A is a hermitian, single-particle operator, chosen to give cluster-embedding separation for the modified Fock eigenstates, and the Fock–Dirac density ρ_1 from equation (12.57) can be expressed in terms of the modified Fock eigenstates, as follows:

$$\rho_1 = \sum_{j'(\mathrm{occ})} |j'\rangle\langle j'| = \sum_{j'(\mathrm{occ})} |\tilde{j}'\rangle\langle\tilde{j}'|. \tag{12.151}$$

We can prove equation (12.151) as follows. For a given arbitrary vector $|\beta\rangle$, write

$$|\beta\rangle = \{|\beta\rangle_{\mathrm{occ}} + |\beta\rangle_{\perp}\}, \tag{12.152}$$

where $|\beta\rangle_{\rm occ}$ and $|\beta\rangle_\perp$ are components in the occupied manifold, and orthogonal to it, respectively. Then

$$\sum_{j'(\rm occ)} |j'\rangle\langle j'|\beta\rangle = |\beta\rangle_{\rm occ},$$

and

$$\sum_{j'(\rm occ)} |\tilde{j}'\rangle\langle \tilde{j}'|\beta\rangle = |\beta\rangle_{\rm occ}.$$

Thus

$$\left\{\sum_{j'(\rm occ)} |j'\rangle\langle j'| - \sum_{j'(\rm occ)} |\tilde{j}'\rangle\langle \tilde{j}'|\right\}|\beta\rangle = 0.$$

Since $|\beta\rangle$ is arbitrary, this proves equation (12.151). Because the modified Fock equation (12.113) has approximately localized eigenstates, a strictly localized approximate solution in terms of LCAO-MOs will be more accurate than such a solution for the Fock equation based on the Fock operator $F = (F' - A_F)$, equation (12.147), where F does not, in general, have eigenstates that are as well localized.

There remains the central problem of determining a suitable localizing operator A, equation (12.125). While the author has published a proposed form for A [Vail (2001)], it will not be described here, because better ideas will surely emerge before long.

12.6   Correlation

We have discussed in detail the Hartree–Fock approximation, in which the many-fermion wave function is approximated by a single Slater determinant, equation (12.19). We have also described a procedure for solving the Fock equation in the approximation of a finite AO basis, section 12.3.3, thereby determining the Slater determinant. When, as is usual, the atomic orbital basis set $\{|k\rangle\}_{N'}$, consisting of N' elements, has $N' > N$, then N' eigenvectors $\underline{a}_j$ are generated by the Fock equation (12.106). The AO basis functions $|k\rangle$ are given in position representation in equation (12.110). This solution for the Fock equation provides us with a means of improving on the Hartree–Fock approximation, as we shall see.

Recall the hamiltonian for the N-electron system, equation (12.2), and the Fock operator, equation (12.94) with equation (12.91). Comparing these two, we are led to write

$$H = \sum_{j=1}^{N}\left\{\tilde{F}(\vec{r}_j, \underline{r}) - \frac{1}{2}\sum_{j'=1}^{N}{}' |\vec{r}_j - \vec{r}_{j'}|^{-1}\right\}, \qquad (12.153)$$

where

$$\tilde{F}(\vec{r}, \underline{r}) = \left\{ -\frac{1}{2}\nabla^2 - \sum_J Z_J |\vec{r} - \vec{R}_J|^{-1} + \sum_{j'=1}^{N} {}' |\vec{r} - \vec{r}_{j'}|^{-1} \right\}, \qquad (12.154)$$

where we recall that the notation $\underline{r}$ stands for $(\underline{r}_1, \underline{r}_2, \dots, \underline{r}_N)$. Examining equation (12.153), we see that the hamiltonian H can be broken into two parts, such that some eigenstates of one part are already known from the solution of the Fock equation. Specifically, let

$$H = (H_0 + H_1), \qquad (12.155)$$

where

$$H_0 = \sum_j \tilde{F}(\vec{r}_j, \underline{r}), \qquad (12.156)$$

$$H_1 = \left\{ -\frac{1}{2}\sum_{j'} {}' |\vec{r}_j - \vec{r}_{j'}|^{-1} \right\}. \qquad (12.157)$$

If *all* the eigenstates of H_0 were known, then H_0 would form the basis of a perturbation-theory approach to the Schrödinger equation (12.3). Thus, consider the eigenvalue equation,

$$H_0|\psi_\lambda^{(0)}\rangle = E_\lambda^{(0)}|\psi_\lambda^{(0)}\rangle, \qquad (12.158)$$

or

$$\begin{aligned} E_\lambda^{(0)} &= \langle\psi_\lambda^{(0)}|H_0|\psi_\lambda^{(0)}\rangle \\ &= \sum_j \langle\psi_\lambda^{(0)}|\tilde{F}(\vec{r}_j, \underline{r})|\psi_\lambda^{(0)}\rangle. \end{aligned} \qquad (12.159)$$

We now recall that the total energy in Hartree–Fock approximation looks somewhat like the right-hand side of equation (12.159), with equation (12.154): see equation (12.60) with equations (12.30), (12.33) and (12.37). In that case, $|\psi\rangle$ is a Slater determinant, equation (12.29). Thus, as in equation (12.60), but without the factor $\frac{1}{2}$, if we consider $|\psi_\lambda^{(0)}\rangle$ to be a Slater determinant, then we have

$$E_\lambda^{(0)} = \sum_{j=1}^{N} \langle j| \left\{ \left[h_1 + \sum_{j'} \langle j'|h_2(1 - P_{jj'})|j'\rangle \right] \right\} |j\rangle. \qquad (12.160)$$

But the quantity in $\{ \}$ brackets in equation (12.160) is precisely the Fock operator, equation (12.94), so that with the Fock equation (12.93), we have, from equation (12.160),

$$E_\lambda^{(0)} = \sum_{j=1}^{N} \langle j|\varepsilon_j|j\rangle = \left(\sum_{j=1}^{N} \varepsilon_j \right). \qquad (12.161)$$

It follows that in equation (12.158), $|\psi_\lambda^{(0)}\rangle$ are Slater determinants formed from distinct selections of N eigenstates of the Fock operator. This of course assumes that the Fock eigenstates are exact. In that case they are a complete orthonormal set, as is the collection of all distinct Slater determinants that can be formed from them. We therefore have, in equations (12.155)–(12.157), along with equations (12.158) and (12.161), the prerequisite for a perturbative treatment of the N-electron system, based on the Hartree–Fock equation.

We now illustrate how many-body perturbation theory generates corrections to the Hartree–Fock approximation. Since correlation is defined as the difference between the exact solution and the Hartree–Fock solution, these corrections are called *correlation corrections*. Thouless (1972, especially Chapter IV, section 1), has given a brilliant critique of many-body perturbation theory, including the issue of system size dependence (N-dependence); see also Davidson and Silver (1977). Thouless describes Rayleigh–Schrödinger perturbation theory. We follow his lead. Let H_0 be a hamiltonian whose eigenvalue problem has been solved, as above:

$$H_0|\psi_\lambda^{(0)}\rangle = E_\lambda^{(0)}|\psi_\lambda^{(0)}\rangle. \tag{12.158}$$

Consider the exact solution,

$$H|\psi_\lambda\rangle = (H_0 + H_1)|\psi_\lambda\rangle = E_\lambda|\psi_\lambda\rangle, \tag{12.162}$$

from equations (12.155), (12.156) and (12.157). Expand $|\psi_\lambda\rangle$ in terms of the complete orthonormal set $\{|\psi_\lambda^{(0)}\rangle\}$:

$$|\psi_\lambda\rangle = \sum_{\lambda=0}^{\infty} c_{\lambda'\lambda}|\psi_{\lambda'}^{(0)}\rangle, \tag{12.163}$$

where

$$c_{\lambda'\lambda} = \langle\psi_{\lambda'}^{(0)}|\psi_\lambda\rangle. \tag{12.164}$$

Thus

$$|\psi_\lambda\rangle = \left\{c_{\lambda\lambda}|\psi_\lambda^{(0)}\rangle + \sum_{\lambda'\neq\lambda} c_{\lambda'\lambda}|\psi_{\lambda'}^{(0)}\rangle\right\}, \tag{12.165}$$

where the second term in $\{\}$ in equation (12.165) gives the perturbative correction to $|\psi_\lambda^{(0)}\rangle$, for given λ. Substituting for $c_{\lambda'\lambda}$ from equation (12.164) into equation (12.165):

$$\{1 - |\psi_\lambda^{(0)}\rangle\langle\psi_\lambda^{(0)}|\}|\psi_\lambda\rangle = \sum_{\lambda'\neq\lambda} c_{\lambda'\lambda}|\psi_{\lambda'}^{(0)}\rangle. \tag{12.166}$$

The right-hand side of equation (12.166) is clearly orthogonal to $|\psi_\lambda^{(0)}\rangle$. Thus the operator P_λ projects *orthogonal* to $|\psi_\lambda^{(0)}\rangle$, where

$$P_\lambda = \{1 - |\psi_\lambda^{(0)}\rangle\langle\psi_\lambda^{(0)}|\}. \tag{12.167}$$

From equation (12.167), it follows that

$$|\psi_\lambda\rangle = \{|\psi_\lambda^{(0)}\rangle\langle\psi_\lambda^{(0)}|\psi_\lambda\rangle + P_\lambda|\psi_\lambda\rangle\}. \tag{12.168}$$

The perturbative correction, $P_\lambda|\psi_\lambda\rangle$ in equation (12.168), would be clearer if it were seen as a correction to $|\psi_\lambda^{(0)}\rangle$. This can be accomplished by applying an unconventional normalization to the exact eigenstate $|\psi_\lambda\rangle$, namely requiring

$$\langle\psi_\lambda^{(0)}|\psi_\lambda\rangle = 1. \tag{12.169}$$

With such a normalization, the interpretation of $|\psi_\lambda\rangle$ as a probability amplitude is not valid. We shall return to this point in a moment. With equation (12.169) we have, from equation (12.168),

$$|\psi_\lambda\rangle = \{|\psi_\lambda^{(0)}\rangle + P_\lambda|\psi_\lambda\rangle\}. \tag{12.170}$$

We can now construct a probability amplitude, with unit norm. From equation (12.170),

$$(1 - P_\lambda)|\psi_\lambda\rangle = |\psi_\lambda^{(0)}\rangle. \tag{12.171}$$

Applying equation (12.169) to equation (12.171),

$$\langle\psi_\lambda|(1 - P_\lambda)|\psi_\lambda\rangle = 1. \tag{12.172}$$

Thus, the state vector $\langle\psi_\lambda'|$ is normalized, where, from equation (12.172),

$$|\psi_\lambda'\rangle = (1 - P_\lambda)^{1/2}|\psi_\lambda\rangle, \tag{12.173}$$

with the normalization of equation (12.169).

We now return to the exact eigenvalue equation (12.162), in the spirit of perturbation theory, using equation (12.170):

$$(H_0 - E_\lambda)|\psi_\lambda\rangle = -H_1|\psi_\lambda\rangle. \tag{12.174}$$

We solve this formally for E_λ, using equations (12.158) and (12.169),

$$\langle\psi_\lambda^{(0)}|(H_0 - E_\lambda)|\psi_\lambda\rangle = \langle\psi_\lambda^{(0)}|(E_\lambda^{(0)} - E_\lambda)|\psi_\lambda\rangle$$
$$= (E_\lambda^{(0)} - E_\lambda) = -\langle\psi_\lambda^{(0)}|H_1|\psi_\lambda\rangle,$$

or

$$E_\lambda = \{E_\lambda^{(0)} + \langle\psi_\lambda^{(0)}|H_1|\psi_\lambda\rangle\}. \tag{12.175}$$

Of course, since we do not know $|\psi_\lambda\rangle$ explicitly, equation (12.175) is an implicit equation for the perturbative correction to the energy $E_\lambda^{(0)}$.

We similarly obtain a formal, perturbative solution for the state vector. From equation (12.170),

$$|\psi_\lambda\rangle = \{|\psi_\lambda^{(0)}\rangle + P_\lambda|\psi_\lambda\rangle\}. \tag{12.170}$$

We formally solve equation (12.174) for $|\psi_\lambda\rangle$,

$$|\psi_\lambda\rangle = -(H_0 - E_\lambda)^{-1} . H_1 |\psi_\lambda\rangle,$$

and substitute this for $|\psi_\lambda\rangle$ in the right-hand side of equation (12.170),

$$|\psi_\lambda\rangle = \{|\psi_\lambda^{(0)}\rangle + P_\lambda (E_\lambda - H_0)^{-1} . H_1 |\psi_\lambda\rangle\}. \qquad (12.176)$$

We see here the perturbative correction, linear in H_1, but with H_1 involved implicitly to all higher orders in E_λ and $|\psi_\lambda\rangle$ in the second term in {} brackets.

   Equations (12.175) and (12.176) can be proliferated iteratively. First consider $|\psi_\lambda\rangle$, equation (12.176). For the first-order correction, replace

$$|\psi_\lambda\rangle \approx |\psi_\lambda^{(0)}\rangle$$

in the right-hand side, obtaining

$$|\psi_\lambda^{(1)}\rangle \approx \{|\psi_\lambda^{(0)}\rangle + P_\lambda (E_\lambda - H_0)^{-1} . H_1 |\psi_\lambda^{(0)}\rangle\}. \qquad (12.177)$$

Substitute this for $|\psi_\lambda\rangle$ to get the second-order correction:

$$
\begin{aligned}
|\psi_\lambda^{(2)}\rangle &\approx \{|\psi_\lambda^{(0)}\rangle + P_\lambda (E_\lambda - H_0)^{-1} . H_1 |\psi_\lambda^{(1)}\rangle\} \\
&= \{|\psi_\lambda^{(0)}\rangle + P_\lambda (E_\lambda - H_0)^{-1} . H_1 |\psi_\lambda^{(0)}\rangle \\
&\quad + P_\lambda (E_\lambda - H_0)^{-1} . H_1 . P_\lambda (E_\lambda - H_0)^{-1} . H_1 |\psi_\lambda^{(0)}\rangle\}. \quad (12.178)
\end{aligned}
$$

In equations (12.177) and (12.178) the orders of H_1 are still partly implicit to all orders through the terms in E_λ. This procedure generates a power series in the operator O defined by

$$O_\lambda = P_\lambda (E_\lambda - H_0)^{-1} . H_1. \qquad (12.179)$$

In fact, we obtain, formally,

$$|\psi_\lambda\rangle = \sum_{n=0}^{\infty} O_\lambda^n |\psi_\lambda^{(0)}\rangle = (1 - O_\lambda)^{-1} . |\psi_\lambda^{(0)}\rangle. \qquad (12.180)$$

   If we now substitute equation (12.180) into equation (12.175), we obtain, for the energy, the formal expression

$$E_\lambda = E_\lambda^{(0)} + \langle \psi_\lambda^{(0)} | H_1 (1 - O_\lambda)^{-1} | \psi_\lambda^{(0)} \rangle. \qquad (12.181)$$

Equations (12.180) and (12.181) represent the *Wigner–Brillouin perturbation theory*. We can eliminate the implicit dependence on E_λ by approaching equation (12.170) differently, through the Schrödinger equation (12.162). Since we do not like having $(H_0 - E_\lambda)^{-1}$ in our formulae, we manipulate equation (12.162) to avoid it. Consider

$$H |\psi_\lambda\rangle = (H_0 + H_1) |\psi_\lambda\rangle = E_\lambda |\psi_\lambda\rangle, \qquad (12.162)$$

whence

$$(H_0 - E_\lambda^{(0)})|\psi_\lambda\rangle = (E_\lambda - H_1 - E_\lambda^{(0)})|\psi_\lambda\rangle. \tag{12.182}$$

Now in equation (12.182), $(E_\lambda - E_\lambda^{(0)})$ is a first-order small quantity, as is H_1, although it contains all higher small orders as well. Thus, we rewrite equation (12.170), replacing $|\psi_\lambda\rangle$ on the right-hand side by using equation (12.182) in the form

$$|\psi_\lambda\rangle = (H_0 - E_\lambda^{(0)})^{-1} \cdot (E_\lambda - E_\lambda^{(0)} - H_1)|\psi_\lambda\rangle, \tag{12.183}$$

obtaining

$$|\psi_\lambda\rangle = \{|\psi_\lambda^{(0)}\rangle + P_\lambda(E_\lambda^{(0)} - H_0)^{-1} \cdot (H_1 - E_\lambda + E_\lambda^{(0)})|\psi_\lambda\rangle\}. \tag{12.184}$$

This in turn can be used with equation (12.175) to generate a fully explicit power series in H_1:

$$E_\lambda = \{E_\lambda^{(0)} + \langle\psi_\lambda^{(0)}|H_1|\psi_\lambda\rangle\}. \tag{12.185}$$

The result is called *Rayleigh–Schrödinger perturbation theory*.

From equations (12.184) and (12.185) we obtain first the zeroth-order solutions:

$$|\psi_\lambda\rangle \approx |\psi_\lambda^{(0)}\rangle, \tag{12.186}$$

$$E_\lambda \approx E_\lambda^{(0)}. \tag{12.187}$$

In equations (12.186) and (12.187), the solution $|\psi_\lambda^{(0)}\rangle$ and $E_\lambda^{(0)}$ are determined by the choice of H_0; see equations (12.155)–(12.158). If the system is such that its exact solution $|\psi_\lambda\rangle$ and E_λ do not satisfy equations (12.186) and (12.187) approximately, then the perturbative method will not give physically correct results. An example of such a case is given in Chapter 14.

We now define the first order solution to come from substituting the zeroth order solutions into equations (12.184) and (12.185), and the nth order to come from substituting the $(n-1)$st order. Thus, in first order,

$$|\psi_\lambda\rangle \approx \{|\psi_\lambda^{(0)}\rangle + P_\lambda(E_\lambda^{(0)} - H_0)^{-1} \cdot H_1|\psi_\lambda^{(0)}\rangle\}, \tag{12.188}$$

$$E_\lambda \approx \{E_\lambda^{(0)} + \langle\psi_\lambda^{(0)}|H_1|\psi_\lambda^{(0)}\rangle\}. \tag{12.189}$$

From equations (12.161), (12.157) and (12.99), we see that this first-order energy, equation (12.189), is just the Hartree–Fock approximation. Proceeding to second order, we substitute equations (12.188) and (12.189) into equations (12.184) and (12.185):

$$|\psi_\lambda\rangle \approx \{|\psi_\lambda^{(0)}\rangle + P_\lambda(E_\lambda^{(0)} - H_0)^{-1}(H_1 - \langle\psi_\lambda^{(0)}|H_1|\psi_\lambda^{(0)}\rangle)$$
$$\times [|\psi_\lambda^{(0)}\rangle + P_\lambda(E_\lambda^{(0)} - H_0)^{-1} \cdot H_1|\psi_\lambda^{(0)}\rangle]\}, \tag{12.190}$$

$$E_\lambda \approx \{E_\lambda^{(0)} + \langle\psi_\lambda^{(0)}|H_1|\psi_\lambda^{(0)}\rangle + \langle\psi_\lambda^{(0)}|H_1 \cdot P_\lambda(E_\lambda^{(0)} - H_0)^{-1} \cdot H_1|\psi_\lambda^{(0)}\rangle\}. \tag{12.191}$$

We note parenthetically at this point that the present formulation is not satisfactory for $|\psi_\lambda\rangle$, because of the term in equation (12.190) of the form

$$-P_\lambda \cdot (E_\lambda^{(0)} - H_0)^{-1} \cdot \langle \psi_\lambda^{(0)}|H_1|\psi_\lambda^{(0)}\rangle \cdot |\psi_\lambda^{(0)}\rangle. \qquad (12.192)$$

Since $\langle \psi_\lambda^{(0)}|H_1|\psi_\lambda^{(0)}\rangle$ is a real number, the operator $(E_\lambda^{(0)} - H_0)|\psi_\lambda^{(0)}\rangle$ gives zero, from equation (12.158), and thus the operator $(E_\lambda^{(0)} - H_0)^{-1}$ is singular. Formally, this can be avoided by replacing $E_\lambda^{(0)}$ by $(E_\lambda^{(0)} + i\gamma)$, with γ real, and then taking the limit $\gamma \to 0$ after all other operations have been performed. An explicit expression for $|\psi_\lambda\rangle$ in second order can be obtained by the usual power-series treatment of perturbation theory: see Szabo and Ostlund (1989), Chapter 6.

We now return to equation (12.191) to obtain the second-order perturbative term in the energy, the lowest-order correction to the Hartree–Fock approximation. It is

$$\langle \psi_\lambda^{(0)}|H_1 \cdot P_\lambda (E_\lambda^{(0)} - H_0)^{-1} \cdot H_1|\psi_\lambda^{(0)}\rangle. \qquad (12.193)$$

If we use the representation of equation (12.167) for P_λ, along with the identity I in the form

$$I = \sum_n |\psi_n^{(0)}\rangle\langle\psi_n^{(0)}|, \qquad (12.194)$$

and the fact that $|\psi_n^{(0)}\rangle$ is an eigenstate of H_0 belonging to eigenvalue $E_n^{(0)}$, we obtain from equation (12.193)

$$\sum_{n(\neq\lambda)} \frac{|\langle \psi_\lambda^{(0)}|H_1|\psi_n^{(0)}\rangle|^2}{(E_\lambda^{(0)} - E_n^{(0)})}. \qquad (12.195)$$

Equation (12.195) can be expressed entirely in terms of eigenstates and eigenvalues of the Fock equation. Since $n \neq \lambda$ in the summation, there must be at least one Fock eigenstate, say $|c\rangle$, in the occupied manifold of state $\psi_n^{(0)}$ that is outside the occupied manifold of state $\psi_\lambda^{(0)}$: see equation (12.93), with j replaced by c. Furthermore, the manifolds λ and n cannot differ by more than two Fock eigenstates, by an extension of the discussion following equation (12.38). Orthogonality among the Fock eigenstates requires that $(n-2)$ of the Fock eigenstates that make up the Slater determinant $|\psi_\lambda^{(0)}\rangle$ be the same as $(n-2)$ of those that make up $|\psi_n^{(0)}\rangle$, for the matrix element of each two-particle term in H_1, equation (12.195), involves integration over two single-particle variables.

In fact, the matrix element of H_1 in equation (12.195) is zero when λ and n differ by only a single Fock eigenstate. We can see this by evaluating the matrix element for such a case. The proof in its most elegant form depends on *Brillouin's theorem* [Brillouin (1934)], see also Slater (1963, Appendix 4). This states that the total hamiltonian H has zero matrix elements between such eigenstates. But $H_1 = (H - H_0)$, equations (12.155)–(12.157), and from equation (12.158)

$$H_0|\psi_n^{(0)}\rangle = E_n^{(0)}|\psi_n^{(0)}\rangle, \tag{12.196}$$

so

$$\langle \psi_\lambda^{(0)}|H_0|\psi_n^{(0)}\rangle = E_n^{(0)}\langle \psi_\lambda^{(0)}|\psi_n^{(0)}\rangle. \tag{12.197}$$

Thus, through the orthogonality of $|\psi_\lambda^{(0)}\rangle$ and $|\psi_n^{(0)}\rangle$ with $n \neq \lambda$, the matrix elements of H_0 are also zero for such states, and so therefore are the matrix elements of H_1.

We now prove Brillouin's theorem. Suppose, as in equation (12.29),

$$\psi_\lambda^{(0)} = A\varphi_1(\underline{r}_1)\varphi_2(\underline{r}_2)\cdots\varphi_N(\underline{r}_N), \tag{12.29}$$

and

$$\psi_n^{(0)} = A\varphi_c(\underline{r}_1)\varphi_2(\underline{r}_2)\cdots\varphi_N(\underline{r}_N). \tag{12.198}$$

Then, from equations (12.29) and (12.198), with equation (12.2),

$$\langle \psi_\lambda^{(0)}|\sum_{j=1}^N h_1(\vec{r}_j)|\psi_n^{(0)}\rangle = \int dr_1 \cdots dr_N [\varphi_1(\underline{r}_1)\varphi_2(\underline{r}_2)\cdots\varphi_N(\underline{r}_N)]^*$$

$$\times \sum_{j=1}^N h_1(\vec{r}_j) . A[\varphi_c(\underline{r}_1)\varphi_2(\underline{r}_2)\cdots\varphi_N(\underline{r}_N)]$$

$$= \int dr_1 \, \varphi_1^*(\underline{r}_1)h_1(\underline{r}_1)\varphi_c(\underline{r}_1)$$

$$= \langle \varphi_1|h_1|\varphi_c\rangle, \tag{12.199}$$

where h_1 is defined in equation (12.33). Similarly,

$$\langle \psi_\lambda^{(0)}|\frac{1}{2}\sum_{j,j'}' h_2(\vec{r}_j,\vec{r}_{j'})|\psi_n^{(0)}\rangle$$

$$= \int dr_1 \cdots dr_n [\varphi_1(\underline{r}_1)\varphi_2(\underline{r}_2)\cdots\varphi_N(\underline{r}_N)]$$

$$\times \frac{1}{2}\sum_{j,j'}' h_2(\vec{r}_j,\vec{r}_{j'}) . A[\varphi_c(\underline{r}_1)\varphi_2(\underline{r}_2)\cdots\varphi_N(\underline{r}_N)]. \tag{12.200}$$

This can only be non-zero if j is equal to one, or if j' is equal to one. Thus equation (12.200) becomes

$$\frac{1}{2}\sum_{j'(\neq 1)} \int dr_1 \, dr_{j'} \, \varphi_1^*(\underline{r}_1)\varphi_{j'}^*(\underline{r}_{j'})h_2(\vec{r}_1,\vec{r}_{j'})[1 - P(\underline{r}_1,\underline{r}_{j'})]\varphi_c(\underline{r}_1)\varphi_{j'}(\underline{r}_{j'})$$

$$+ \frac{1}{2}\sum_{j(\neq 1)} \int dr_1 \, dr_j \, \varphi_1^*(\underline{r}_1)\varphi_j^*(\underline{r}_j)h_2(\vec{r}_j,\vec{r}_1)[1 - P(\underline{r}_j,\underline{r}_1)]\varphi_c(\underline{r}_1)\varphi_j(\underline{r}_j). \tag{12.201}$$

Since $h_2(\vec{r}, \vec{r}') = h_2(\vec{r}', \vec{r})$ and similarly for P, we have for (12.200)

$$\sum_{j(\neq 1)} \langle \varphi_1 \varphi_j | h_2(1 - P) | \varphi_c \varphi_j \rangle = \langle \varphi_1 | \left\{ \sum_{j(\neq 1)} \langle \varphi_j | h_2(1 - P) | \varphi_j \rangle \right\} | \varphi_c \rangle. \quad (12.202)$$

Combining equations (12.199) with (12.200) and (12.202),

$$\langle \psi_\lambda^{(0)} | H | \psi_n^{(0)} \rangle = \langle \varphi_1 | \left\{ h_1 + \sum_{j(\neq 1)} \langle \varphi_j | h_2(1 - P) | \varphi_j \rangle \right\} | \varphi_c \rangle. \quad (12.203)$$

But the operator in { } brackets in equation (12.203), with the redundant restriction $j(\neq 1)$ removed [see discussion following equation (12.43)] is just the Fock operator, equation (12.94) with equations (12.33) and (12.63). Furthermore, $|\varphi_c\rangle$ is a Fock eigenstate with eigenvalue ε_c. Thus equation (12.203) is

$$\langle \psi_\lambda^{(0)} | H | \psi_n^{(0)} \rangle = \varepsilon_c \langle \varphi_1 | \varphi_c \rangle = 0, \quad (12.204)$$

from the orthogonality of Fock eigenstates $|\varphi_1\rangle$ and $|\varphi_c\rangle$. This, with equations (12.29) and (12.198), proves Brillouin's theorem. We combine equation (12.204) with the similar result in equation (12.197),

$$\langle \varphi_\lambda^{(0)} | H_0 | \psi_n^{(0)} \rangle = 0, \quad (12.205)$$

to obtain

$$\langle \psi_\lambda^{(0)} | (H - H_0) | \psi_n^{(0)} \rangle = \langle \psi_\lambda^{(0)} | H_1 | \psi_n^{(0)} \rangle = 0. \quad (12.206)$$

From the discussion following equation (12.195) and the result of equation (12.206) we see that the second-order perturbative correction to the Hartree–Fock approximation involves states $|\psi_\lambda^{(0)}\rangle$ and $|\psi_n^{(0)}\rangle$ that differ by exactly two single-particle Fock eigenstates. Thus, in general, we can get $|\psi_n^{(0)}\rangle$ from $|\psi_\lambda^{(0)}\rangle$ by de-occupying two Fock eigenstates, say $|\varphi_a\rangle$ and $|\varphi_b\rangle$, from the occupied manifold of $|\psi_\lambda^{(0)}\rangle$, and occupying two of its virtual manifold eigenstates, say $|\varphi_c\rangle$ and $|\varphi_d\rangle$. The matrix element in equation (12.195) is now

$$\langle \psi_\lambda^{(0)} | H_1 | \psi_n^{(0)} \rangle = \langle ab | H_1 (1 - P) | cd \rangle, \quad (12.207)$$

with $|a\rangle$ and $|b\rangle$ from the occupied manifold of $|\psi_\lambda^{(0)}\rangle$, and $|c\rangle$ and $|d\rangle$ from its virtual, or unoccupied manifold. In equation (12.207),

$$|a\rangle \equiv |\varphi_a\rangle \quad (12.208)$$

and similarly for $|b\rangle$, $|c\rangle$ and $|d\rangle$, and $|\psi_n^{(0)}\rangle$ is *not* given by equation (12.198), but rather

$$|\psi_n^{(0)}\rangle = \{ A\varphi_1(\underline{r}_1) \cdots \varphi_c(\underline{r}_a) \cdots \varphi_d(\underline{r}_b) \cdots \varphi_N(\underline{r}_N) \}. \quad (12.209)$$

Recalling the definition of $E_\lambda^{(0)}$, equation (12.161), we find the energy denominator in equation (12.195) to be

$$(E_\lambda^{(0)} - E_n^{(0)}) = (\varepsilon_a + \varepsilon_b - \varepsilon_c - \varepsilon_d). \qquad (12.210)$$

Thus combining equations (12.207) and (12.210) we obtain for equation (12.195):

$$\sum_{n(\neq\lambda)} \frac{|\langle\psi_\lambda^{(0)}|H_1|\psi_n^{(0)}\rangle|^2}{(E_\lambda^{(0)} - E_n^{(0)})} = \sum_{\substack{a<b(\text{occ})\\c<d(\text{virt})}} \frac{|\langle ab|H_1(1-P)|cd\rangle|^2}{(\varepsilon_a + \varepsilon_b - \varepsilon_c - \varepsilon_d)}. \qquad (12.211)$$

We note that the second-order perturbative correlation correction, equation (12.211), is negative, because ε_c and ε_d are greater than ε_a and ε_b. Higher-order corrections quickly become much more complicated, and require special methods, as discussed lucidly by Thouless (1972). Second-, and even third- and fourth-order Rayleigh–Schrödinger many-body perturbation theory correlation corrections are practicable, and are incorporated in quantum-chemistry-type computations routinely. In many problems, including some optical absorption processes, for example, correlation must be included in order to get quantitative accuracy.

12.7　One-, two- and N-particle density functionals

12.7.1　Introduction

The Hartree–Fock method has two features, at least, that one might wish to improve on. First, the Slater determinant wave function, equation (12.29), must have N molecular orbitals φ_j. For most problems, this means that computations are large, perhaps larger than one might like. Consider, for example, a molecular cluster $(\text{Ba}^{2+})_1(\text{F}^-)_8$ embedded in a BaF_2 crystal. In that case $N = [54 + (10 \times 8)] = 134$. Matrices of at least this dimensionality need to be manipulated in obtaining the solution. On the other hand, rather than knowing all that detail, one might be content to have a rough picture of the nine ions, representable as a combination of, at minimum, nine localized functions. Something closer to nine than to 134 might give a perfectly adequate picture of the molecular cluster, in terms of particle density. For the analysis of physical properties and processes, however, accurate total energies are required. One is therefore led to ask whether a rigorous relationship exists between particle density and total energy and, further, whether the variational principle can be applied to determine the density, as is done, for example, in solving the Fock equation. The other unsatisfactory feature of the Hartree–Fock method is that it is not useful for problems for which correlation is dominant, and not very accurate where correlation is significant. One would like to have a rigorous analytical approach in which correlation is not treated as a separate, small correction.

In section 12.7.2 we discuss the so-called density functional formulation of Hohenberg and Kohn (1964), in which it is shown that the total energy is a unique functional of the particle density. This is obviously a very powerful result. The fact that the functional is not explicitly known, however, is a drawback. In section 12.7.3 we show how to write the total energy of an arbitrary system, fermions or bosons, with n-body interactions, *explicitly* as a functional of a so-called reduced n-particle density matrix. In section 12.7.4 we examine the applicability of this formulation to the many-fermion system. It turns out not to lead to a simple variational method, but its attractive features encourage continuing efforts to use it as the basis of a computational scheme. In section 12.7.5 we show the relationship between the reduced density matrix (two-particle formulation) and the single-particle density Hohenberg and Kohn formulation.

12.7.2 Density functional of Hohenberg and Kohn

The many-electron hamiltonian, equation (12.2), possesses eigenstates for which a particle density $\rho(\vec{r})$ can be identified, equation (12.45):

$$\rho(\vec{r}) = \langle\psi| \sum_j \delta(\vec{r} - \vec{r}_j)|\psi\rangle. \tag{12.212}$$

We concentrate on the ground state $|\psi\rangle$. The specific atomic system, and its ground state, depend solely on the nuclear charges Z_J and positions $\vec{R}_J$, which determine the potential external to the electronic system, $V_{\text{ext}}(\vec{r})$, where

$$V_{\text{ext}}(\vec{r}) = -\sum_J Z_J|\vec{r} - \vec{R}_J|^{-1}, \tag{12.213}$$

see equation (12.2). The ground-state wave function ψ, and thus the density ρ, are therefore functionals of V_{ext}. The converse is also true, namely, the external potential V_{ext} is a functional of the density ρ. Furthermore, the total ground-state energy E_0, is also therefore a functional of ρ. If it is a *unique* functional, then one can consider the possibility of determining the ground state by minimizing the energy with respect to variation of ρ. We shall show that V_{ext}, and therefore the total ground-state energy E_0, are unique functionals of the density ρ. This is the Hohenberg–Kohn theorem.

The proof is as follows [Hohenberg and Kohn (1964)]. Consider two physically distinct N-electron systems, corresponding to V_{ext} and V'_{ext}, differing by more than a constant. The corresponding Schrödinger equations are

$$H|\psi\rangle = E_0|\psi\rangle, \tag{12.214}$$

$$H'|\psi'\rangle = E'_0|\psi'\rangle, \tag{12.215}$$

where

$$(H' - H) = (V' - V), \qquad (12.216)$$

where V is the potential energy of the system due to V_{ext}:

$$V = \sum_{j=1}^{N} V_{\text{ext}}(\vec{r}_j). \qquad (12.217)$$

Note then that in equation (12.216), V and V' differ by having different sets of values for (Z_J, R_J): see equation (12.213). Since $|\psi'\rangle$ is the ground state for H', equation (12.215), we have, from equations (12.214)–(12.216),

$$\langle \psi' | H' | \psi' \rangle < \langle \psi | H' | \psi \rangle = \langle \psi | H | \psi \rangle + \langle \psi | (V' - V) | \psi \rangle, \qquad (12.218)$$

or

$$E_0' < \{ E_0 + \langle \psi | (V' - V) | \psi \rangle \}. \qquad (12.219)$$

Now we can write

$$V = \sum_j V_{\text{ext}}(\vec{r}_j) = \sum_j \int d\vec{r}\, \delta(\vec{r} - \vec{r}_j) V_{\text{ext}}(\vec{r}). \qquad (12.220)$$

Thus, from equation (12.220),

$$\langle \psi | V | \psi \rangle = \int d\vec{r}\, V_{\text{ext}}(\vec{r}) \langle \psi | \sum_j \delta(\vec{r} - \vec{r}_j) | \psi \rangle. \qquad (12.221)$$

From equation (12.212), this becomes

$$\langle \psi | V | \psi \rangle = \int d\vec{r}\, V_{\text{ext}}(\vec{r}) \rho(\vec{r}). \qquad (12.222)$$

In turn, using equation (12.222), equation (12.219) is now

$$E_0' < \left\{ E_0 + \int d\vec{r}\, [V_{\text{ext}}'(\vec{r}) - V_{\text{ext}}(\vec{r})] \rho(\vec{r}) \right\}. \qquad (12.223)$$

Similarly to equation (12.218), we can write

$$\langle \psi | H | \psi \rangle < \langle \psi' | H | \psi' \rangle = \{ \langle \psi' | H' | \psi' \rangle - \langle \psi' | (V' - V) | \psi' \rangle \} \qquad (12.224)$$

or

$$E_0 < \{ E_0' - \langle \psi' | V' - V | \psi' \rangle \} \qquad (12.225)$$

which, with equation (12.222), becomes

$$E_0 < \left\{ E_0' + \int d\vec{r}\, [V_{\text{ext}}(\vec{r}) - V_{\text{ext}}'(\vec{r})] \rho'(\vec{r}) \right\}. \qquad (12.226)$$

Now assume that the ground state density $\rho'(\vec{r})$ associated with $V_{\text{ext}}'(\vec{r})$ is the same as $\rho(\vec{r})$ associated with the distinct potential $V_{\text{ext}}(\vec{r})$; i.e. assume

$$\rho'(\vec{r}) = \rho(\vec{r}). \qquad (12.227)$$

Then equation (12.226) becomes

$$E_0 < \left\{ E_0' + \int \mathrm{d}\vec{r}\, [V_{\mathrm{ext}}(\vec{r}) - V'_{\mathrm{ext}}(\vec{r})]\rho(\vec{r}) \right\}, \tag{12.228}$$

and combining equations (12.223) and (12.228) we conclude that

$$(E_0' + E_0) < (E_0' + E_0). \tag{12.229}$$

The contradiction expressed by equation (12.229) shows that the assumption of equation (12.227) is invalid. Thus, a unique density $\rho(\vec{r})$, equation (12.212), for the ground-state ψ, cannot come from two distinct external potentials, $V_{\mathrm{ext}}(\vec{r})$, equation (12.213). It means that $\rho(\vec{r})$ is a unique functional of $V_{\mathrm{ext}}(\vec{r})$, and vice versa, $V_{\mathrm{ext}}(\vec{r})$ is a unique functional of $\rho(\vec{r})$. Consequently V, equation (12.217), and therefore the ground-state energy E_0 and wavefunction ψ, are unique functionals of the density ρ, which we express as

$$E_0 = E_0[\rho]. \tag{12.230}$$

So far, the seminal result of equation (12.230) has not led to the explicit, general functional relationship $E_0[\rho]$. Nevertheless, it has led to computational procedures that are, in practice, widely useful in determining the ground-state properties of a variety of solid, molecular, and atomic systems. So much so that, in 1998, Walter Kohn shared the Nobel Prize in Chemistry for the progress that this discovery had made possible. Kohn shared the Nobel Prize with J A Pople, whose contribution had been in Hartree–Fock-based computational methods. In the field of computational quantum chemistry, methods based on the density functional theorem rival if not surpass in popularity those based on the Hartree–Fock approximation. In the field of quantum electronic structure computations in physics, particularly in the US, density functional methods have been dominant. Much theoretical work has been done, elucidating and refining such methods. The qualitative difference between Hartree–Fock and density functional methods is that the explicit form used for part of the density functional is essentially empirical. Pedagogically, the work by Parr and Yang (1989) serves as an introduction to both the basic theory and the applications of the density functional method. Beyond that, the relevant literature is extremely voluminous, and growing. In section 12.7.5 we show a formal implicit relationship between the particle density ρ and the total energy, involving the two-particle density matrix. The utility of this relationship, however, has yet to be demonstrated.

12.7.3   Reduced density matrices

Let us re-examine the many-particle total energy, with a view to developing a computational method that does not rely on one, or a few, Slater determinants. Let $(E, |\psi\rangle)$ be the total energy and its eigenstate,

$$E = \langle \psi | H | \psi \rangle, \tag{12.231}$$

where

$$\psi \equiv \psi(\underline{r}_1, \underline{r}_2, \ldots, \underline{r}_N). \tag{12.232}$$

For H, we limit ourselves to the non-relativistic, field-free, spin-independent form with two-particle interactions only, as in equation (12.2). We consider N identical particles, fermions or bosons, for which the symmetry conditions are, respectively,

$$P_{ij}\psi(\underline{r}_1, \ldots, \underline{r}_N) = \mp\psi(\underline{r}_1, \ldots, \underline{r}_N), \tag{12.233}$$

where P_{ij} is the particle pairwise interchange operator, as in equation (12.6). For fermions, equation (12.233) is the general statement of the Pauli principle.

Consider the external, single-particle potential contribution V to the energy E, from equation (12.217). It is

$$\langle \psi | V | \psi \rangle = \sum_{j=1}^{N} \int d\underline{r}_1, \ldots, d\underline{r}_N \psi^*(\underline{r}_1, \ldots, \underline{r}_N) V_{\text{ext}}(\vec{r}_j)\psi(\underline{r}_1, \ldots, \underline{r}_N), \tag{12.234}$$

where

$$V_{\text{ext}}(\vec{r}) = -\sum_J Z_J |\vec{r} - \vec{R}_J|^{-1}. \tag{12.235}$$

Consider an arbitrary term j in equation (12.234). Interchange variables $\underline{r}_j$ and $\underline{r}_1$ in the integration. Then we have $V_{\text{ext}}(\vec{r}_j) \rightarrow V_{\text{ext}}(\vec{r}_1)$. We can thereafter interchange $\underline{r}_j$ and $\underline{r}_1$ in both ψ^* and ψ, introducing the factor $(\mp 1)^2 = +1$ for either fermions or bosons. It follows that equation (12.234) reduces to

$$\langle \psi | V | \psi \rangle = N \int d\underline{r}_1, \ldots, d\underline{r}_N \psi^*(\underline{r}_1, \ldots, \underline{r}_N) V_{\text{ext}}(\vec{r}_1)\psi(\underline{r}_1, \ldots, \underline{r}_N)$$

$$= N \int d\underline{r} \cdot d\underline{r}_2, \ldots, d\underline{r}_N \psi^*(\underline{r}, \underline{r}_2, \ldots, \underline{r}_N) V_{\text{ext}}(\vec{r})\psi(\underline{r}, \underline{r}_2, \ldots, \underline{r}_N). \tag{12.236}$$

Note that in this section the notation $\underline{r}$ refers to a *single-particle* space and spin coordinate, unlike its meaning in earlier sections. The same kind of result occurs for the kinetic energy, and a similar one for the pairwise interactions. The net result is that the energy can be written

$$E = N \int d\underline{r} \cdot d\underline{r}' \cdot d\underline{r}_3, \ldots, d\underline{r}_N \psi^*(\underline{r}, \underline{r}', \underline{r}_3, \ldots, \underline{r}_N)$$

$$\times \{-\tfrac{1}{2}\nabla^2 + V_{\text{ext}}(\underline{r}) + \tfrac{1}{2}(N-1)v(\underline{r}, \underline{r}')\} \cdot \psi(\underline{r}, \underline{r}', \underline{r}_3, \ldots, \underline{r}_N), \tag{12.237}$$

where

$$v(\underline{r},\underline{r}') = |\vec{r} - \vec{r}'|^{-1}. \tag{12.238}$$

Equation (12.236), in comparison with equation (12.234), shows that the total external potential energy is simply N times a single-particle potential energy. Given the indistinguishability of identical particles, this result should be obvious: in a many-body system, since all particles are identical, all must contribute equally to the energy, be it single-particle potential energy, or kinetic energy, as seen in equation (12.237). Similarly, the two-body energy is contributed equally by all pairs of particles in the system. Thus the result of equation (12.237) supports the idea that indistinguishability has been correctly incorporated into the formulation.

The potential energy terms in equation (12.237) simplify further, as follows:

$$N \int d\underline{r} \cdot d\underline{r}' \{V_{\text{ext}}(\vec{r}) + \tfrac{1}{2}(N-1)v(\underline{r},\underline{r}')\}$$

$$\times \int d\underline{r}_3, \ldots, d\underline{r}_N \psi^*(\underline{r},\underline{r}',\underline{r}_3,\ldots,\underline{r}_N) \cdot \psi(\underline{r},\underline{r}',\underline{r}_3,\ldots,\underline{r}_N). \tag{12.239}$$

The integral over $(d\underline{r}_3,\ldots,d\underline{r}_N)$ is simply the probability density of finding a specific particle at $\underline{r}$ and another specific particle at $\underline{r}'$. Since the identical particles are indistinguishable, this is not an observable. Rather, we speak of the probability density of finding any particle (a particle) at $\vec{r}$, and any other at $\vec{r}'$. For N particles there are $N(N-1)$ specific pairs. We therefore define the two-particle density $\rho_2(\underline{r},\underline{r}';\underline{r},\underline{r}')$ as follows:

$$\rho_2(\underline{r},\underline{r}';\underline{r},\underline{r}') = N(N-1) \int d\underline{r}_3,\ldots,d\underline{r}_N \psi^*(\underline{r},\underline{r}',\underline{r}_3,\ldots,\underline{r}_N)$$

$$\times \psi(\underline{r},\underline{r}',\underline{r}_3,\ldots,\underline{r}_N). \tag{12.240}$$

The notation for ρ_2 is more cumbersome than needed because we want to generalize it as follows:

$$\rho_2(\underline{r},\underline{r}';\underline{q};\underline{q}') = N(N-1) \int d\underline{r}_3,\ldots,d\underline{r}_N \psi^*(\underline{r},\underline{r}',\underline{r}_3,\ldots,\underline{r}_N)$$

$$\times \psi(\underline{q},\underline{q}',\underline{r}_3,\ldots,\underline{r}_N). \tag{12.241}$$

Equation (12.241) defines the two-particle density *operator* $\rho_2(\underline{r},\underline{r}',\underline{q},\underline{q}')$. It is referred to as a *reduced density matrix* for the following reason. The probability density for finding specific particles at $(\underline{r}_1,\ldots,\underline{r}_N)$ is

$$|\psi|^2 = \psi^*(\underline{r}_1,\ldots,\underline{r}_N)\psi(\underline{r}_1,\ldots,\underline{r}_N)$$

$$= \langle \underline{r}_1,\ldots,\underline{r}_N|\psi\rangle\langle\psi|\underline{r}_1,\ldots,\underline{r}_N\rangle. \tag{12.242}$$

From equation (12.242) ($|\psi\rangle\langle\psi|$) is identified as the *N-particle density operator*. In a matrix representation, it is referred to as the *N-particle density matrix*. If it is averaged (integrated) over $(N - 2)$ particles, as in equation (12.241), it becomes the *two particle reduced density matrix*, or, more simply, the two-particle density matrix. The two-particle density matrix is sometimes defined without the normalization $N(N - 1)$ given in equation (12.241).

Returning to equations (12.237) and (12.239), we see that the total energy E can be expressed in terms of ρ_2, equation (12.241), as follows:

$$E = \int d\underline{r} \cdot d\underline{r}' \int d\underline{q} \cdot d\underline{q}' \cdot \delta(\underline{r} - \underline{q})\delta(\underline{r}' - \underline{q}')$$
$$\times \{(N - 1)^{-1}[-\tfrac{1}{2}\nabla_q^2 + V_{ext}(\underline{q})] + \tfrac{1}{2}v(\underline{q}, \underline{q}')\}\rho_2(\underline{r}, \underline{r}'; \underline{q}, \underline{q}'). \quad (12.243)$$

We thus see that the total energy is an explicit linear functional of ρ_2. This remarkable result was apparently first discovered by Husimi (1940).

The complexity of having a function of four variables, $\rho_2(\underline{r}, \underline{r}'; \underline{q}, \underline{q}')$, is only apparent. In fact, equation (12.243) simplifies as follows:

$$E = \left\{ \int d\underline{r} \cdot d\underline{q}(N - 1)^{-1}\delta(\underline{r} - \underline{q})(-\tfrac{1}{2}\nabla_q^2) \int d\underline{r}' \, \rho_2(\underline{r}, \underline{r}'; \underline{q}, \underline{r}') \right.$$
$$+ \int d\underline{r}(N - 1)^{-1} V_{ext}(\underline{r}) \int d\underline{r}' \, \rho_2(\underline{r}, \underline{r}'; \underline{r}, \underline{r}')$$
$$\left. + \frac{1}{2}\int d\underline{r} \cdot d\underline{r}' \, v(\underline{r}, \underline{r}')\rho_2(\underline{r}, \underline{r}'; \underline{r}, \underline{r}') \right\}. \quad (12.244)$$

Thus the term in v has ρ_2 dependent on only two variables, $\underline{r}$ and $\underline{r}'$. As we saw earlier, in equation (12.222), the term in V_{ext} is

$$\int d\underline{r} \, V_{ext}(\underline{r})(N - 1)^{-1} \int d\underline{r}' \, \rho_2(\underline{r}, \underline{r}'; \underline{r}, \underline{r}') = \int d\underline{r} \, V_{ext}(\vec{r})\rho(\underline{r}), \quad (12.245)$$

where $\rho(\underline{r})$ is the density of particles. Thus,

$$\rho(\underline{r}) = (N - 1)^{-1} \int d\underline{r}' \, \rho_2(\underline{r}, \underline{r}'; \underline{r}, \underline{r}')$$
$$= N \int d\underline{r}_2 \, d\underline{r}_3, \ldots, d\underline{r}_N \psi^*(\underline{r}, \underline{r}_2, \underline{r}_3, \ldots, \underline{r}_N) \cdot \psi(\underline{r}, \underline{r}_2, \underline{r}_3, \ldots, \underline{r}_N). \quad (12.246)$$

The second line of equation (12.246) comes from the definition of ρ_2, equation (12.241). Correspondingly, we define the first-order, or single-particle reduced density matrix, $\rho_1(\underline{r}, \underline{q})$, as follows:

$$\rho_1(\underline{r}; \underline{q}) = N \int d\underline{r}_2, \ldots, d\underline{r}_N \, \psi^*(\underline{r}, \underline{r}_2, \ldots, \underline{r}_N)\psi(\underline{q}, \underline{r}_2, \ldots, \underline{r}_N)$$
$$= (N - 1)^{-1} \int d\underline{r}' \, \rho_2(\underline{r}, \underline{r}'; \underline{q}, \underline{r}'). \quad (12.247)$$

From equations (12.247) and (12.246), we see

$$\rho(\underline{r}) = \rho_1(\underline{r}; \underline{r}). \tag{12.248}$$

It follows from equation (12.246) that

$$\int d\underline{r}\, \rho(\underline{r}) = N, \tag{12.249}$$

from the normalization of ψ. Equation (12.249) is, of course, required from the definition of particle density in an N-particle system. Finally, combining equations (12.244), (12.245) and (12.247),

$$E = \int d\underline{r} \left\{ \int d\underline{q}\, \delta(\underline{r} - \underline{q})(-\tfrac{1}{2}\nabla_q^2)\rho_1(\underline{r}; \underline{q}) + V_{\text{ext}}(\underline{r})\rho(\underline{r}) \right.$$
$$\left. + \frac{1}{2}\int d\underline{r}'\, v(\underline{r}, \underline{r}')\rho_2(\underline{r}, \underline{r}'; \underline{r}, \underline{r}') \right\}. \tag{12.250}$$

Thus E is a linear combination of explicit linear functionals of ρ_1 and ρ_2 [see equation (12.248)]. Basically, there is only one physical constraint on this system, namely that the number of particles should be N. From equations (12.247)–(12.249),

$$N(N - 1) = \int d\underline{r}\, d\underline{r}'\, \rho_2(\underline{r}, \underline{r}'; \underline{r}, \underline{r}'). \tag{12.251}$$

This expresses the constraint in terms of ρ_2, compatible with the energy functional, equation (12.243). Other constraints, such as given total spin or given total angular momentum, can be expressed in terms of ρ_2, analogous to equation (12.251).

The derivation of equation (12.243), leading to the two-particle density functional for a system of N particles with two-body interactions, generalizes to a similar linear functional of the n-particle density for systems with n-body interactions. We reiterate these results are equally valid for bosons and fermions.

12.7.4 The many-fermion system

Since we have an explicit energy functional in terms of ρ_2, equation (12.143), and a corresponding constraint, equation (12.151), we might consider the possibility of applying the minimum-energy variational principle. In practical terms, we might consider also an expansion in terms of a complete set of atomic-like orbitals, as in equations (12.103) and (12.110) for the Hartree–Fock approximation. In practice, an incomplete set would have to be endured. Denote such a set $\{f_\lambda(\underline{r})\}$. Then

$$\rho_2(\underline{r}, \underline{r}'; \underline{q}, \underline{q}') = \sum_{\lambda\lambda'\mu\mu'} \tilde{c}_{\lambda\lambda'\mu\mu'}\, f_\lambda(\underline{r})f_{\lambda'}(\underline{r}')f_\mu(\underline{q})f_{\mu'}(\underline{q}'), \tag{12.252}$$

for an arbitrary function of four such variables. In our case, for fermions, the Pauli principle requires the following constraints:

$$\rho_2(\underline{r},\underline{r}';\underline{q},\underline{q}') = -\rho_2(\underline{r}',\underline{r};\underline{q},\underline{q}') = -\rho_2(\underline{r},\underline{r}';\underline{q}',\underline{q}) = \rho_2(\underline{r}',\underline{r};\underline{q}',\underline{q}). \quad (12.253)$$

See the definition of ρ_2, equation (12.241), and the requirement of the Pauli principle, equation (12.233). The constraints of equation (12.253) can be satisfied by introducing antisymmetrical pairs of single-particle functions:

$$f_{\lambda\lambda'}(\underline{r},\underline{r}') = 2^{-1/2}[f_\lambda(\underline{r})f_{\lambda'}(\underline{r}') - f_\lambda(\underline{r}')f_{\lambda'}(\underline{r})]. \quad (12.254)$$

Such pair functions are sometimes called *geminals*. From equations (12.253) and (12.254) we now have

$$\rho_2(\underline{r},\underline{r}';\underline{q},\underline{q}') = \sum_{\lambda\lambda'\mu\mu'} \tilde{c}_{\lambda\lambda'\mu\mu'} f_{\lambda\lambda'}(\underline{r},\underline{r}')f_{\mu\mu'}(\underline{q},\underline{q}'). \quad (12.255)$$

In fact, since

$$f_{\lambda\lambda'}(\underline{r},\underline{r}') = -f_{\lambda'\lambda}(\underline{r},\underline{r}'), \quad (12.256)$$

we can, without loss of generality, write equation (12.255) in the form

$$\rho_2(\underline{r},\underline{r}';\underline{q},\underline{q}') = \sum_{\lambda > \lambda',\mu > \mu'} c_{\lambda\lambda'\mu\mu'} f_{\lambda\lambda'}(\underline{r},\underline{r}')f_{\mu\mu'}(\underline{q},\underline{q}'). \quad (12.257)$$

In equations (12.252) and (12.257), note the distinction between $\tilde{c}_{\lambda\lambda'\mu\mu'}$ and $c_{\lambda\lambda'\mu\mu'}$.

We now substitute the Fourier expansion for ρ_2, equation (12.257), into the total-energy and normalization equations (12.243) and (12.251) respectively. The result is:

$$E = \sum_{\lambda > \lambda',\mu > \mu'} c_{\lambda\lambda'\mu\mu'} h_{\lambda\lambda'\mu\mu'}, \quad (12.258)$$

$$N(N-1) = \sum_{\lambda > \lambda',\mu > \mu'} c_{\lambda\lambda'\mu\mu'} n_{\lambda\lambda'\mu\mu'}, \quad (12.259)$$

where

$$h_{\lambda\lambda'\mu\mu'} = \int d\underline{r}\,d\underline{r}'\, f_{\lambda\lambda'}(\underline{r},\underline{r}')\{(N-1)^{-1}[-\tfrac{1}{2}\nabla^2 + V_{\text{ext}}(\underline{r})] + v(\underline{r},\underline{r}')\}f_{\mu\mu'}(\underline{r},\underline{r}'), \quad (12.260)$$

$$n_{\lambda\lambda'\mu\mu'} = \int d\underline{r}\,d\underline{r}'\, f_{\lambda\lambda'}(\underline{r},\underline{r}')f_{\mu\mu'}(\underline{r},\underline{r}'). \quad (12.261)$$

Several features of the formalism developed here are of interest. First, there is no self-consistent field problem. Second, there does not appear to be a requirement that an approximation scheme must be based on any minimum number (greater than two) of basis functions $f_\lambda(\underline{r})$. Third, correlation is not identified separately. However, upon examining equations

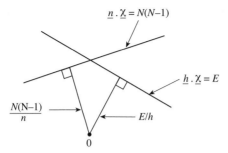

Figure 12.1. Illustrating hyperplanes for many-electron energy and normalization in Fourier coefficients $\underline{x}$ of the two-particle density: see equations (12.262) and (12.263) and associated discussion.

(12.258) and (12.259), insuperable difficulties emerge. We naturally think of applying the variational principle to the energy E, by varying the coefficients $c_{\lambda\lambda'\mu\mu'}$, to obtain minimum energy, subject to the constraint equation (12.259). This does not work out, because, in the space of variables $c_{\lambda\lambda'\mu\mu'}$, to be taken of finite dimension for practical computation, equations (12.258) and (12.259) are hyperplanes. If we consider the set of independent variables $\{c_{\lambda\lambda'\mu\mu'}\}$ as defining a hyperspace, denote the vector $\{c_{\lambda\lambda'\mu\mu'}\}$ by $\underline{x}$. Then equations (12.258) and (12.259) take the form

$$\underline{h} \cdot \underline{x} = E, \qquad (12.262)$$

$$\underline{n} \cdot \underline{x} = N(N-1). \qquad (12.263)$$

Equations (12.262) and (12.263) are represented schematically in figure 12.1, in which the hyperplanes are represented as planes in three-dimensional space and are viewed edge-on. For fixed $\underline{n}$, the plane defined by its non-unit normal $\underline{n}$ and by $N(N-1)$, is fixed. We have

$$\left(\frac{1}{n}\right)\underline{n} \cdot \underline{x} = \frac{N(N-1)}{n} = d_1, \qquad (12.264)$$

where d_1 is the perpendicular distance of the plane from the origin O. Similarly, for given E the plane defined by its non-unit normal $\underline{h}$ and by E is shown. In general, the two planes intersect in a line perpendicular to the plane of the page, i.e. perpendicular to the plane of $\underline{h}$ and $\underline{n}$, and this intersection fails to determine $\underline{x}$ uniquely: $\underline{x}$ may lie anywhere on the line. Furthermore, there is no minimum value of E required to satisfy these equations: E may become negative more than any finite negative number, and still have the planes intersecting. In two dimensions, the value of $\underline{x}$ is unique, but possible values of E are still unbounded below.

The question of appropriate additional constraints is still unresolved, after more than sixty years, although it is still very much a question of lively investigation. This is most eloquently illustrated by the recent volumes by Coleman and Yukalov (2000) and by Cioslowski (2000). An important earlier source of detailed information on reduced density matrices is the book by Davidson (1976). We can see the nature of the outstanding problem from one point of view by re-examining the definition of ρ_2, equation (12.240), and the role of the Pauli principle as we have introduced it so far, equation (12.253); see also equations (12.5) and (12.6). We have

$$\rho_2(\underline{r}, \underline{r}'; \underline{q}, \underline{q}') = N(N-1) \int d\underline{r}_3, \ldots, d\underline{r}_N \, \psi^*(\underline{r}, \underline{r}', \underline{r}_3, \ldots, \underline{r}_N)$$

$$\times \psi(\underline{q}, \underline{q}', \underline{r}_3, \ldots, \underline{r}_N), \qquad (12.265)$$

$$\rho_2(\underline{r}, \underline{r}'; \underline{q}, \underline{q}') = -\rho_2(\underline{r}'\underline{r}; \underline{q}, \underline{q}') = \rho_2(\underline{r}', \underline{r}; \underline{q}', \underline{q}). \qquad (12.266)$$

The problem is that equation (12.266) would be satisfied by a wave function $\psi(\underline{r}, \underline{r}', \underline{r}_3, \ldots, \underline{r}_N)$ that is not antisymmetric in the variables $\underline{r}_3, \ldots, \underline{r}_N$. The problem of analytically representing the additional constraints needed to fully satisfy the Pauli principle is called *the N-representability problem*. The further practical problem is to correctly include Pauli-principle effects in a viable, efficient computational approximation.

The formulation presented in this section has been presented because of its elegance, and because of the hope that the N-representability problem can be managed in such a way that the apparent advantages of the formulation can be realized in a computationally efficient manner.

12.7.5 The density functional and the two-particle density operator

In section 12.7.2 we discussed the theorem of Hohenberg and Kohn, that the ground-state energy E_0 of a many-body system is a unique, unspecified functional $E_0[\rho]$ of the single-particle density $\rho(\vec{r})$, equation (12.212); see also equation (12.230). In section 12.7.4 we saw that the total energy is an explicit linear functional of the two-particle density ρ_2, as in equation (12.243):

$$E = W[\rho_2]. \qquad (12.267)$$

Furthermore, the single-particle density ρ is an explicit linear functional of ρ_2, from equations (12.247) and (12.248):

$$\rho = \rho[\rho_2]. \qquad (12.268)$$

The relationships of equation (12.267) and (12.268) may provide a framework within which a fruitful re-examination of the Hohenberg–Kohn theorem can be conducted. equation (12.243) is of the form

$$E = \int d\underline{r} \, d\underline{r}' \, d\underline{q} \, d\underline{q}' \, h(\underline{r}, \underline{r}'; \underline{q}, \underline{q}') \rho_2(\underline{r}, \underline{r}'; \underline{q}, \underline{q}'), \qquad (12.269)$$

and equations (12.248) and (12.247) can be combined in the form

$$\rho(\underline{r}) = (N-1)^{-1} \int d\underline{r}' \, d\underline{q} \, d\underline{q}' \, \delta(\underline{q} - \underline{r})\delta(\underline{q}' - \underline{r}')\rho_2(\underline{r}, \underline{r}', \underline{q}, \underline{q}'). \qquad (12.270)$$

Equations (12.269) and (12.270) show the formal, implicit integral relationship between the total energy E and the single-particle density ρ, both as explicit linear functionals of ρ_2. This lets us see clearly why, to date, no explicit form of the Hohenberg–Kohn density functional has been found.

Chapter 13

Paramagnetism and diamagnetism in the electron gas

13.1 Introduction

Magnetism is a subject of great scope, both in variety of phenomena and in theoretical challenges. In this chapter we limit ourselves to two of the latter: paramagnetism and diamagnetism in the electron gas. The electron gas is an oversimplified model of conduction electrons in a metal. Paramagnetism represents the tendency of electron spin magnetic moments to align with an applied magnetic induction field, making a characteristic contribution to the magnetic susceptibility of the system. Diamagnetism represents the fact that electrons in such a field have orbits whose perpendicular projections are circular, and the resultant current loops make a different contribution to the magnetic susceptibility. The treatment of electron paramagnetism is presented here in section 13.2 as a particularly instructive application of quantum statistical thermodynamics. In section 13.3, diamagnetism is discussed, where the quantization of electron orbits in a magnetic induction field, section 13.3.2, results in a spectacular change in the topology of the Fermi surface, sections 13.3.3 and 13.3.4, attended by periodic variation of the magnetic susceptibility as a function of magnetic induction field: the De Haas–van Alphen effect (section 13.3.5). A well-known but nonetheless surprisingly simple relationship between low-temperature paramagnetic and diamagnetic susceptibilities is arrived at in section 13.3.6. For paramagnetism, I have expanded some details of Huang's presentation (1967, especially sections 8.1, 8.3, 9.6, 11.1 and 11.5) in his excellent work on statistical mechanics, and for diamagnetism I have similarly adopted Pippard's elegant discussion at the 1961 Les Houches Summer School [Pippard (1962, p. 11, section IIA)].

To study the subject of magnetism in a more extensive and balanced way, the reader might consider, among others, the following: van Vleck's (1932) timeless, lucid exposition of susceptibilities; Morrish's (1965) still relevant textbook; White's (1983) theoretical outline; Mattis's (1987) coverage of some modern topics; and Majlis's (2000) contemporary textbook.

13.2 Paramagnetism of the electron gas

13.2.1 The total energy

Consider a metal where the conduction electrons, in their interaction with the periodic field of the rest of the crystal, may be approximated by free non-interacting electrons with effective band mass m_b [see for example Grosso and Pastori Parravicini (2000, Chapter I, section 6)]. The electrons have intrinsic magnetic moment $\vec{\mu}$, according to Dirac electron theory [see for example Sakurai (1967, section 3.3)] given by

$$\vec{\mu} = -\left(\frac{e\hbar}{2m}\right)\vec{\sigma}, \qquad e > 0, \tag{13.1}$$

where $\vec{\sigma}$ are Pauli spin matrices. Consider a uniform, time-independent magnetic induction field $\vec{B}$ in the z direction,

$$\vec{B} = B\hat{e}_3, \tag{13.2}$$

where $B = |\vec{B}|$ and $\hat{e}_3$ is a unit vector in the z direction. Then

$$\vec{\sigma} \cdot \vec{B} = \sigma_3 B = \begin{pmatrix} 1 & 0 \\ 0 & -1 \end{pmatrix} \cdot B, \tag{13.3}$$

using a representation of the Pauli matrices in which σ_3 is diagonal [see for example Sakurai (1967, p. 80, footnote)]. Thus the interaction energy of the electron's spin magnetic moment with the $\vec{B}$ field [see for example, Greiner (1989, chapter 12)] is

$$-\vec{\mu} \cdot \vec{B} = \mu_B B \sigma_3, \tag{13.4}$$

where $\mu_B = [e\hbar/2m]$ is the *Bohr magneton*. The hamiltonian for the N-electron system in such a $\vec{B}$-field is

$$H = \sum_{j=1}^{N} \left\{ \frac{1}{2m_b} (\vec{p}_j + e\vec{A}(\vec{r}_j))^2 + \mu_B B \sigma_{3j} \right\}, \tag{13.5}$$

where $(\vec{p}_j, \vec{r}_j)$ are canonical momentum and position vector operators of the jth electron, and $\vec{A}(\vec{r})$ is the vector potential [see for example Goldstein (1980, p. 346)]. For a sufficiently weak field, B 'small', the term in A^2 in equation (13.5) may be neglected.

Apart from the terms $[p_j^2/(2m_b)]$, consider the remaining terms, in expectation value,

$$\left\{ \frac{1}{2m_b} \cdot e \cdot \langle \vec{p} \cdot \vec{A} + \vec{A} \cdot \vec{p} \rangle + \mu_B B \langle \sigma_3 \rangle \right\}. \tag{13.6}$$

We take $\langle \sigma_3 \rangle = 1$. Also,

$$(\vec{\nabla} \times \vec{A}) = \vec{B}. \tag{13.7}$$

Let us estimate $\langle \vec{p} \cdot \vec{A} + \vec{A} \cdot \vec{p} \rangle$ as follows:

$$\langle \vec{p} \cdot \vec{A} + \vec{A} \cdot \vec{p} \rangle \sim 2(m_\mathrm{b} v_{\mathrm{av}}) \cdot B\left(\frac{\hbar}{m_\mathrm{b} v_{\mathrm{av}}}\right) \approx 2B\hbar. \tag{13.8}$$

This is done on the basis that the partial derivatives in $(\vec{\nabla} \times \vec{A})$, equation (13.7), render $\vec{A}$ in the expectation value to be of order $B/(\vec{p}/\hbar)$, on the basis that $\vec{p} = -i\hbar\vec{\nabla}$, and assuming that $\vec{p}$ in the expectation value contributes of order $m_\mathrm{b}|\vec{v}_{\mathrm{av}}|$, where $|\vec{v}_{\mathrm{av}}| = v_{\mathrm{av}}$ is some average value of the speed of an electron in the N-electron system. Actually, of course, we know that

$$\vec{v}_j = \frac{1}{m_\mathrm{b}}(p_j + e\vec{A}(\vec{r}_j)). \tag{13.9}$$

Substituting equation (13.8) into equation (13.6), we have

$$(13.6) \sim \left\{\frac{e}{2m_\mathrm{b}} \cdot 2B\hbar + \mu_\mathrm{B} B\right\} = \frac{Be\hbar}{m_\mathrm{b}}\left\{1 + \frac{m_\mathrm{b}}{2m}\right\}, \tag{13.10}$$

where we have used equation (13.1) for μ_B. Thus the term in $\langle \vec{p} \cdot \vec{A} + \vec{A} \cdot \vec{p} \rangle$ will be negligible relative to the term in $\langle \sigma_3 \rangle$ in equation (13.6) if

$$m_\mathrm{b} \gg 2m. \tag{13.11}$$

We assume this to be the case, and so consider the effective hamiltonian

$$H = \sum_{j=1}^{N}\left\{-\frac{\hbar^2}{2m_\mathrm{b}}\nabla_j^2 + \mu_\mathrm{B} B\sigma_{3j}\right\}. \tag{13.12}$$

The magnetic property of such a many-electron system is called *paramagnetic*. As we shall see, it simply represents the tendency of free electron spin magnetic moments to align with an applied $\vec{B}$-field. The discussion of section 13.2 is largely based on that of Huang (1967), especially his sections 8.3, 9.6, 11.1 and 11.5.

We can easily determine eigenstates of this paramagnetic hamiltonian, equation (13.12). They are Slater determinants [see section 12.2.1, equation (12.19)] consisting of products of single-particle states $\psi_{\vec{k},s}$ that are simultaneous eigenstates of p^2 and of σ_3. Let the eigenstates of σ_3 be denoted η_s. Then, from equation (13.3),

$$\sigma_3 \eta_s = s\eta_s \rightarrow s = \pm 1 \tag{13.13}$$

and

$$\eta_1 = \begin{pmatrix} 1 \\ 0 \end{pmatrix}, \qquad \eta_{-1} = \begin{pmatrix} 0 \\ 1 \end{pmatrix}. \tag{13.14}$$

Let the eigenstates $\phi_k(\vec{r})$ of p^2 be determined subject to Born–von Karmann boundary conditions (see section 7.5.1). Then

$$\varphi_{\vec{k}}(\vec{r}) = \Omega^{-1/2}\, e^{i\vec{k}\cdot\vec{r}}, \tag{13.15}$$

where

$$\Omega = L^3, \qquad k_\alpha = \frac{2\pi n_\alpha}{L}, \qquad n_\alpha = 0, \pm 1, \pm 2, \ldots, \qquad \alpha = 1, 2, 3. \qquad (13.16)$$

The single-particle eigenstates of equation (13.12) are then

$$\psi_{\vec{k},s} = \varphi_{\vec{k}} \cdot \eta_s, \qquad (13.17)$$

or in Dirac notation $|\vec{k}, s\rangle$, where

$$p^2|k, s\rangle = \hbar^2 k^2 |\vec{k}, s\rangle \qquad (13.18)$$

and

$$\sigma_3 |\vec{k}, s\rangle = s|\vec{k}, s\rangle. \qquad (13.19)$$

Correspondingly, the single-particle eigenvalues $\varepsilon_{\vec{k},s}$ are

$$\varepsilon_{\vec{k},s} = \left(\frac{\hbar^2 k^2}{2m_b} + \mu_B B s \right). \qquad (13.20)$$

An N-electron eigenstate and the corresponding N-electron eigenvalues contain a set of N or fewer distinct $\vec{k}$-values: N if all are of the same spin, fewer if some $\vec{k}$ values occur with both $s = \pm 1$. We can write the total energy E_N simply in terms of the occupancy, one or zero, in each single-particle state $|\vec{k}, s\rangle$:

$$E_N = \sum_{\vec{k},s} n_{\vec{k}s} \varepsilon_{\vec{k},s} = \sum_{\vec{k}} \left\{ n_{\vec{k}}^{(+)} \left(\frac{\hbar^2 k^2}{2m_b} + \mu_B B \right) + n_{\vec{k}}^{(-)} \left(\frac{\hbar^2 k^2}{2m_b} - \mu_B B \right) \right\}, \qquad (13.21)$$

where $n_{\vec{k}}^{(\pm)} \equiv n_{\vec{k},\pm 1}$ is the number (zero or one) of electrons in $|\vec{k}, \pm 1\rangle$. We define the total number $N^{(\pm)}$ of spin-up or spin-down,

$$N^{(\pm)} = \sum_{\vec{k}} n_{\vec{k}}^{(\pm)}, \qquad (13.22)$$

and require the constraint upon equation (13.21) of total number of electrons N,

$$N = (N^{(+)} + N^{(-)}). \qquad (13.23)$$

Then equation (13.21) simplifies to

$$E_N = \left\{ \sum_{\vec{k}} (n_{\vec{k}}^{(+)} + n_{\vec{k}}^{(-)}) \frac{\hbar^2 k^2}{2m_b} + \mu_B B (N^{(+)} - N^{(-)}) \right\}. \qquad (13.24)$$

From equation (13.24), we can see that a many-electron eigenfunction can be specified by a set of occupation numbers $\{n_{\vec{k}}^{(\pm)}\}$, subject to the constraint of equation (13.23) with equation (13.22). There are degeneracies among the possible eigenvalues:

$$E_N = E_N(\{n_{\vec{k}}^{(\pm)}\}). \tag{13.25}$$

13.2.2 The magnetic susceptibility

The magnetic susceptibility is a measure of the tendency of a material to acquire a magnetic moment in the presence of an applied magnetic induction field. We define the magnetic susceptibility as

$$\chi_{\alpha\beta} = \mu_0 \left(\frac{\partial M_\alpha}{\partial B^\beta} \right),$$

a second rank tensor where μ_0 is the permeability of free space, and $\vec{M}$ is the magnetization, i.e. the induced magnetic dipole moment per unit volume. For a linear isotropic material,

$$\vec{M} = \frac{\chi_m}{\mu_0} \vec{B},$$

where χ_m is a scalar, and more generally:

$$\chi_m = \mu_0 \frac{\partial M}{\partial B}. \tag{13.26}$$

These matters are reviewed, for example, in Reitz *et al.* (1979), sections 9.5 and 9.6. For a specific state, characterized by $\{n_{\vec{k}}^{(\pm)}\}$, as in equation (13.25), the magnetization is

$$M = -\frac{\mu_B}{\Omega}(N^{(+)} - N^{(-)}) = -\frac{\mu_B}{\Omega}(2N^{(+)} - N), \tag{13.27}$$

where we have used equation (13.23).

We wish to apply equilibrium statistical thermodynamics to determine the dependence of M on B [see equation (13.26)] as a function of temperature. For this purpose, we shall need to know the thermal average values $\overline{N}^{(+)}$ and $\overline{N}^{(-)}$. We begin with the partition function Q_N of the system,

$$Q_N = \sum_{\{n_{\vec{k}}^{(\pm)}\}}' \exp[-\beta E_N(\{n_{\vec{k}}^{(\pm)}\})], \tag{13.28}$$

where $\beta = (k_B T)^{-1}$, k_B is Boltzmann's constant, and T is the Kelvin temperature. We shall use equation (13.24) for $E_N(\{n_{\vec{k}}^{(\pm)}\})$. In equation (13.28), the prime on the sum limits it to the cases

$$n_{\vec{k}}^{(\pm)} = 0 \text{ or } 1, \qquad \sum_{\vec{k}}(n_{\vec{k}}^{(+)} + n_k^{(-)}) = N, \tag{13.29}$$

where

$$\sum_{\vec{k}} n_{\vec{k}}^{(+)} = N^{(+)}, \qquad \sum_{\vec{k}} n_k^{(-)} = N^{(-)} = (N - N^{(+)}). \tag{13.30}$$

Using equation (13.24) in equation (13.28), we have

$$Q_N = \exp(\beta\mu_B BN) \sideset{}{'}\sum_{N^{(+)}} \exp(-2\beta\mu_B BN^{(+)})$$

$$\times \sideset{}{''}\sum_{n_k^{(+)}} \exp\left(-\beta\sum_{\vec{k}} \frac{\hbar^2 k^2}{2m_b} n_{\vec{k}}^{(+)}\right) \sideset{}{'''}\sum_{n_k^{(-)}} \exp\left(-\beta\sum_{\vec{k}} \frac{\hbar^2 k^2}{2m_b} n_{\vec{k}}^{(-)}\right). \quad (13.31)$$

In equation (13.31) $\sum'$ ranges over $0 \le N^{(+)} \le N$, and $\sum''$ and $\sum'''$ are subject to the two constraints of equations (13.30). The last two factors in equation (13.31) are of the same form $Q_{N'}^{(0)}$, where

$$Q_{N'}^{(0)} = \sum_{\{n_{\vec{k}}\}} \exp\left(-\beta\sum_{\vec{k}} \frac{\hbar^2 k^2}{2m_b} n_{\vec{k}}\right). \quad (13.32)$$

In equation (13.32), the symbol $\{n_{\vec{k}}\}$ refers to a particular set of values of $n_{\vec{k}}$ for all $\vec{k}$ that satisfy the restrictions

$$\sum_{\vec{k}} n_{\vec{k}} = N', \qquad n_{\vec{k}} = 0 \text{ or } 1. \quad (13.33)$$

In terms of the notation of equation (13.32), the partition function, equation (13.31) becomes

$$Q_N = \exp(\beta\mu_B BN) \sum_{N^{(+)}=1}^{N} \exp(-2\beta\mu_B BN^{(+)}) Q_{N^{(+)}}^{(0)} Q_{N-N^{(+)}}^{(0)}. \quad (13.34)$$

Now Q_N, the partition function, is related to the Helmholtz free energy $A_N(\Omega, T)$ by the relationship

$$\ln Q_N = -\beta A_N(\Omega, T); \quad (13.35)$$

see Huang (1967), section 8.1. Let us define $A^{(0)}(N)$ by

$$A_N^{(0)} = -\frac{1}{\beta} \ln Q_N^{(0)} \equiv A^{(0)}(N). \quad (13.36)$$

Then, from equations (13.34) to (13.36), the Helmholtz free energy per particle is

$$\frac{1}{N} A(\Omega, T) = -\frac{1}{N\beta} \ln Q_N$$

$$= \left\{-\mu_B B - \frac{1}{N\beta} \ln \sum_{N^{(+)}} \exp[\beta g(N^{(+)})]\right\}, \quad (13.37)$$

where

$$g(N^{(+)}) = \{-2\mu_B BN^{(+)} - A^{(0)}(N^{(+)}) - A^{(0)}(N - N^{(+)})\}. \quad (13.38)$$

Now in equation (13.37), the $(\ln \sum)$ term is of the form

$$\ln \left(\sum_{n=0}^{N} x_n \right). \tag{13.39}$$

Let x_M be the largest of the values of x_n. Then

$$\ln \left(\sum_{n=0}^{N} x_n \right) = \ln \left\{ x_M \left(1 + \sum_n' \frac{x_n}{x_M} \right) \right\}$$

$$= \left\{ \ln(x_M) + \ln \left(1 + \sum_n' \frac{x_n}{x_M} \right) \right\}. \tag{13.40}$$

Now all of the terms summed in the second term of equation (13.40) are less than or equal to one. In the term $\sum_n'$, the prime indicates a sum over only N terms. Thus, for large N

$$\ln \left(1 + \sum_n' \frac{x_n}{x_M} \right) \lesssim \ln(N + 1) \sim \ln(N). \tag{13.41}$$

It follows from equations (13.40) and (13.41) that

$$\ln \left(\sum_{n=0}^{N} x_n \right) = \ln(x_M) + O(\ln N), \tag{13.42}$$

where $O(N)$ indicates 'of the order of N'. We now return to equation (13.37), and identify $\overline{N}^{(+)}$ as the value of $N^{(+)}$ that maximizes $g(N^{(+)})$. Applying the results of equations (13.39)–(13.42), we have

$$\frac{1}{N} A(\Omega, T) = \left\{ -\mu_B B - \frac{1}{N\beta} \ln[\exp(\beta g(\overline{N}^{(+)}))] - \frac{1}{N\beta} O(\ln N) \right\}. \tag{13.43}$$

From the definition of g, equation (13.38), we see that equation (13.43) reduces to

$$\frac{1}{N} A(\Omega, T) \approx -\mu_B B - \frac{1}{N\beta} \beta \left\{ -2\mu_B B \overline{N}^{(+)} - A^{(0)}(\overline{N}^{(+)}) - A^{(0)}(N - \overline{N}^{(+)}) \right\}$$

$$= \left\{ -\mu_B B \left(1 - \frac{2\overline{N}^{(+)}}{N} \right) + \frac{1}{N} [A^{(0)}(\overline{N}^{(+)}) + A^{(0)}(N - \overline{N}^{(+)})] \right\}, \tag{13.44}$$

where in equation (13.44) we have neglected the small term $O(N^{-1} \ln N)$ from equation (13.43). In equation (13.44) we see contributions to $(N^{-1}A)$ from spin-up and spin-down parts of the system in the absence of magnetic induction field B, plus the B-dependent contribution.

We now observe that the equilibrium value of $\overline{N}^{(+)}$ will minimize the Helmholtz free energy A under the given conditions of temperature and field. Thus $\overline{N}^{(+)}$ is determined by

$$\frac{1}{N}\frac{\partial A}{\partial \overline{N}^{(+)}} = \frac{2\mu_{\mathrm{B}}B}{N} + \frac{1}{N}\left\{\frac{\partial A^{(0)}(\overline{N}^{(+)})}{\partial \overline{N}^{(+)}} + \frac{\partial}{\partial \overline{N}^{(+)}}A^{(0)}(N - \overline{N}^{(+)})\right\} = 0,$$

(13.45)

having used equation (13.44) for A. From equation (13.45) we have $\overline{N}^{(+)}$ determined by

$$\left\{\frac{\partial A^{(0)}(\overline{N}^{(+)})}{\partial \overline{N}^{(+)}} - \frac{\partial A^{(0)}(N - \overline{N}^{(+)})}{\partial(N - \overline{N}^{(+)})}\right\} = -2\mu_{\mathrm{B}}B. \qquad (13.46)$$

We recall from equations (13.36) and (13.32) that $A^{(0)}(n)$ is the Helmholtz free energy of a gas of n fermions all of the same spin. From classical thermodynamics, we then have

$$\frac{\partial A^{(0)}(n)}{\partial(n)} = \mu^{(0)}(n), \qquad (13.47)$$

where $\mu^{(0)}(n)$ is the chemical potential per particle. From equations (13.46) and (13.47) we then have

$$\{\mu^{(0)}(\overline{N}^{(+)}) - \mu^{(0)}(N - \overline{N}^{(+)})\} = -2\mu_{\mathrm{B}}B. \qquad (13.48)$$

The left-hand side of equation (13.48) is simply the difference in chemical potential between spin-up and spin-down particles in our original Fermi gas. When the solution for $\overline{N}^{(+)}$ is substituted for $N^{(+)}$ in equation (13.27), we obtain the magnetization as a function of B, whence the magnetic susceptibility, equation (13.26).

13.2.3 Solution at low temperature

It remains to determine the explicit functional dependence of the chemical potential $\mu^{(0)}$ on $\overline{N}^{(+)}$ in equation (13.48). We approach it from the viewpoint of the grand partition function, bearing in mind that the system in question is a Fermi gas of particles all of the same spin. According to Huang (1967), section 8.3, the grand partition function Q is

$$Q = \sum_{N=0}^{\infty} Q_N^{(0)} \exp(N\beta\mu^{(0)}). \qquad (13.49)$$

With equation (13.32), this can be written as

$$Q = \sum_{N=0}^{\infty} z^N \sum_{\{n_{\vec{k}}\}}' \exp\left(-\beta \sum_{\vec{k}} n_{\vec{k}} \varepsilon_{\vec{k}}\right). \qquad (13.50)$$

In equation (13.50), we have substituted

$$z = \exp(\beta\mu^{(0)}),\tag{13.51}$$

$$\varepsilon_{\vec{k}} = \frac{\hbar^2 k^2}{2m_b}.\tag{13.52}$$

Also, the prime on the sum indicates that the possible values of $n_{\vec{k}}$ are limited to zero and one, such that

$$\sum_{\vec{k}} n_{\vec{k}} = N.\tag{13.53}$$

Equation (13.50) can therefore be written as

$$Q = \sum_{N=0}^{\infty} \sum_{\{n_{\vec{k}}\}}{}' \prod_{\vec{k}} [z\exp(-\beta\varepsilon_{\vec{k}})]^{n_{\vec{k}}}.\tag{13.54}$$

Now the double summation contains all sets of $n_{\vec{k}}$ adding up to $N = 0$, $N = 1, \ldots$ to infinity. The product is of the form

$$\prod_{\vec{k}} (x_{\vec{k}})^{n_{\vec{k}}}.\tag{13.55}$$

Thus in the double sum in equation (13.54), each term contains as many factors as there are $\vec{k}$ values, see equation (13.16), but the powers of $x_{\vec{k}}$ that occur span all possibilities $n = 0, 1, 2, \ldots, \infty$. Thus equation (13.54) is

$$Q = \prod_{\vec{k}} [(x_{\vec{k}})^0 + (x_{\vec{k}})^1] = \prod_{\vec{k}} \{1 + z\exp(-\beta\varepsilon_{\vec{k}})\}.\tag{13.56}$$

We now obtain the thermal average $\overline{N}^{(+)}$ of the number of particles. From equation (13.50) with equation (13.32), the thermal average is

$$\overline{N}^{(+)} = \frac{1}{Q}\sum_{N=0}^{\infty} Nz^N Q_N^{(0)} = z\frac{\partial}{\partial z}\ln Q.\tag{13.57}$$

From equation (13.56) this becomes

$$\overline{N}^{(+)} = z\frac{\partial}{\partial z}\ln\left\{\prod_{\vec{k}} [1 + z\exp(-\beta\varepsilon_{\vec{k}})]\right\}$$

$$= z\frac{\partial}{\partial z}\left\{\sum_{\vec{k}} \ln[1 + z\exp(-\beta\varepsilon_{\vec{k}})]\right\}$$

$$= z\sum_{\vec{k}} \frac{\exp(-\beta\varepsilon_{\vec{k}})}{[1 + z\exp(-\beta\varepsilon_{\vec{k}})]}$$

$$= \sum_{\vec{k}} \frac{1}{[z^{-1}\exp(\beta\varepsilon_k) + 1]}.\tag{13.58}$$

We convert the sum to an integral,

$$\overline{N}^{(+)} = \sum_{\vec{k}} \overline{n}_{\vec{k}} = \int d^3 k \, d(\vec{k}) \overline{n}(\vec{k}) \qquad (13.59)$$

where the density $d(\vec{k})$ of points in $\vec{k}$ space is

$$d(\vec{k}) = \left[\frac{(2\pi)^3}{\Omega} \right]^{-1} = \frac{\Omega}{(2\pi)^3}. \qquad (13.60)$$

From equation (13.58) we identify the thermal average $\overline{n}(\vec{k})$ of the particle distribution,

$$\overline{n}(\vec{k}) = \left[z^{-1} \exp\left(\frac{\beta \hbar^2 k^2}{2 m_{\mathrm{b}}} \right) + 1 \right]^{-1}. \qquad (13.61)$$

From equations (13.59)–(13.61) we have

$$\overline{N}^{(+)} = \frac{\Omega}{(2\pi)^3} 4\pi \int_0^\infty dk \, k^2 \left[z^{-1} \exp\left(\frac{\beta \hbar^2 k^2}{2 m_{\mathrm{b}}} \right) + 1 \right]^{-1}$$

$$= \frac{\Omega}{2\pi^2} \left(\frac{2 m_{\mathrm{b}}}{\beta \hbar^2} \right)^{3/2} \int_0^\infty du \, u^2 [z^{-1} \exp(u^2) + 1]^{-1}, \qquad (13.62)$$

where

$$u^2 = \frac{\beta \hbar^2 k^2}{2 m_{\mathrm{b}}}. \qquad (13.63)$$

Equation (13.62) gives $\overline{N}^{(+)}$ as a function of z and β, or from equations (13.51) and the definitions

$$z = \exp(\beta \mu^{(0)}), \qquad \beta = (k_{\mathrm{B}} T)^{-1}, \qquad (13.64)$$

we have $\overline{N}^{(+)}$ as a function of T and chemical potential $\mu^{(0)}$. Inverting that relationship would give $\mu^{(0)}$ as a function of $\overline{N}^{(+)}$ and T, from which equation (13.48) could, in principle, be solved for $\overline{N}^{(+)}$ as a function of B and T. We proceed to carry out this process in the low-temperature approximation in this section, and at high temperature in the next section.

First we rewrite equation (13.62),

$$\frac{\lambda^3 \overline{N}^{(+)}}{\Omega} = \frac{4}{\sqrt{\pi}} \int_0^\infty du \, u^2 [z^{-1} \exp(u^2) + 1]^{-1}, \qquad (13.65)$$

where

$$\lambda = \left(\frac{2\pi \hbar^2 \beta}{m_{\mathrm{b}}} \right)^{1/2}. \qquad (13.66)$$

From equation (13.64), define

$$\frac{\mu^{(0)}}{k_B T} = \nu, \tag{13.67}$$

so that $z = e^{\nu}$, and the integral in equation (13.65) becomes

$$\int_0^{\infty} du\, u^2 [\exp(u^2 - \nu) + 1]^{-1}, \qquad t = (u^2 - \nu), \tag{13.68}$$

$$= \int_{-\nu}^{\infty} dt\, \frac{1}{2}(t + \nu)^{1/2}[\exp(t) + 1]^{-1}, \qquad \text{integrate by parts}$$

$$= \frac{1}{2} \left\{ \frac{2}{3}(t + \nu)^{3/2}[\exp(t) + 1]^{-1} \Big|_{-\nu}^{\infty} \right.$$

$$\left. - \frac{2}{3}\int_{-\nu}^{\infty} dt(t + \nu)^{3/2}(-1)[\exp(t) + 1]^{-2}\exp(t) \right\}$$

$$= \frac{1}{3}\int_{-\nu}^{\infty} dt\, \frac{(t + \nu)^{3/2}}{(e^t + 1)^2} e^t. \tag{13.69}$$

In equation (13.69) we have used the result, verifiable by using L'Hôpital's rule twice, that

$$\lim_{t \to \infty} \frac{(t + \nu)^{3/2}}{(e^t + 1)} = 0. \tag{13.70}$$

Relating to equation (13.69), consider

$$\int_{-\infty}^{-\nu} dt\, \frac{(t + \nu)^{3/2} e^t}{(e^t + 1)^2} = \int_{-\infty}^{0} dy\, \frac{y^{3/2} e^y e^{-\nu}}{[e^{(y-\nu)} + 1]^2} \sim e^{-\nu}, \qquad y = (t + \nu). \tag{13.71}$$

Now for small T, $\nu = \beta\mu^{(0)}$ is large, so the integral in equation (13.71) is negligible. Thus equation (13.69) is approximately

$$\frac{1}{3}\int_{-\infty}^{\infty} dt\, \frac{(t + \nu)^{3/2} e^t}{(e^t + 1)^2}. \tag{13.72}$$

Now, for large ν,

$$(t + \nu)^{3/2} = \nu^{3/2}\left(1 + \frac{t}{\nu}\right)^{3/2} \approx \nu^{3/2}\left\{1 + \frac{3}{2}\frac{t}{\nu} + \frac{3}{8}\frac{t^2}{\nu^2} + \cdots\right\} \tag{13.73}$$

for $t < \nu$. Furthermore the integrand in equation (13.72) goes to zero as $t \to \pm\infty$. Thus we may write, approximately for equations (13.69) and (13.72),

$$\frac{1}{3}\left\{\nu^{3/2}\int_{-\infty}^{\infty} \frac{dt\, e^t}{(e^t + 1)^2} + \nu^{-1/2}\frac{3}{8}\int_{-\infty}^{\infty} \frac{dt\, t^2 e^t}{(e^t + 1)^2}\right\}, \tag{13.74}$$

to second order, the first-order term being zero because its integrand has odd parity, and the region of integration is symmetrical. The first integral is

$$\int_{-\infty}^{\infty} \frac{dt\, e^t}{(e^t + 1)^2} = \int_{-\infty}^{\infty} d\left[\frac{-1}{(e^t + 1)}\right] = 1. \tag{13.75}$$

The second integral may be transformed into a Riemann zeta function, as

$$\int_{-\infty}^{\infty} \frac{dt\, t^2 e^t}{(e^t + 1)^2} = \lim_{\lambda \to 1} \frac{\partial}{\partial\lambda} \left\{ 2\int_0^{\infty} dt\, t\left[\frac{-1}{(e^{\lambda t} + 1)}\right]\right\}$$

$$(u = \lambda t) \to = \lim_{\lambda \to 1} \frac{\partial}{\partial\lambda} \left\{ -\frac{2}{\lambda^2}\int_0^{\infty} \frac{du\, u}{(e^u + 1)}\right\}. \tag{13.76}$$

Now, from Jahnke and Emde (1945), p. 269, the Riemann zeta function $\zeta(n)$ is

$$\zeta(n) = \frac{1}{(1 - 2^{1-n})} \frac{1}{(n-1)!} \int_0^{\infty} \frac{du\, u^{n-1}}{(e^u + 1)}, \tag{13.77}$$

whence

$$\zeta(2) = 2\int_0^{\infty} \frac{du\, u}{(e^u + 1)}, \tag{13.78}$$

and equation (13.76) reduces to

$$\int_{-\infty}^{\infty} \frac{dt\, t^2 e^t}{(e^t + 1)^2} = 2\zeta(2). \tag{13.79}$$

From Jahnke and Emde (1945), p. 272,

$$\zeta(2) = \frac{2\pi^2}{(2!)} \frac{1}{6} = \frac{\pi^2}{6}. \tag{13.80}$$

Thus combining equations (13.65), (13.69), (13.74), (13.75), (13.79) and (13.80), we have

$$\frac{\lambda^3 \overline{N}^{(+)}}{\Omega} = \frac{4}{\sqrt{\pi}} \frac{1}{3}\left\{ \nu^{3/2} + \frac{\pi^2}{8}\nu^{-1/2}\right\}. \tag{13.81}$$

We solve equation (13.81) first in lowest order,

$$\frac{\lambda^3 \overline{N}^{(+)}}{\Omega} = \frac{4}{3\sqrt{\pi}} \nu^{3/2}. \tag{13.82}$$

Now in equations (13.58) and (13.59) we introduced the quantity

$$\overline{n}_{\vec{k}} = [z^{-1}\exp(\beta\varepsilon_{\vec{k}}) + 1]^{-1} \tag{13.83}$$

which from equation (13.59) has the meaning of the thermal average of $n_{\vec{k}}$, the number of particles in state $\vec{k}$. We also recall from equations (13.64) and (13.67) that

$$z = e^{\nu}. \tag{13.84}$$

Thus

$$\bar{n}_{\vec{k}} = [\exp(\beta\varepsilon_{\vec{k}} - \nu) + 1]^{-1}. \tag{13.85}$$

From equation (13.67),

$$\nu = \beta\mu^{(0)}, \tag{13.86}$$

so equation (13.85) is

$$\bar{n}_{\vec{k}} = \{\exp[\beta(\varepsilon_{\vec{k}} - \mu^{(0)})] + 1\}^{-1}. \tag{13.87}$$

From equation (13.82) with equation (13.86), we evaluate $\mu^{(0)}$:

$$
\begin{aligned}
\mu^{(0)} &= \frac{1}{\beta}\left(\frac{\lambda^3 \overline{N}^{(+)} 3\sqrt{\pi}}{4\Omega}\right)^{2/3} \\
&= \left(\frac{2\pi\hbar^2}{m_{\mathrm{b}}}\right)\left(\frac{3\sqrt{\pi}\,\overline{N}^{(+)}}{4\Omega}\right)^{2/3} \\
&= \frac{\hbar^2}{2m_{\mathrm{b}}}\left(\frac{6\pi^2\overline{N}^{(+)}}{\Omega}\right)^{2/3},
\end{aligned}
\tag{13.88}
$$

where we have used equation (13.66), the definition of λ, in the first step of equations (13.88). In the limit of low temperature, $T \to 0$, $\beta \to \infty$, we denote $\mu^{(0)}$ by $\varepsilon_{\mathrm{F}}^{(0)}$, the Fermi energy,

$$\varepsilon_{\mathrm{F}}^{(0)} = \frac{\hbar^2}{2m_{\mathrm{b}}}(6\pi^2\rho^{(0)})^{2/3} \tag{13.89}$$

where

$$\lim_{T \to 0}\left(\frac{\overline{N}^{(+)}}{\Omega}\right) = \frac{N}{\Omega} \equiv \rho^{(0)}, \tag{13.90}$$

the particle density. Then equation (13.87) for the occupation numbers $n_{\vec{k}}$ at $T = 0$ becomes

$$
\begin{aligned}
n_{\vec{k}} &= \lim_{\beta \to \infty}\{\exp[\beta(\varepsilon_{\vec{k}} - \varepsilon_{\mathrm{F}}^{(0)})] + 1\}^{-1} \\
&= 0 \quad \text{for } \varepsilon_{\vec{k}} > \varepsilon_{\mathrm{F}}^{(0)}, \\
&= 1 \quad \text{for } \varepsilon_k \le \varepsilon_{\mathrm{F}}^{(0)}.
\end{aligned}
\tag{13.91}
$$

From equation (13.89), we denote the quantity

$$(6\pi^2\rho^{(0)})^{1/3} = k_{\mathrm{F}}^{(0)}, \tag{13.92}$$

the Fermi wave number. For particles all of one spin, it is the largest occupied value of $|\vec{k}|$. We can see this as,

$$\int_0^{k_F^{(0)}} d^3k \, d^{(0)}(\vec{k}) = N, \tag{13.93}$$

whence from equation (13.60),

$$N = 4\pi \int_0^{k_F^{(0)}} dk \, k^2 \left[\frac{\Omega}{(2\pi)^3}\right] = \Omega \frac{1}{2\pi^2} \frac{k_F^{(0)3}}{3}, \tag{13.94}$$

in agreement with equation (13.92).

Now returning to equation (13.86), in zeroth order, in the limit $T = 0$,

$$\mu^{(0)} = k_B T \nu = \varepsilon_F^{(0)} \qquad (T = 0). \tag{13.95}$$

This shows clearly that for low temperature, ν is large. Referring to equation (13.81), we had, to second order,

$$\frac{\lambda^3 \overline{N}^{(+)}}{\Omega} \approx \frac{4}{3\sqrt{\pi}} \nu^{3/2} \left(1 + \frac{\pi^2}{8\nu^2}\right). \tag{13.96}$$

Now $\overline{N}^{(+)}$ is determined, in non-zero field, from equation (13.48). To get ν, or $\mu^{(0)}$, from equation (13.96) in terms of $\overline{N}^{(+)}$, we approximate the second order term, in ν^{-2}, at $T = 0$,

$$\frac{\pi^2}{8\nu^2} \approx \frac{\pi^2}{8} \left(\frac{k_B T}{\varepsilon_F^{(0)}}\right)^2, \tag{13.97}$$

having used equation (13.95). We then solve equation (13.96) with equation (13.97):

$$\nu^{3/2} \approx \frac{3\sqrt{\pi}\lambda^3 \overline{N}^{(+)}}{4\Omega \left[1 + \dfrac{\pi^2}{8} \left(\dfrac{k_B T}{\varepsilon_F^{(0)}}\right)^2\right]} \tag{13.98}$$

or

$$\nu \approx \left(\frac{3\sqrt{\pi}\lambda^3 \overline{N}^{(+)}}{4\Omega}\right)^{2/3} \left[1 - \frac{2}{3}\frac{\pi^2}{8} \left(\frac{k_B T}{\varepsilon_F^{(0)}}\right)^2\right]. \tag{13.99}$$

having used

$$(1 + \alpha)^{-2/3} \approx (1 - \tfrac{2}{3}\alpha), \qquad \alpha \ll 1. \tag{13.100}$$

Now, we are considering an electron gas of N electrons that may be either spin-up or spin-down, whereas $\varepsilon_F^{(0)}$ is determined for electrons all of the same spin. The density of states $d(\vec{k})$ for electrons of either spin is double that for a single spin, $d^{(0)}(\vec{k})$, preceding equation (13.94), so the Fermi wave number k_F, from equation (13.94) is

$$k_F = (3\pi^2 N/\Omega)^{1/3} \tag{13.101}$$

and the Fermi energy, from equation (13.89) becomes

$$\varepsilon_F = \frac{\hbar^2}{2m_b}(3\pi^2 N/\Omega)^{2/3}. \tag{13.102}$$

Thus, from equation (13.89),

$$\varepsilon_F^{(0)}(N) = \frac{\hbar^2}{2m_b}(3\pi^2 2N/\Omega)^{2/3} = \varepsilon_F(2N), \tag{13.103}$$

see equation (13.102). In equation (13.99), with equation (13.66),

$$\left(\frac{3\sqrt{\pi}\lambda^3 \overline{N}^{(+)}}{4\Omega}\right)^{2/3} = \left(\frac{3\sqrt{\pi}\,\overline{N}^{(+)}}{4\Omega}\right)^{2/3} \frac{2\pi\hbar^2\beta}{m_b}$$

$$= \frac{\beta\hbar^2}{2m_b}(6\pi^2 \overline{N}^{(+)}/\Omega)^{2/3} = \beta\varepsilon_F(2\overline{N}^{(+)}), \tag{13.104}$$

having also used equation (13.103). Thus equation (13.99) takes the form

$$\nu(\overline{N}^{(+)}) \approx \beta\varepsilon_F(2\overline{N}^{(+)})\left\{1 - \frac{1}{12}\left[\frac{\pi}{\beta\varepsilon_F(2\overline{N}^{(+)})}\right]^2\right\}. \tag{13.105}$$

In order to determine $\overline{N}^{(+)}$, we need to solve equation (13.48), where the second term is

$$\mu^{(0)}(N - \overline{N}^{(+)}) = \mu^{(0)}(\overline{N}^{(-)}). \tag{13.106}$$

In equation (13.106), we therefore have again the chemical potential for particles of one spin, namely, spin down. Thus, equation (13.105) can be adapted, as follows:

$$\nu(N - \overline{N}^{(+)}) = \beta\varepsilon_F(2(N - \overline{N}^{(+)}))\left\{1 - \frac{1}{12}\left[\frac{\pi}{\beta\varepsilon_F(2(N - \overline{N}^{(+)}))}\right]^2\right\}. \tag{13.107}$$

Then from equation (13.86), with equations (13.105), (13.107) and (13.48), we have

$$\left\{[\varepsilon_F(2\overline{N}^{(+)}) - \varepsilon_F(2N - 2\overline{N}^{(+)})]\right.$$

$$\left. - \frac{1}{12}\left(\frac{\pi}{\beta}\right)^2\left[\frac{1}{\varepsilon_F(2\overline{N}^{(+)})} - \frac{1}{\varepsilon_F(2N - 2\overline{N}^{(+)})}\right]\right\} = -2\mu_B B. \tag{13.108}$$

We can make equation (13.108) appear more malleable by introducing

$$r = \left(\frac{2\overline{N}^{(+)}}{N} - 1\right), \qquad 2\overline{N}^{(+)} = N(1 + r) \qquad (13.109)$$

where we note that, since $0 \le N^{(+)} \le N$,

$$-1 \le r \le 1. \qquad (13.110)$$

Now referring to the definition of ε_F, equation (13.103), we see

$$\varepsilon_F(N(1 \pm r)) = \varepsilon_F(N)(1 \pm r)^{2/3}. \qquad (13.111)$$

Thus, substituting equations (13.109) and (13.111) into equation (13.108), we have

$$\left\{ [(1 + r)^{2/3} - (1 - r)^{2/3}] - \frac{1}{12}\left[\frac{\pi}{\beta\varepsilon_F(N)}\right]^2 \right.$$
$$\left. \times [(1 + r)^{-2/3} - (1 - r)^{-2/3}] \right\} = \frac{-2\mu_B B}{\varepsilon_F(N)}. \qquad (13.112)$$

In equation (13.112), consider first the low-temperature limit, $T \to 0$, $\beta \to \infty$. Now from equation (13.110), namely the requirement that $\overline{N}^{(+)} \le N$, we see that $|r| \le 1$. Thus, we expand as follows:

$$[(1 + r)^{2/3} - (1 - r)^{2/3}] \approx [(1 + \tfrac{2}{3}r + \cdots) - (1 - \tfrac{2}{3}r + \cdots)] \approx \tfrac{4}{3}r. \quad (13.113)$$

Then for equations (13.112) and (13.113) to be consistent with $|r| \le 1$ we require

$$-1 \le r \approx \left(\frac{-3\mu_B B}{2\varepsilon_F(N)}\right) \le 0. \qquad (13.114)$$

The present results will therefore only be valid for relatively weak field, in the sense of equation (13.114). For still weaker field,

$$B \ll \frac{2\varepsilon_F(N)}{3\mu_B}, \qquad (13.115)$$

we have, from equation (13.109),

$$r = \left(\frac{2\overline{N}^{(+)}}{N} - 1\right) \approx \left(\frac{-3\mu_B B}{2\varepsilon_F(N)}\right) \qquad (13.116)$$

or

$$\overline{N}^{(\pm)} \approx \frac{N}{2}\left(1 \mp \frac{3\mu_B B}{2\varepsilon_F(N)}\right). \qquad (13.117)$$

We note that this gives the reasonable result that at zero field, $\overline{N}^{(+)} = \overline{N}^{(-)} = (N/2)$.

For the first-order correction to equation (13.116), we expand

$$[(1+r)^{-2/3} - (1-r)^{-2/3}] \approx -\tfrac{4}{3}r. \tag{13.118}$$

With equations (13.118) and (13.113), equation (13.112) becomes

$$\frac{4}{3}r\left\{1 + \frac{1}{12}\left[\frac{\pi}{\beta\varepsilon_F(N)}\right]^2\right\} \approx \frac{-2\mu_B B}{\varepsilon_F(N)}, \tag{13.119}$$

or

$$r \approx \frac{-3\mu_B B}{2\varepsilon_F(N)}\left\{1 - \frac{1}{12}\left[\frac{\pi}{\beta\varepsilon_F(N)}\right]^2\right\}. \tag{13.120}$$

We are finally in a position to evaluate the magnetic susceptibility χ, from equation (13.26) with equations (13.27), (13.109) and (13.120). We have

$$\overline{M} = \frac{-\mu_B}{\Omega}(2\overline{N}^{(+)} - N) = -\mu_B\rho^{(0)}r$$

$$\approx \frac{3\mu_B^2\rho^{(0)}B}{2\varepsilon_F(N)}\left\{1 - \frac{1}{12}\left[\frac{\pi k_B T}{\varepsilon_F(N)}\right]^2\right\} \tag{13.121}$$

where recall that μ_B, the Bohr magneton is

$$\mu_B = \left(\frac{e\hbar}{2m}\right),$$

and

$$\rho^{(0)} = \frac{N}{\Omega}, \qquad \beta = (k_B T)^{-1}.$$

It follows from equation (13.121) with equation (13.26) that the paramagnetic susceptibility $\overline{\chi}_p$ at low temperature is

$$\overline{\chi}_p = \frac{\mu_0\overline{M}}{B} = \frac{3\mu_B^2\rho^{(0)}\mu_0}{2\varepsilon_F(N)}\left\{1 - \frac{1}{12}\left[\frac{\pi k_B T}{\varepsilon_F(N)}\right]^2\right\}. \tag{13.122}$$

This is a parabola, opening downward, as shown schematically in figure 13.1. In equations (13.121) and (13.122), $\overline{M}$ and $\overline{\chi}_p$ are thermal averages in the sense of statistical thermodynamics. Note also that, in equation (13.122),

$$\varepsilon_F(N) = \frac{\hbar^2}{2m_b}(3\pi^2\rho^{(0)})^{2/3}, \tag{13.123}$$

from equation (13.103).

13.2.4 Solution at high temperature

We return to the general relationship, for all temperatures T, for $\overline{N}^{(+)}$, equation (13.65). From the definition of λ, equation (13.66), we see that

$$\frac{\lambda^3\overline{N}^{(+)}}{\Omega} \approx \frac{1}{(k_B T)^{3/2}}, \tag{13.124}$$

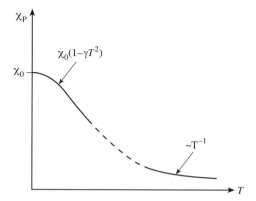

Figure 13.1. Paramagnetic susceptibility χ_P of the electron gas (schematic) as a function of temperature T, for weak field $B < (2\varepsilon_F)/(3\mu_B)$, equation (13.115), showing parabolic behaviour $\overline{\chi}_P = \chi_0(1 - \gamma T^2)$ at low temperature and $\overline{\chi}_P \approx T^{-1}$ at high temperature: see equations (13.122) and (13.136) respectively.

provided, as we expect and as we shall show, $\overline{N}^{(+)}$ is approximately independent of T at high temperature. Thus the integral in equation (13.65) must go to zero as $T \to \infty$. Now the integral goes to zero (from above) as $z \to 0^+$. Furthermore, the integral increases monotonically with $z > 0$, and therefore does not approach zero for any other value of z.

    Proof:

$$\frac{\partial}{\partial z} \int_0^\infty \frac{du\, u^2}{(z^{-1} e^{u^2} + 1)} = \int_0^\infty \frac{du\, u^2 e^{u^2}}{(e^{u^2} + z)^2} > 0. \tag{13.125}$$

Thus the Taylor series expansion of the integral about $z = 0$ must apply for high temperature:

$$\int_0^\infty \frac{du\, u^2}{(z^{-1} e^{u^2} + 1)} \approx z \lim_{z \to 0^+} \int_0^\infty \frac{du\, u^2 e^{u^2}}{(e^{u^2} + z)^2}$$

$$= z \int_0^\infty du\, u^2 e^{-u^2} = \frac{\sqrt{\pi}}{4} z. \tag{13.126}$$

Thus, from equations (13.65) and (13.126),

$$\frac{\lambda^3 \overline{N}^{(+)}}{\Omega} \approx \exp(\beta \mu^{(0)}), \tag{13.127}$$

having used equation (13.51). Inverting equation (13.127), we have

$$\mu^{(0)}(\overline{N}^{(+)}) \approx \frac{1}{\beta} \ln\left(\frac{\lambda^3 \overline{N}^{(+)}}{\Omega}\right). \tag{13.128}$$

Substituting equation (13.128) into equation (13.48), and using the notation of equations (13.109), we have

$$\left\{ \ln\left[\frac{\lambda^3}{\Omega}\frac{N}{2}(1+r)\right] - \ln\left[\frac{\lambda^3}{\Omega}\frac{N}{2}(1-r)\right] \right\} = -2\mu_B\beta B, \qquad (13.129)$$

or

$$\ln\left(\frac{1+r}{1-r}\right) = -2\mu_B\beta B, \qquad (13.130)$$

whose solution is

$$\left(\frac{1+r}{1-r}\right) = \exp(-2\mu_B\beta B), \qquad (13.131)$$

$$r = \frac{[\exp(-2\mu_B\beta B) - 1]}{[\exp(-2\mu_B\beta B) + 1]} = -\tanh(\mu_B\beta B), \qquad (13.132)$$

Now from the first line of equations (13.121), this gives, for the magnetization,

$$\overline{M} = -\mu_B\rho^{(0)}r = \mu_B\rho^{(0)}\tanh(\mu_B\beta B), \qquad (13.133)$$

for high temperature. Thus the high-temperature, paramagnetic suscepti-bility, $\overline{\chi}_p$ from equation (13.26) is

$$\overline{\chi}_p = \mu_0\frac{\partial \overline{M}}{\partial B} = \mu_0\mu_B^2\rho^{(0)}\beta[\mathrm{sech}(\mu_B\beta B)]^2 \sim \frac{1}{(k_B T)}. \qquad (13.134)$$

As in section 13.2.3, consider weak field, equation (13.115). Then denote $x = (\mu_B\beta B)$, and consider

$$(\mathrm{sech}\, x)^2 = \frac{4}{(e^x + e^{-x})^2}$$

$$\approx \frac{4}{[(1 + x + \frac{1}{2}x^2 + \cdots) + (1 - x + \frac{1}{2}x^2 \cdots)]^2}$$

$$\approx \frac{4}{(2 + x^2)^2} \approx \frac{4}{(4 + 4x^2)} = \frac{1}{(1 + x^2)} \approx (1 - x^2). \qquad (13.135)$$

Thus, in lowest order of weak-field approximation, equations (13.134) and (13.135) give

$$\overline{\chi}_p \approx \frac{\mu_0\mu_B^2\rho^{(0)}}{(k_B T)}. \qquad (13.136)$$

This high-temperature behavior is shown in figure 13.1, along with the low-temperature case, equation (13.122). From this figure we can see the qualitative dependence of the paramagnetic susceptibility of the electron gas over the whole range $0 \leq T < \infty$.

13.3   Diamagnetism of the electron gas

13.3.1   Introduction

We refer back to section 13.2.1, where, in weak magnetic induction field $\vec{B}$, the effective hamiltonian for a system of metallic electrons of band mass m_b is given in equation (13.6). In section 13.2 we discussed paramagnetism based on the approximation of equation (13.11). Paramagnetism arises from the tendency of the intrinsic magnetic dipole moment of electrons to align with an applied magnetic induction field, producing magnetization parallel to the field, and a characteristic magnetic susceptibility, shown schematically in figure 13.1. We now consider the opposite case:

$$m_b \ll m. \tag{13.137}$$

This will lead us to *diamagnetic* properties. If neither equation (13.11) or (13.137) applies, then the magnetic properties of the approximately free metallic electrons will be a combination of paramagnetic and diamagnetic features.

We are familiar, presumably, with the fact that the Lorentz force upon a charged particle in a magnetic induction field causes the particle to follow an orbit centered on an axis parallel to the field. The orbit is that of a circular helix if the particle has a velocity component parallel to the field, and simply circular if not. The Lorentz force for an electron is

$$\vec{F} = -e(\vec{v} \times \vec{B}), \tag{13.138}$$

where $\vec{v}$ is the electron's velocity, and $\vec{B}$ is the magnetic induction field. Looking along the direction of $\vec{B}$, we would see the electron circling clockwise, from the right-hand rule for the vector cross product. This constitutes a *counterclockwise* electrical current, because of the electron's negative charge $(-e)$. Such a counterclockwise current loop is associated with a magnetic dipole moment oriented in the direction *opposite* to $\vec{B}$. Thus, the contribution of electrons in an electron gas to the magnetization due to the Lorentz force is expected to be in the direction opposite to $\vec{B}$, and the corresponding magnetic susceptibility is expected to be *negative*. It can be shown that for *classical* charged particles in thermal equilibrium this, and all other magnetic effects, are zero, a result known as van Leeuwen's theorem. [For a succinct proof, see Ashcroft and Mermin (1976, p. 646, fn. 7.)] In quantum systems, however, the result is not zero, and the effect is called *diamagnetism*. In this chapter we shall discuss only the low temperature limit: $T = 0$. This leads us, in section 13.3.2, to the de Haas–van Alphen effect, an oscillatory dependence of magnetization on $\vec{B}$ field. In section 13.3.3 we obtain the low-temperature susceptibility, and compare it with the corresponding paramagnetic susceptibility. Section 13.3 is largely based on Pippard (1962). Other insightful discussions are given by Madelung (1978, section 2.1.2) and by Huang (1967, sections 11.3–11.4).

13.3.2 The Landau levels

We now consider the high-field case, for the electron gas, equation (13.5), omitting the paramagnetic term:

$$H \approx \sum_{j=1}^{N} \left(\frac{1}{2m_b} \right) (\vec{p}_j + e\vec{A}(\vec{r}_j))^2. \tag{13.139}$$

We choose a specific gauge, in which

$$\vec{A} = Bx\hat{\varepsilon}_2, \tag{13.140}$$

where position vector $\vec{r}$ has cartesian components (x, y, z), and $\hat{\varepsilon}_j$ $(j = 1, 2, 3)$ are unit vectors in the (x, y, z) directions. This ensures that, from equation (13.140),

$$(\vec{\nabla} \times \vec{A}) = B\hat{\varepsilon}_3 = \vec{B}, \tag{13.141}$$

with $\vec{B}$ in the z direction. From equations (13.139) and (13.140),

$$H = \frac{1}{2m_b} \sum_j (\vec{p}_j + eBx_j\hat{\varepsilon}_2)^2. \tag{13.142}$$

Because there are no electron–electron interactions in our effective hamiltonian equation (13.142), the total energy at the Hartree–Fock level will be the sum of N eigenvalues of the single particle hamiltonian (Fock operator):

$$H_1 = \frac{1}{2m_b} (\vec{p} + eBx\hat{\varepsilon}_2)^2. \tag{13.143}$$

We note that, from equation (13.143),

$$[p_3, H_1] = [p_2, H_1] = 0, \tag{13.144}$$

where $[p_\lambda, H_1]$ is the commutator,

$$[p_\lambda, H_1] \equiv (p_\lambda H_1 - H_1 p_\lambda), \tag{13.145}$$

and λ labels cartesian components. Equation (13.144) says that p_3 and p_2 are conserved quantities. Initially, this seems counterintuitive, because the physical system is cylindrically symmetrical about the direction of $\vec{B}$, i.e.: the z or $\hat{\varepsilon}_3$ direction, and so why is p_2 conserved? We defer this question, remarking only that $\vec{p}$ is the *canonical* momentum, not to be confused with $m\vec{v}$, where $\vec{v}$ is the electron's velocity.

In position representation,

$$\vec{p} = -i\hbar\vec{\nabla}, \tag{13.146}$$

and the eigenstates of p_3 and p_2 satisfy

$$p_\lambda \varphi_\lambda = \hbar k_\lambda \varphi_\lambda, \tag{13.147}$$

whence, from equations (13.146) and (13.147),

$$\varphi_3 \sim \exp(ik_3 z), \qquad \varphi_2 \sim \exp(ik_2 y). \tag{13.148}$$

The full single-particle eigenfunctions are thus

$$\varphi(\vec{r}) = \exp[i(k_2 y + k_3 z)] f(x). \tag{13.149}$$

From equations (13.143) and (13.149), the single-particle Schrödinger (or Fock) equation becomes

$$H_1 \varphi = \frac{1}{2m_b} [p^2 + eB(p_2 x + x p_2) + (eBx)^2] \varphi$$

$$= \frac{1}{2m_b} [p_1^2 + \hbar^2(k_2^2 + k_3^2) + 2eBx\hbar k_2 + (eBx)^2] \varphi = \varepsilon \varphi, \tag{13.150}$$

where ε is the single-particle energy eigenvalue. With equation (13.149), this reduces to

$$\left\{ \frac{1}{2m_b} p_1^2 + \left(\frac{m}{m_b}\right)\left(\frac{eB}{m}\right) \hbar k_2 x + \frac{\hbar^2}{2m_b}(k_2^2 + k_3^2) \right.$$

$$\left. + \frac{1}{2}\left(\frac{m^2}{m_b}\right)\left(\frac{eB}{m}\right)^2 x^2 \right\} f(x) = \varepsilon f(x). \tag{13.151}$$

We now introduce the cyclotron frequency ω_c (actually, angular frequency) for a free electron:

$$\omega_c = \left(\frac{eB}{m}\right), \tag{13.152}$$

Then equation (13.151) becomes

$$\left\{ \frac{1}{2m_b} p_1^2 + \frac{1}{2} m_b \left(\frac{m\omega_c}{m_b}\right)^2 x^2 + \left(\frac{m\omega_c}{m_b}\right) \hbar k_2 x \right\} f$$

$$= \left[\varepsilon - \frac{\hbar^2}{2m_b}(k_2^2 + k_3^2) \right] f. \tag{13.153}$$

For the terms in x^2 and x on the left-hand side of equation (13.153), complete the square, as follows:

$$\left(\frac{1}{2} m_b \tilde{\omega}_c^2 x^2 + \tilde{\omega}_c \hbar k_2 x \right)$$

$$= \left\{ \frac{1}{2} m_b \tilde{\omega}_c^2 \left[x^2 + 2\left(\frac{\hbar k_2}{m_b \tilde{\omega}_c}\right) x + \left(\frac{\hbar k_2}{m_b \tilde{\omega}_c}\right)^2 \right] - \left(\frac{\hbar^2 k_2^2}{2m_b}\right) \right\}$$

$$= \left\{ \frac{1}{2} m_b \tilde{\omega}_c^2 (x')^2 - \left(\frac{\hbar^2 k_2^2}{2m_b}\right) \right\}. \tag{13.154}$$

In equation (13.154), we have introduced x' in terms of x_0, and $\tilde{\omega}_c$, where

$$\tilde{\omega}_c = \frac{m}{m_b}\omega_c, \qquad x' = (x + x_0), \qquad x_0 = \left(\frac{\hbar k_2}{m_b\tilde{\omega}_c}\right). \tag{13.155}$$

Now, in equation (13.153),

$$p_1 = -i\hbar\frac{\partial}{\partial x} = -i\hbar\frac{\partial}{\partial x'} = p_1'. \tag{13.156}$$

Thus equation (13.153) with equations (13.154)–(13.156) becomes

$$\left\{\frac{1}{2m_b}(p_1')^2 + \frac{1}{2}m_b\tilde{\omega}_c^2(x')^2\right\}g(x') = \left(\varepsilon - \frac{\hbar^2k_3^2}{2m_b}\right)g(x'), \tag{13.157}$$

where, through equation (13.155),

$$g(x') = f(x). \tag{13.158}$$

The left-hand side of equation (13.157) contains the hamiltonian of a harmonic oscillator on the x-axis about the center $x = -x_0$, whose eigenvalues are

$$\left(\varepsilon - \frac{\hbar^2k_3^2}{2m_b}\right) = \left(n + \frac{1}{2}\right)\hbar\tilde{\omega}_c, \qquad n = 0, 1, 2, \ldots, \tag{13.159}$$

whence the single particle energies ε are

$$\varepsilon(n, k_3) = \left\{\left(n + \frac{1}{2}\right)\hbar\tilde{\omega}_c + \frac{\hbar^2k_3^2}{2m_b}\right\}. \tag{13.160}$$

Referring to equation (13.160), we see that the $\vec{k}$-space geometry only enters the solution through k_3, even though p_2 is conserved: see equation (13.144) and discussion following it. The momentum components $\hbar k_1$ and $\hbar k_2$ associated with a free electron only enter the motion through the harmonic oscillator states $g(x')$, equations (13.157) and (13.158).

We can now examine the relationship between the corresponding classical oscillator and our expectation that the classical motion is circular when projected on the x–y plane, with angular frequency $\tilde{\omega}_c$, even though the gauge which we have chosen, equation (13.140), gives a hamiltonian that appears to be biased in the y-direction, equation (13.143). For uniform circular motion in the x–y plane with radius R about the point $x = -x_0$ [see equation (13.155)], $y = 0$, with angular frequency $\tilde{\omega}_c$, we have

$$x = -x_0 + R\cos(\tilde{\omega}_c t); \qquad v_1 = -\tilde{\omega}_c R\sin(\tilde{\omega}_c t);$$
$$y = R\sin(\tilde{\omega}_c t); \qquad v_2 = \tilde{\omega}_c R\cos(\tilde{\omega}_c t). \tag{13.161}$$

From equations (13.161) we see that

$$v_2 = \tilde{\omega}_c(x + x_0). \tag{13.162}$$

Now, from the hamiltonian formulation,

$$v_2 = \frac{1}{m_b}(p_2 + eBx); \qquad (13.163)$$

see equation (13.143). In equation (13.163), p_2 is constant [from equation (13.144)], and from equation (13.147) its value has been taken to be $(\hbar k_2)$. Thus, with equations (13.155) and (13.152), equation (13.163) becomes

$$v_2 = \frac{1}{m_b}(\hbar k_2 + eBx) = \left(\frac{\hbar k_2}{m_b} + \tilde{\omega}_c x\right) = \tilde{\omega}_c(x + x_0). \qquad (13.164)$$

What we have shown is that v_2, as determined from our original hamiltonian, equation (13.143), with the gauge given in equation (13.140), which appears to destroy the cylindrical symmetry of the system about the z-axis, in fact does not do so, but leaves v_2 (distinct from (p_2/m)) in the form required for uniform *circular* motion in the x–y plane about $x = -x_0, y = 0$, expressed by the *one-dimensional* harmonic oscillator part of equation (13.157). For the *free* electron gas, the energy of an electron is given by

$$\frac{\hbar^2}{2m_b}(k_1^2 + k_2^2 + k_3^2). \qquad (13.165)$$

In a magnetic induction field in the z-direction, the term in k_3^2 is maintained where k_α ($\alpha = 1, 2, 3$) are quasi-continuous variables:

$$k_\alpha = \frac{2\pi n_\alpha}{L}, \qquad n_\alpha = 0, \pm 1, \pm 2, \ldots, \qquad (13.166)$$

where $L^2 = \Omega$ is the Born–von Karmann volume: see section 7.5.1. The other two terms, in $(k_1^2 + k_2^2)$, however, are replaced by quantized harmonic oscillator levels, the so-called *Landau levels*:

$$\left(n + \frac{1}{2}\right)\hbar\tilde{\omega}_c = \left(n + \frac{1}{2}\right)\frac{e\hbar B}{m_b}; \qquad (13.167)$$

see equation (13.159). The level spacing of the transverse part of the motion, instead of being of order L^{-2} with $L \to \infty$, is of the order of the field strength B, and the new levels are highly degenerate, as we shall see.

13.3.3 The Fermi distribution

We now introduce the Fermi energy ε_F, the highest occupied energy level at absolute zero, $T = 0$. It depends primarily on the number of particles (for given total volume Ω). The quantum numbers (n, k_3) for the occupied levels must satisfy the condition

$$\varepsilon(n, k_3) = \left[\left(n + \frac{1}{2}\right)\hbar\tilde{\omega}_c + \frac{\hbar k_3^2}{2m_b}\right] \leq \varepsilon_F; \qquad (13.168)$$

see equation (13.160). Since $\tilde{\omega}_c \approx B$, we see that, for a given value of B, n must satisfy the relationship

$$\left(n + \frac{1}{2}\right) \leq \frac{\varepsilon_F}{\hbar\tilde{\omega}_c} \sim \frac{1}{B}. \tag{13.169}$$

Thus, for small B, many oscillator levels will be occupied (each with high degeneracy), and for large B, only a small number. For a given value of n, at given field, the values of k_3 must satisfy

$$|k_3| \leq (k_3)_{F,n} \equiv \left[\frac{2m_b}{\hbar^2}\left\{\varepsilon_F - \left(n + \frac{1}{2}\right)\hbar\tilde{\omega}_c\right\}\right]^{1/2}, \tag{13.170}$$

thus defining the position $(k_3)_{F,n}$ of the Fermi surface in k-space in the z-direction for given n. Clearly, for large enough field $\vec{B}$ (large enough $\tilde{\omega}_c$: see equations (13.155) and (13.152)), ε_F must depend on B in equation (13.170) in such a way that the square root is real. We consider here the approximation of $\vec{B}$-field weak enough so that ε_F is approximately equal to the zero-field value, equation (13.102): see Pippard (1961), p. 23.

Now we know that applying a $\vec{B}$-field does not change the energy of a charged particle: classically, it only drives the particle into a circular (two dimensions) or helical (three dimensions) orbit. Thus, comparing equations (13.160) and (13.165), we might think that

$$\frac{\hbar^2}{2m_b}(k_1^2 + k_2^2) = \left(n + \frac{1}{2}\right)\hbar\tilde{\omega}_c. \tag{13.171}$$

This cannot be taken literally, because in the free electron gas, most electrons do not have values of (k_1, k_2) that can satisfy equation (13.171), for any value of n within the occupied range given by equation (13.169). In fact, it is quantization of the oscillation associated with the B-field that drives the transverse motion into the discrete oscillator levels labelled by n. The situation regarding equation (13.171) is that the sum over all electrons, or all occupied states (k_1, k_2) for the free electrons, and all occupied n in the presence of the $\vec{B}$-field, must be equal in total energy in the sense of equation (13.171). The total number of occupied states in the two cases must be equal. Since k_1 and k_2 are quasi-continuous for a large Born–von Karmann volume, and n is not very large for B not too small (equation (13.169)), the degeneracy of states in n must be high. What is this degeneracy? Well, from equation (13.155) we see that the classical orbits in a non-zero B-field are centered on points $x = -x_0$, given by

$$x_0 = \left(\frac{\hbar k_2}{m_b\tilde{\omega}_c}\right), \tag{13.172}$$

where, from equation (13.166),

$$k_2 = \frac{2\pi n_2}{L}, \qquad n_2 = 0, \pm 1, \pm 2, \ldots \tag{13.173}$$

Possible values of x_0 are therefore

$$x_0 = \left(\frac{\hbar}{m_b\tilde{\omega}_c}\right)\left(\frac{2\pi n_2}{L}\right), \tag{13.174}$$

The spacing S of centers is therefore

$$S = \left(\frac{2\pi\hbar}{m_b\tilde{\omega}_c L}\right). \tag{13.175}$$

The range of values of x_0 within the Born–von Karmann volume is L. The number of centers is therefore L/S,

$$\frac{L}{S} = \left(\frac{m_b\tilde{\omega}_c L^2}{2\pi\hbar}\right). \tag{13.176}$$

For spin-$\frac{1}{2}$ the degeneracy, for each oscillator level n, is twice this, namely,

$$(\text{degeneracy}) = \left(\frac{m_b\tilde{\omega}_c L^2}{\pi\hbar}\right). \tag{13.177}$$

Since this many electrons, equation (13.177), all have the same energy for a state with given n, we can represent them in $\vec{k}$-space as all having values of (k_1, k_2) satisfying equation (13.171) for a given value of k_3. Such points in $\vec{k}$-space lie on a circular cylinder with axis corresponding to the k_3 axis and, with radius R_n (see figure 13.2),

$$R_n = \left[\frac{2m_b\tilde{\omega}_c}{\hbar}\left(n + \frac{1}{2}\right)\right]^{1/2}. \tag{13.178}$$

Note that R_n increases with n. Since it is a radius in $\vec{k}$-space, its dimensionality is inverse length. According to equation (13.170), these concentric cylinders all have lengths $2(k_3)_{F,n}$, where the cylinder with $n = 0$ is longest, with length decreasing as n increases. For large enough B (large $\tilde{\omega}_c$), only $n = 0$ is occupied (see equation (13.170)), and the cylindrical occupied area in $\vec{k}$-space, from equation (13.178) has radius

$$R_{n=0} = \left(\frac{m_b\tilde{\omega}_c}{\hbar}\right)^{1/2}. \tag{13.179}$$

For N electrons in volume $\Omega = L^3$, we can determine $(k_3)_{F,n=0}$,

$$N = (\text{degeneracy}) \times \int_{-(k_3)_{F,0}}^{+(k_3)_{F,0}} dk_3 \left(\frac{L}{2\pi}\right), \tag{13.180}$$

where '(degeneracy)' is that of the oscillator $n = 0$ (independent of n) from equation (13.177), and $L/(2\pi)$ is the density of points in the k_3 direction. Thus, equation (13.180) gives

$$N = \left(\frac{m_b\tilde{\omega}_c L^2}{\pi\hbar}\right)\frac{L}{2\pi}2(k_3)_{F,0}, \tag{13.181}$$

whence

$$(k_3)_{F,0} = \pi^2 \left(\frac{\hbar}{m_b \tilde{\omega}_c}\right) \rho^{(0)} = \frac{\pi^2}{R_0^2} \rho^{(0)}, \tag{13.182}$$

where $\rho^{(0)}$ is the density of electrons,

$$\rho^{(0)} = \frac{N}{L^3} = \frac{N}{\Omega}, \tag{13.183}$$

and R_0 is given by equation (13.179). The result of equation (13.182) is to be compared with the Fermi radius $k_F^{(0)}$ for the free electron gas,

$$N = 2\left(\frac{L}{2\pi}\right)^3 4\pi \int_0^{k_F^{(0)}} dk\, k^2$$

$$= \frac{L^3}{\pi^2} \frac{k_F^{(0)^3}}{3} \tag{13.184}$$

whence

$$k_F^{(0)} = (3\pi^2 \rho^{(0)})^{1/3}. \tag{13.185}$$

In the literature [e.g. Madelung (1978, figure 2.8, p. 34)] it is often indicated that for 'strong field', the cylinder $n = 0$, holding all the occupied states on its surface, has the condition

$$(k_3)_{F,0} < k_F^{(0)}. \tag{13.186}$$

Let us examine the ratio

$$\frac{(k_3)_{F,0}}{k_F^{(0)}} = \frac{\left(\dfrac{\pi^2 \hbar \rho^{(0)}}{m_b \tilde{\omega}_c}\right)}{(3\pi^2 \rho^{(0)})^{1/3}} = \left(\frac{\pi^4}{3}\right)^{1/3} \left(\frac{\hbar}{m_b \tilde{\omega}_c}\right) (\rho^{(0)})^{2/3}$$

$$= \left(\frac{\pi^4}{3}\right)^{1/3} \left(\frac{\hbar}{eB}\right) (\rho^{(0)})^{2/3}, \tag{13.187}$$

having used equations (13.152) and (13.155) for $\tilde{\omega}_c$. Now, for typical metals, $k_F^{(0)} \sim 10^{10}\ \mathrm{m}^{-1}$ [see Ashcroft and Mermin (1976, table 2.1, p. 38)]. With this value, equation (13.185) gives

$$\rho^{(0)} = \frac{(k_F^{(0)})^3}{3\pi^2} \sim 3 \times 10^{28}\ \mathrm{m}^{-3}. \tag{13.188}$$

Using this in equation (13.187), we have

$$\frac{(k_3)_{F,0}}{k_F^{(0)}} \sim \left(\frac{\pi^4}{3}\right)^{1/3} \left(\frac{1}{2\pi \times 2.4 \times 10^{14}}\right) \frac{(3 \times 10^{28})^{2/3}}{B}$$

$$\sim \frac{2 \times 10^4}{B}. \tag{13.189}$$

Thus, for equation (13.186) to be satisfied, we should need a magnetic induction field B, with

$$B > 10^4 \text{ tesla}. \qquad (13.190)$$

Such a field strength is orders of magnitude greater than those that are usual for studies of magnetic properties. Nevertheless, it is what must be understood by 'strong magnetic field' in the present context. It shows that essentially all studies of diamagnetism will be performed at 'low fields', such that the largest n-value in equation (13.169) will be much larger than unity.

Let us further discuss the 'weak field' case. First, rewrite equation (13.170) as

$$\varepsilon_F = \left\{ \frac{\hbar^2 (k_3)_{F,n}^2}{2m_b} + \hbar \tilde{\omega}_c \left(n + \frac{1}{2} \right) \right\}. \qquad (13.191)$$

For very weak field $B \geq 0$, the Fermi energy ε_F must be approximately equal to the free-electron Fermi energy $\varepsilon_F^{(0)}$:

$$\varepsilon_F^{(0)} = \frac{\hbar^2 k_F^{(0)^2}}{2m_b}; \qquad (13.192)$$

see equations (13.184) and (13.185). Thus, from equation (13.191) with equations (13.152) and (13.155),

$$\left\{ \frac{\hbar^2 (k_3)_{F,n}^2}{2m_b} + \frac{e\hbar B}{m_b} \left(n + \frac{1}{2} \right) \right\} \approx \varepsilon_F^{(0)}. \qquad (13.193)$$

Recalling our expression for the radius of the nth cylinder in k-space, equation (13.178), we have, from equation (13.193),

$$\frac{\hbar^2}{2m_b} \{ (k_3)_{F,n}^2 + R_n^2 \} \approx \varepsilon_F^{(0)}. \qquad (13.194)$$

From equation (13.194) with equation (13.192), we see that, even for $n = 0$,

$$(k_3)_{F,0} \lesssim k_F^{(0)}, \qquad (13.195)$$

and in general

$$[(k_3)_{F,n}^2 + R_n^2] \approx (k_F^{(0)})^2. \qquad (13.196)$$

This situation is illustrated in figure 13.2.

The topology of the Fermi surface, as illustrated in figure 13.2, changes discontinuously as the strength of the B-field is varied. Specifically, as B increases, we see that the number of cylinders (the number of occupied oscillator levels) decreases in integer steps. We now begin the study of this effect. Consider an arbitrary value of $k_3 < (k_3)_{F,0}$ which is overlapped by

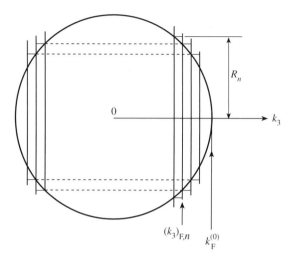

Figure 13.2. The Fermi surface of the electron gas under diamagnetic conditions, with weak magnetic induction field $\vec{B}$ in the z-direction. The k_3 component of $\vec{k}$ vectors is the z-component. Illustrated are the field-free Fermi sphere of radius $k_F^{(0)}$, and the coaxial cylindrical surfaces for $B \gtrsim 0$, of length $2(k_3)_{F,n}$ and radius R_n: see equations (13.196) and (13.178) respectively.

one or more cylinders. Consider a slab of k-space of thickness δk_3 at k_3. Let $\delta N(k_3, B)$ be the number of states in the slab. From the degeneracy given in equation (13.177), this number, for a given value of n, is

$$\left(\frac{m_b \tilde{\omega}_c L^2}{\pi \hbar}\right)\left(\frac{L}{2\pi}\right)\delta k_3 = \left(\frac{m_b \tilde{\omega}_c \Omega}{2\pi^2 \hbar}\right)\delta k_3, \tag{13.197}$$

where $[L/(2\pi)]$ is the density of points on k_3 and $\Omega = L^3$ is the Born–von Karmann volume. From equations (13.152) and (13.155), this is

$$\frac{\Omega e B}{2\pi^2 \hbar}\delta k_3 \sim B, \tag{13.198}$$

independent of n. Now let n' be the largest integer satisfying equation (13.191). Then from equation (13.198), we have

$$\delta N(k_3, B) = (n' + 1)\left(\frac{e}{2\pi^2 \hbar}\right)B\Omega\delta k_3. \tag{13.199}$$

The factor $(n' + 1)$ accounts for the oscillator level $n = 0$. Now consider decreasing B continuously from infinity. At some value, n' increases by unity from zero to one [see equation (13.169)], and $\delta N(k_3, B)$, for given value of k_3, increases by

$$\left(\frac{e}{2\pi^2\hbar}\right)B\Omega\delta k_3 \sim B. \tag{13.200}$$

This discontinuous increase occurs when the end of the cylinder for $n' = 0$ coincides with the chosen value of k_3. In order to visualize this better, let us rewrite equation (13.191) as

$$\varepsilon_F \equiv \left(\frac{\hbar^2 k_F^2}{2m_b}\right) = \left\{\frac{\hbar^2 (k_3)_{F,n'}^2}{2m_b} + \frac{e\hbar B}{m_b}\left(n' + \frac{1}{2}\right)\right\}. \tag{13.201}$$

Equation (13.201) defines k_F. Now with $(k_3)_{F,n'}$ equal to the chosen value of k_3, we can see from equation (13.201) that the value of B for which the discontinuities occur is given by

$$B_{n'} = \frac{\hbar}{2e}\frac{(k_F^2 - k_3^2)}{(n' + \frac{1}{2})}. \tag{13.202}$$

We can now plot δN, equation (13.199), as a function of B. This is commonly done in terms of the variable B^{-1}, increasing to the left, as in figure 13.3. From equation (13.199) we have, in between discontinuities,

$$\delta N(k_3, B^{-1}) = (n' + 1)\left(\frac{e}{2\pi^2\hbar}\right)\frac{\Omega}{B^{-1}}\delta k_3. \tag{13.203}$$

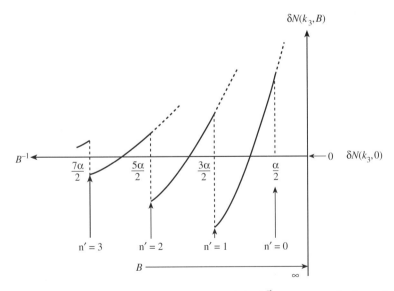

Figure 13.3. The number of occupied states δN in a slab of $\vec{k}$-space perpendicular to the k_3 axis, as a function of B^{-1}, for the diamagnetic electron gas in a magnetic induction field $\vec{B}$: see equations (13.203)–(13.205).

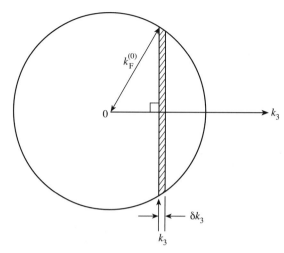

Figure 13.4. The region of the field-free Fermi sphere corresponding to a slab of thickness δk_3 at k_3 on the k_3-axis, for the electron gas: see equation (13.206).

The discontinuities occur, according to equation (13.202) at

$$B_{n'}^{-1} = \alpha(n' + \tfrac{1}{2}), \qquad n' = 0, 1, 2, \ldots, \tag{13.204}$$

where

$$\alpha = \frac{2e}{\hbar(k_F^2 - k_3^2)}. \tag{13.205}$$

We note that, with $B = 0$, i.e. for the free electron gas,

$$\delta N(k_3, B = 0) = \pi(k_F^{(0)^2} - k_3^2)2\left(\frac{L}{2\pi}\right)^3 \delta k_3; \tag{13.206}$$

see figure 13.4. Now from equations (13.203)–(13.205), approaching the discontinuity from values of B^{-1} above it, we have

$$\delta N(k_3, B_{n'}^{-1}) = \left[\left(\frac{B_{n'}^{-1}}{\alpha}\right) + \frac{1}{2}\right]\left(\frac{e}{2\pi^2\hbar}\right)\frac{\Omega}{B_{n'}^{-1}}\delta k_3$$

$$= \frac{\Omega}{\alpha}\left(\frac{e}{2\pi^2\hbar}\right)\delta k_3 + \frac{1}{2}\left(\frac{e}{2\pi^2\hbar}\right)\frac{\Omega}{B_{n'}^{-1}}\delta k_3. \tag{13.207}$$

The second term in equation (13.207) is one half of the discontinuity: see equation (13.200). Furthermore, the first term is, from equation (13.205),

$$\frac{\Omega(k_F^2 - k_3^2)}{4\pi^2}\delta k_3. \tag{13.208}$$

For weak field, where $k_F \approx k_F^{(0)}$, this is $\delta N(k_3, B = 0)$: see equation (13.206). Thus, equation (13.207) reads

$$\delta N(k_3, B^{-1}) = \{\delta N(k_3, B = 0) + \tfrac{1}{2}(\text{discontinuity})\}, \qquad (13.209)$$

for a value of B^{-1} just above that at the discontinuity. This shows that the discontinuities in figure 13.3 are symmetrical about $\delta N(k_3, B = 0)$, as shown.

13.3.4 Energy considerations

For finite, non-zero field, the cylinders, figure 13.2, do not coincide exactly with the free-electron Fermi sphere. Let us consider the deviation of the energy for electrons in the slab of thickness δk_3 at k_3, relative to the free-electron $(B = 0)$ case. For this, we equate the number of electrons in a small region of $\vec{k}$-space within the Fermi surface with the number of $\vec{k}$-space points (times 2 for spin) in the same region. That is, we view the effect of the $\vec{B}$-field as squeezing electrons out of the free-electron Fermi sphere in the (k_1, k_2) directions, into states of energy $\hbar^2 k^2/(2m_b)$. Still considering a slab of $\vec{k}$-space, of thickness δk_3 at k_3, we have an annulus of thickness δk_3, width $\delta k'$, and inner radius $(k_F^{(0)^2} - k_3^2)^{1/2}$, outside the free-electron Fermi sphere: see figure 13.5. Then the number of occupied states

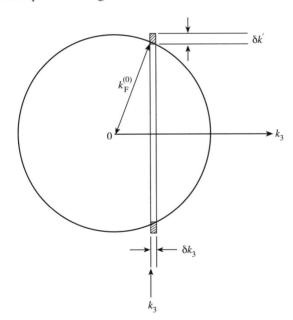

Figure 13.5. Annulus of $\vec{k}$-space points outside the field-free Fermi sphere that come onto segments of cylinders (figure 13.2) when a weak $\vec{B}$-field is applied to the electron gas. Width of annulus is $\delta k'$, thickness is δk_3, with the annulus perpendicular to the k_3 axis at k_3: see equation (13.210).

in the annulus is

$$\{\delta N_{max}(k_3, B^{-1}) - \delta N(k_3, B = 0)\}$$

$$= 2\pi(k_F^{(0)^2} - k_3^2)^{1/2} \delta k_3 \, \delta k' \, 2 \frac{\Omega}{(2\pi)^3}$$

$$= (k_F^{(0)^2} - k_3^2)^{1/2} \delta k_3 \, \delta k' \, \frac{\Omega}{(2\pi^2)}. \tag{13.210}$$

Now the maximum energy, E_{max}, of any state in the annulus is

$$E_{max} \approx \frac{\hbar^2}{2m_b} \{[(k_F^{(0)^2} - k_3^2)^{1/2} + \delta k']^2 + k_3^2\}, \tag{13.211}$$

and the minimum energy, E_{min}, is

$$E_{min} = \frac{\hbar^2}{2m_b} k_F^{(0)^2}. \tag{13.212}$$

Thus the maximum excess energy of the states in the annulus is $(E_{max} - E_{min})$, and we take the mean excess energy of these states to be half this,

$$\frac{1}{2}(E_{max} - E_{min}) \approx \frac{1}{2}\frac{\hbar^2}{2m_b} \{[(k_F^{(0)^2} - k_3^2) + 2(k_F^{(0)^2} - k_3^2)^{1/2}\delta k'] + k_3^2 - k_F^{(0)^2}\}$$

$$= \frac{\hbar^2}{2m_b}(k_F^{(0)^2} - k_3^2)^{1/2}\delta k', \tag{13.213}$$

to lowest order in $\delta k'$. From equation (13.210), equation (13.213) is

$$\tfrac{1}{2}(E_{max} - E_{min}) = \text{(mean excess energy per state)}$$

$$= \frac{\hbar^2}{2m_B}\left(\frac{2\pi^2}{\Omega}\right)\frac{[\delta N(k_3, B^{-1}) - \delta N(k_3, B = 0)]}{\delta k_3}. \tag{13.214}$$

Now denote the total excess energy associated with the slab δk_3 by δE. Then, from equation (13.214) and equation (13.210),

$$\delta E = \text{(mean excess energy per state)} \times \text{(number of states)}$$

$$= \frac{\hbar^2}{2m_b}\left(\frac{2\pi^2}{\Omega}\right)\frac{[\delta N - \delta N_0]^2}{\delta k_3}, \tag{13.215}$$

where

$$\delta N_0 \equiv \delta N(k_3, B = 0). \tag{13.216}$$

To the extent that $\delta N(k_3, B^{-1})$ in figure 13.3 is approximately linear in between discontinuities, δE, equation (13.215) is approximately parabolic.

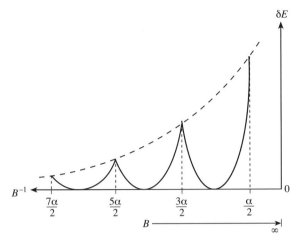

Figure 13.6. Excess energy δE as a function of B^{-1} for the electron gas due to interaction of orbital magnetic moment with applied $\vec{B}$-field for occupied states in a slab δk_3 of $\vec{k}$-space: see equations (13.215)–(13.219) and associated discussion.

This is illustrated in figure 13.6. As a function of B^{-1}, the maximum value δE_{max} of δE occurs at the discontinuities of δN. The maximum value of $(\delta N - \delta N_0)$, we have shown, equation (13.209), is half the discontinuity, namely, from equation (13.200),

$$(\delta N - \delta N_0)_{max} = \frac{1}{2}\left(\frac{eB\Omega}{2\pi^2\hbar}\delta k_3\right). \tag{13.217}$$

From equations (13.215) and (13.217),

$$\delta E_{max} = \frac{\hbar^2}{2m_b}\left(\frac{2\pi^2}{\Omega}\right)\left(\frac{eB\Omega}{4\pi^2\hbar}\right)^2\delta k_3 = \left(\frac{eB}{4\pi}\right)^2\frac{\Omega\delta k_3}{m_b}. \tag{13.218}$$

In terms of B^{-1}, we have, from (13.218),

$$\delta E_{max} \approx \frac{1}{(B^{-1})^2}. \tag{13.219}$$

Thus δE_{max} drops off as $1/x^2$ as a function of $x = B^{-1}$: this is the locus of maxima shown in figure 13.6.

13.3.5 Magnetization: the de Haas–van Alphen effect

For a linear, isotropic material, we had preceding equation (13.26):

$$\vec{M} = \frac{\chi_m}{\mu_0}\cdot\vec{B}, \tag{13.220}$$

where χ_m, the magnetic susceptibility, is a scalar. The potential energy density ε of magnetization in a material due to its interaction with the $\vec{B}$-field is

$$\varepsilon = -\vec{M} \cdot \vec{B}, \tag{13.221}$$

see Jackson (1962), p. 150. Thus,

$$M_\alpha = \left(-\frac{\partial \varepsilon}{\partial B_\alpha} \right), \tag{13.222}$$

where α labels cartesian coordinates.

From equation (13.215) we can deduce the contribution $\delta \varepsilon$ to the energy density from the $\vec{k}$-space slab δk_3 at k_3:

$$\delta \varepsilon = \frac{\delta E}{\Omega} = \frac{\hbar^2}{2m_b} \left(\frac{2\pi^2}{\Omega^2} \right) \frac{(\delta N - \delta N_0)^2}{\delta k_3}. \tag{13.223}$$

Then applying equation (13.222), we determine the corresponding contribution δM to the magnetization,

$$\delta M = -\frac{\hbar^2}{2m_b} \left(\frac{4\pi^2}{\Omega^2} \right) \frac{(\delta N - \delta N_0)}{\delta k_3} \frac{\partial(\delta N)}{\partial B}, \tag{13.224}$$

with $\vec{M}$ in the direction $\vec{B}$. From equation (13.203), this is

$$\delta M = -\frac{e\hbar}{m_b} \frac{(\delta N - \delta N_0)}{\Omega} (n' + 1). \tag{13.225}$$

We recall that in equations (13.203) and (13.225), n' is the largest integer satisfying equation (13.191),

$$n' \lesssim \left\{ \left[\varepsilon_F - \frac{\hbar^2 k_3^2}{2m_b} \right] \frac{m_b}{e\hbar B} - \frac{1}{2} \right\}. \tag{13.226}$$

For small B, we neglect the number $(1/2)$ relative to the other term in equation (13.226), and replace ε_F by $\varepsilon_F^{(0)}$: see equation (13.192):

$$\left(\varepsilon_F - \frac{\hbar^2 k_3^2}{2m_b} \right) \approx \frac{\hbar^2}{2m_b} (k_F^{(0)^2} - k_3^2). \tag{13.227}$$

With these approximations, equations (13.225)–(13.227) give

$$\delta M = -\frac{(\delta N - \delta N_0)}{\Omega B} \frac{\hbar^2}{2m_b} (k_F^{(0)^2} - k_3^2). \tag{13.228}$$

We see from equation (13.228) that δM has the same pattern of discontinuities as those of δN (see figure 13.3) modulated as to amplitude by the factor $(1/B)$.

We shall now concoct an analytical approximation to δM as a function of B^{-1}, similar to figure 13.3 for δN. Due to the negative sign in equation

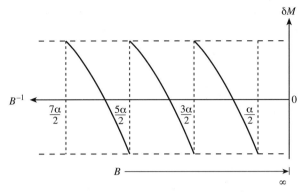

Figure 13.7. Diamagnetic magnetization δM as a function of B^{-1} for the electron gas, for a slab δk_3 of occupied states in $\vec{k}$-space: see equation (13.229) and associated discussion.

(13.228), δM slopes up when δN slopes down. Also from equation (13.228), the maxima of δM are

$$|\delta M_{\max}| = \frac{|\delta N - \delta N_0|_{\max}}{\Omega B} \frac{\hbar^2}{2m_\mathrm{b}} (k_\mathrm{F}^{(0)^2} - k_3^2), \qquad (13.229)$$

where B is evaluated at a discontinuity. From equations (13.207) and (13.209), we have

$$|\delta N - \delta N_0|_{\max} = \frac{e\Omega \delta k_3 B}{4\pi^2 \hbar} \sim B.$$

From this we see that $|\delta M|_{\max}$ in equation (13.229) is independent of B. We therefore deduce that δM versus B^{-1} is qualitatively as shown in figure 13.7. We view this approximately as part of an infinite sawtooth pattern, with straight line segments between discontinuities. For this pattern we have

$$\delta M \approx \frac{(2\delta M_{\max})}{\alpha} B^{-1}, \qquad \text{for } -\frac{\alpha}{2} \le B^{-1} \le \frac{\alpha}{2}. \qquad (13.230)$$

This is periodically repeated with period α in B^{-1} from minus infinity to plus infinity. We now introduce a variable x, proportional to B^{-1}, which ranges from $-\pi$ to π as B^{-1} ranges from $(-\alpha/2)$ to $(+\alpha/2)$, namely,

$$x = (2\pi B^{-1}/\alpha). \qquad (13.231)$$

Substituting this into equation (13.230), we have

$$\delta M = \frac{2\delta M_{\max}}{\alpha} \frac{\alpha}{2\pi} x = \left(\frac{\delta M_{\max}}{\pi}\right) x. \qquad (13.232)$$

Define

$$C = (\delta M_{\max}/\pi). \qquad (13.233)$$

Then

$$\delta M = Cx \qquad \text{for } -\pi \leq x \leq \pi. \tag{13.234}$$

Since δM, equation (13.234), is an odd function, the Fourier series for the periodic repetition of δM in equation (13.234) is

$$\delta M = Cx = \sum_{\mu=1}^{\infty} A_{\mu} \sin(\mu x), \tag{13.235}$$

where μ are integers. We evaluate the linear coefficients A_{μ}, using Fourier's theorem:

$$\int_{-\pi}^{\pi} dx\, Cx \sin(\mu' x) = \sum_{\mu} A_{\mu} \int_{-\pi}^{\pi} dx\, \sin(\mu' x) \sin(\mu x)$$

$$= \sum_{\mu} A_{\mu} \pi \delta_{\mu,\mu'} = \pi A_{\mu'}. \tag{13.236}$$

For the first integral in equation (13.236), we have

$$\int_{-\pi}^{\pi} dx\, x \sin(\mu' x) = \left[\frac{x \cos(\mu' x)}{(-\mu')} \right]_{-\pi}^{\pi} - \int_{-\pi}^{\pi} dx\, \frac{\cos(\mu' x)}{(-\mu')}, \tag{13.237}$$

having integrated by parts. The last integral in equation (13.237) is zero. Thus,

$$\int_{-\pi}^{\pi} dx\, x \sin(\mu' x) = \frac{[\pi(-1)^{\mu'} - (-\pi)(-1)^{\mu'}]}{(-\mu')}$$

$$= \frac{2\pi(-1)^{\mu'+1}}{\mu'}. \tag{13.238}$$

From equations (13.236) and (13.238) we conclude that

$$A_{\mu} = \frac{2(-1)^{\mu+1}}{\mu} C, \tag{13.239}$$

where recall C is defined in equation (13.233).

Let us now write out the expression for δM, equation (13.235), explicitly. From equation (13.239),

$$\delta M = \sum_{\mu=1}^{\infty} A_{\mu} \sin(\mu x)$$

$$= \sum_{\mu} \frac{2(-1)^{\mu+1}}{\mu} C \sin(\mu x). \tag{13.240}$$

Thus, from equations (13.233) and (13.229),

$$\delta M = \sum_{\mu} \frac{2(-1)^{\mu+1}}{\mu} \frac{|\delta N - \delta N_0|_{\text{max}}}{\pi \Omega B} \frac{\hbar^2}{2m_b} (k_F^{(0)^2} - k_3^2) \sin\left[\mu\left(\frac{2\pi}{\alpha B}\right)\right], \quad (13.241)$$

so from equation (13.205), with equations (13.207) and (13.209),

$$\delta M = \sum_{\mu} \frac{2(-1)^{\mu+1}}{\mu} \frac{e}{4\pi^3\hbar} \delta k_3 \frac{\hbar^2}{2m_b} (k_F^{(0)^2} - k_3^2)$$

$$\times \sin\left[\mu \frac{2\pi\hbar}{2eB} (k_F^{(0)^2} - k_3^2)\right]. \quad (13.242)$$

This simplifies to

$$\delta M = \frac{e}{2\pi^3\hbar} \sum_{\mu} \frac{(-1)^{\mu+1}}{\mu} \varepsilon_F^{(0)} \left(1 - \frac{k_3^2}{k_F^{(0)^2}}\right)$$

$$\times \sin\left[\mu 2\pi \frac{\varepsilon_F^{(0)}}{\hbar\tilde{\omega}_c} \left(1 - \frac{k_3^2}{k_F^{(0)^2}}\right)\right] \delta k_3. \quad (13.243)$$

In order to determine the total magnetization $\vec{M}$, we now integrate over all slabs δk_3, for

$$-k_F^{(0)} \lesssim k_3 \lesssim k_F^{(0)}, \quad (13.244)$$

introducing the notation

$$\frac{k_3}{k_F^{(0)}} = y. \quad (13.245)$$

Then equations (13.243)–(13.245) give

$$M = \int \delta M = \frac{ek_F^{(0)} \varepsilon_F^{(0)}}{2\pi^3\hbar} \sum_{\mu} \frac{(-1)^{\mu+1}}{\mu} \int_{-1}^{1} dy\,(1 - y^2)$$

$$\times \sin\left[2\pi\mu \frac{\varepsilon_F^{(0)}}{\hbar\tilde{\omega}_c} (1 - y^2)\right]. \quad (13.246)$$

The nature of the integral in equation (13.246) becomes clearer if we substitute

$$u = (1 - y^2). \quad (13.247)$$

Then we define the integral I by

$$I = \int_{-1}^{1} dy\,(1 - y^2) \sin\left[\frac{2\pi}{\lambda} (1 - y^2)\right]$$

$$= 2 \int_{0}^{1} \frac{du}{2(1 - u)^{1/2}} u \sin\left(\frac{2\pi}{\lambda} u\right). \quad (13.248)$$

where the wavelength λ of the sine in the variable u is

$$\lambda = \left(\frac{\hbar \tilde{\omega}_c}{\mu \varepsilon_F^{(0)}} \right), \tag{13.249}$$

the longest wavelength being

$$\lambda_{\text{max}} = \left(\frac{\hbar \tilde{\omega}_c}{\varepsilon_F^{(0)}} \right) = \left(\frac{e \hbar B}{m_b \varepsilon_F^{(0)}} \right). \tag{13.250}$$

Since the integration, equation (13.248), has u ranging over values $u \leq 1$, we can formulate weak $\vec{B}$-field in terms of equation (13.250),

$$\left(\frac{e \hbar B}{m_b \varepsilon_F^{(0)}} \right) = \lambda_{\text{max}} \ll 1. \tag{13.251}$$

This in turn means that the sine in equation (13.248) undergoes many oscillations within the range $0 \leq u \leq 1$. The integrand of equation (13.248) has two factors: $\sin(2\pi u/\lambda)$ and $\frac{1}{2} u(1-u)^{-1/2}$, which are shown schematically in figure 13.8, along with their product. It is clear that most of the contribution to the integral comes from the vicinity of $u = 1$, or from equation (13.247), $y = 0$. Thus equation (13.248) can be approximated as

$$I = 2 \int_0^1 dy \, (1 - y^2) \sin \left[\frac{2\pi}{\lambda} (1 - y^2) \right]$$

$$\approx 2 \int_0^\beta dy \, (1 - y^2) \sin \left[\frac{2\pi}{\lambda} (1 - y^2) \right]$$

$$\approx 2 \int_0^\beta dy \, \sin \left[\frac{2\pi}{\lambda} (1 - y^2) \right], \tag{13.252}$$

with $\beta \ll 1$. Now, as we shall see, the integral in equation (13.252) can be evaluated exactly if the upper limit β is extended to infinity. For $y > 1$, the phase of the sine increases in magnitude quadratically, giving an increasingly fast oscillation whose integral is negligible. In order to estimate the integral, we shall neglect the discrepancy associated with including the region $\beta < y \leq 1$ in the integration. We therefore consider

$$I \approx 2 \int_0^\infty dy \, \sin \left[\frac{2\pi}{\lambda} (1 - y^2) \right]. \tag{13.253}$$

The right hand side of equation (13.253) can be expanded to

$$I \approx 2 \int_0^\infty dy \left\{ \sin \left(\frac{2\pi}{\lambda} \right) \cos \left(\frac{2\pi y^2}{\lambda} \right) - \cos \left(\frac{2\pi}{\lambda} \right) \sin \left(\frac{2\pi y^2}{\lambda} \right) \right\}. \tag{13.254}$$

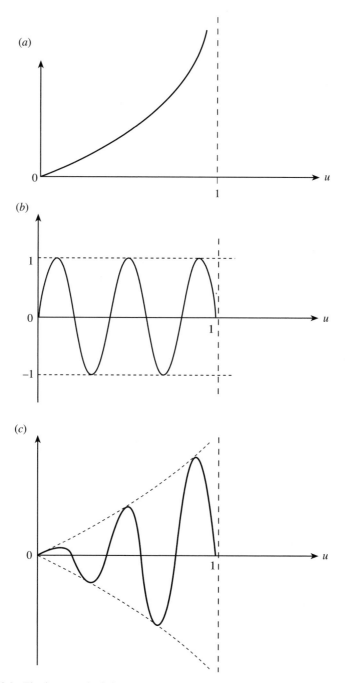

Figure 13.8. The integrand of the integral I in equation (13.248) as a function of u. (*a*) $u(1 - u)^{-1/2}$; (*b*) $\sin(2\pi u/\lambda)$; (*c*) the integrand: the product $u(1 - u)^{-1/2}\sin(2\pi u/\lambda)$.

Now we can evaluate explicitly the integrals in equation (13.254): see Appendix 13.1. The result is

$$\int_0^\infty dy \, \frac{\cos}{\sin} \left(\frac{2\pi y^2}{\lambda} \right) = \left(\frac{\lambda}{2\pi} \right)^{1/2} \frac{1}{2} \left(\frac{\pi}{2} \right)^{1/2}. \tag{13.255}$$

We now return to equation (13.246) for the magnetization M, and assemble the intervening results. From equations (13.246), (13.248), (13.254) and (13.255),

$$
\begin{aligned}
M &= \frac{e k_F^{(0)} \varepsilon_F^{(0)}}{2\pi^3 \hbar} \sum_{\mu=1}^\infty \frac{(-1)^{\mu+1}}{\mu} \frac{\lambda^{1/2}}{2} \left[\sin\left(\frac{2\pi}{\lambda} \right) - \cos\left(\frac{2\pi}{\lambda} \right) \right] \\
&= \frac{e k_F^{(0)} \varepsilon_F^{(0)}}{2\pi^3 \hbar} \sum_\mu \frac{(-1)^{\mu+1}}{\mu} \left(\frac{\hbar \tilde{\omega}_c}{2\mu \varepsilon_F^{(0)}} \right)^{1/2} \left[\frac{1}{\sqrt{2}} \sin\left(\frac{2\pi}{\lambda} \right) - \frac{1}{\sqrt{2}} \cos\left(\frac{2\pi}{\lambda} \right) \right] \\
&= \frac{k_F^{(0)} (\varepsilon_F^{(0)})^{1/2}}{2\pi^3} \left(\frac{e^3 B}{2\hbar m_b} \right)^{1/2} \sum_\mu \frac{(-1)^{\mu+1}}{\mu^{3/2}} \sin\left[2\pi\mu \left(\frac{\varepsilon_F^{(0)} m_b}{e\hbar B} \right) - \frac{\pi}{4} \right] \\
&= -\frac{\varepsilon_F^{(0)}}{2\pi^3} \left(\frac{e}{\hbar} \right)^{3/2} B^{1/2} \sum_\mu \frac{1}{\mu^{3/2}} \sin\left\{ 2\pi\mu \, \frac{\varepsilon_F^{(0)} m_b}{e\hbar} \left(\frac{1}{B} - \frac{e\hbar}{2m_b \varepsilon_F^{(0)}} \right) - \frac{\pi}{4} \right\}.
\end{aligned}
\tag{13.256}
$$

In the last three lines we have used the following results

$$\sin\frac{\pi}{4} = \cos\frac{\pi}{4} = \frac{1}{\sqrt{2}}, \tag{13.257}$$

$$k_F(0) = \frac{(2m_b \varepsilon_F^{(0)})^{1/2}}{\hbar}, \tag{13.258}$$

$$\lambda = \left(\frac{\hbar \tilde{\omega}_c}{\mu \varepsilon_F^{(0)}} \right), \tag{13.259}$$

$$\sin(\theta - \mu\pi) = (-1)^\mu \sin\theta, \tag{13.260}$$

$$\hbar \tilde{\omega}_c = \left(\frac{e\hbar B}{m_b} \right). \tag{13.261}$$

Referring to equation (13.256), we see that the magnetization M is given by a Fourier series, periodic in B^{-1} with period Λ:

$$\Lambda = \left(\frac{e\hbar}{m_b \varepsilon_F^{(0)}} \right). \tag{13.262}$$

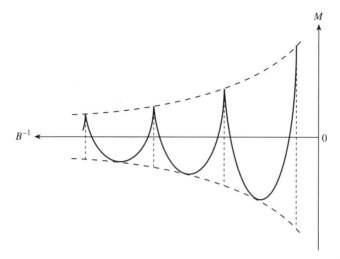

Figure 13.9. Diamagnetic magnetization M (schematically) as a function of B^{-1} for the electron gas, including the entire Fermi distribution: see equation (13.256).

The amplitude as a function of B^{-1} drops off as $(B^{-1})^{-1/2}$. This periodic variation of M with (B^{-1}) is the *de Haas–van Alphen effect*, illustrated in figure 13.9. The periodicity Λ, equation (13.262), varies as $(\varepsilon_F^{(0)})^{-1}$. The observed periodicity therefore gives us the Fermi energy, which is directionally dependent in crystalline metals. This accounts for the fact that the de Haas–van Alphen effect is a powerful tool in mapping out the Fermi surfaces of metals.

13.3.6 Diamagnetism at $T = 0$

We now wish to evaluate a diamagnetic susceptibility χ_D which incorporates an average over the range of B-values within one oscillation of the de Haas–van Alphen effect. According to equation (13.220) this requires an average over $\vec{M}$, which in turn, according to equation (13.221) with equation (13.223) requires an average over the total excess energy δE of the slab δk_3, equation (13.215),

$$\delta E = \frac{\hbar^2}{2m_b}\left(\frac{2\pi^2}{\Omega}\right)\frac{[\delta N - \delta N_0]^2}{\delta k_3}. \tag{13.263}$$

We have remarked, following equation (13.216), that δE is approximately parabolic as a function of B^{-1} between discontinuities. The discontinuities occur at

$$B^{-1} = \left(n' \pm \frac{1}{2}\right)\alpha = \left(n' \pm \frac{1}{2}\right)\frac{2e}{\hbar(k_F^2 - k_3^2)}; \tag{13.264}$$

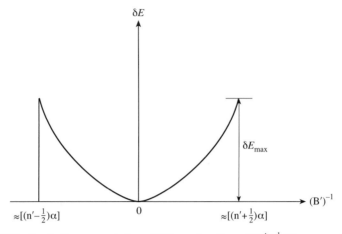

Figure 13.10. Parabolic representation of δE as a function of $(B')^{-1}$ (B measured relative to the minimum of the curve), for a slab δk_3 of occupied states in $\vec{k}$-space: see equations (13.264)–(13.268).

see figure 13.6 and equation (13.205). At the discontinuities,

$$\delta E_{\max} = \left(\frac{eB}{4\pi}\right)^2 \frac{\Omega\,\delta k_3}{m_b} : \tag{13.265}$$

see equation (13.218). We represent δE as a parabola within one oscillation,

$$\delta E = Ax^2, \qquad x = (B')^{-1}, \tag{13.266}$$

where B' is the B field measured relative to the minimum of δE: see figure 13.10. The variable x ranges over values such that $(-\alpha/2) \le x \le (\alpha/2)$: see equations (13.230) and (13.205). At the endpoints of this range, δE has the value $\delta E_{\max}$, equation (13.265). Thus

$$\delta E_{\max} = A(\alpha/2)^2, \tag{13.267}$$

whence

$$A = \left(\frac{2}{\alpha}\right)^2 \delta E_{\max}, \tag{13.268}$$

and so, from equations (13.266) and (13.268),

$$\delta E \approx \delta E_{\max}\left(\frac{2x}{\alpha}\right)^2 \sim \delta k_3; \tag{13.269}$$

see equation (13.265). We now define the average value δE_{av} of δE over one oscillation

$$\delta E_{av} = \frac{1}{\alpha} \int_{(-\alpha/2)}^{(\alpha/2)} dx\, Ax^2 = \frac{1}{3}\delta E_{max}. \tag{13.270}$$

Correspondingly, we define the average value δM_{av} of δM over one oscillation:

$$\delta M_{av} = -\frac{1}{\Omega}\frac{\partial}{\partial B}(\delta E_{av}): \tag{13.271}$$

see equations (13.222). For the average total magnetization M_{av}, we must integrate over k_3 within the Fermi surface

$$M_{av} = \int_{k_3 = -k_F^{(0)}}^{k_F^{(0)}} (\delta M_{av}) = -\frac{1}{\Omega}\frac{\partial}{\partial B}\int_{k_3 = -k_F^{(0)}}^{k_F^{(0)}} \delta E_{av}$$

$$= -\frac{1}{\Omega}\frac{\partial}{\partial B}\int_{k_3 = -k_F^{(0)}}^{k_F^{(0)}} \frac{1}{3}\delta E_{max}$$

$$= -\frac{1}{3\Omega}\frac{\partial}{\partial B}\int_{k_3 = -k_F^{(0)}}^{k_F^{(0)}} \left(\frac{eB}{4\pi}\right)^2 \frac{\Omega\, dk_3}{m_b}$$

$$= -\frac{1}{3}\left(\frac{e}{4\pi}\right)^2 2\left(\frac{B}{m_b}\right) 2k_F^{(0)} = \frac{-e^2 k_F^{(0)} B}{12\pi^2 m_b}. \tag{13.272}$$

In equation (13.272) we have successively used equations (13.271), (13.270), (13.268) and (13.265). It follows that the *diamagnetic susceptibility* χ_D at $T = 0$ is given by

$$\chi_D = \mu_0\left(\frac{\partial M_{av}}{\partial B}\right)_{T=0} = \left(\frac{-e^2 k_F^{(0)} \mu_0}{12\pi^2 m_b}\right). \tag{13.273}$$

We can compare χ_D, equation (13.273), with the *paramagnetic susceptibility* χ_P at $T = 0$, given by equation (13.122),

$$\chi_P = \left(\frac{3\mu_B^2 \rho^{(0)}\mu_0}{2\varepsilon_F}\right). \tag{13.274}$$

where, from equations (13.1) and (13.90),

$$\mu_B = \left(\frac{e\hbar}{2m}\right), \qquad \rho^{(0)} = \frac{N}{\Omega}, \tag{13.275}$$

where m is the free electron mass and, from equations (13.101) and (13.102),

$$\varepsilon_F = \frac{\hbar^2 k_F^2}{2m_b}, \qquad k_F = (3\pi^2 \rho^{(0)})^{1/3}. \tag{13.276}$$

Then, from equation (13.274),

$$\chi_P = \frac{3}{2}\left(\frac{e\hbar}{2m}\right)^2 \frac{\rho^{(0)}\mu_0}{\left(\frac{\hbar^2}{2m_b}\right)(3\pi^2\rho^{(0)})^{2/3}}$$

$$= \frac{3}{2}\frac{e^2}{m}\left(\frac{m_b}{m}\right)\frac{k_F\mu_0}{6\pi^2}. \tag{13.277}$$

We note that in this section, section 13.3, the symbol $k_F^{(0)}$, equation (13.185), where superscript (0) refers to *zero field* ($B = 0$), has the same meaning as k_F in section 13.2, equation (13.101). [In section 13.2, the symbol $k_F^{(0)}$ referred to electrons all of the same spin.] Thus in equation (13.277) we must identify k_F with $k_F^{(0)}$. Comparing equations (13.273) and (13.277) with $k_F \equiv k_F^{(0)}$, we then see that

$$\chi_P = -3\left(\frac{m_b}{m}\right)^2 \chi_D, \tag{13.278}$$

or

$$\chi_D = -\frac{1}{3}\left(\frac{m}{m_b}\right)^2 \chi_P, \qquad T = 0. \tag{13.279}$$

Since from equation (13.122), $\chi_P > 0$ (which is also intuitively obvious, since the intrinsic dipole moments of the electrons align parallel to the $\vec{B}$-field in paramagnetism), we see that from equation (13.279) the diamagnetic susceptibility is negative. As we mentioned in section 13.2.1, this simply reflects the fact that electrons (negatively charged) are induced into orbits that circle the $\vec{B}$-field in the right hand sense (see equation (13.138)), constituting electric current loops with current in the left-hand sense. Such current loops represent, in lowest multipole order, magnetic dipoles oriented *oppositely* to the $\vec{B}$-field, whence the negative susceptibility.

Appendix to Chapter 13

We wish to evaluate

$$\int_0^\infty dy \, \frac{\cos}{\sin}\left(\frac{2\pi y^2}{\lambda}\right) = \left(\frac{\lambda}{2\pi}\right)^{1/2} \frac{\mathrm{Re}}{\mathrm{Im}} \int_0^\infty dx \, e^{ix^2}, \tag{A13.1}$$

where Re and Im refer to real and imaginary parts: see equation (13.255). This leads us to consider an application of the Cauchy integral theorem (see standard textbooks on complex variables) for a contour C in the plane of complex numbers z

$$z = (x + iy); \qquad x, y \text{ both real.} \tag{A13.2}$$

That is, we consider

$$\oint_C dz\, e^{iz^2}. \tag{A13.3}$$

Note that x and y in equation (A13.1) are not the same things as x and y in equation (A13.2). We want to determine that part of the contour integral which lies on the positive real (x) axis, since $x = \text{Re}(z)$. Thus C must contain the positive real axis. The integrand in equation (A13.3) is infinite (diverges), as follows. Let

$$z = (x + iy) = r\, e^{i\theta}, \tag{A13.4}$$

with both r and θ real and r is positive. Then

$$z^2 = r^2\, e^{2i\theta} \tag{A13.5}$$

and since

$$i = \exp(i\pi/2)$$

$$iz^2 = r^2 \exp\left[i\left(2\theta + \frac{\pi}{2} \right) \right]$$

$$= r^2 \left[\cos\left(2\theta + \frac{\pi}{2} \right) + i \sin\left(2\theta + \frac{\pi}{2} \right) \right]. \tag{A13.6}$$

Thus the real part of (iz^2) is positive for

$$-\frac{\pi}{2} < \left(2\theta + \frac{\pi}{2} \right) < \frac{\pi}{2}$$

i.e.

$$-\pi < 2\theta < 0.$$

Thus for $0 \le \theta \le 2\pi$, i.e. $0 \le 2\theta \le 4\pi$, the real part of (iz^2) is positive for

$$-\frac{\pi}{2} < \theta < 0 \qquad \text{and} \qquad \frac{\pi}{2} < \theta < \pi, \tag{A13.7}$$

i.e. in the second and fourth quadrants of the (x, y) plane. As $r \to \infty$, the integrand $\exp(iz^2)$ diverges within these quadrants. We therefore consider the contour C in the first quadrant, shown in figure 13.11. It contains no

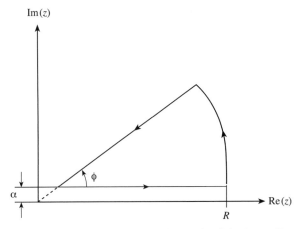

Figure 13.11. Contour for the evaluation of the integrals of the Appendix to Chapter 13: see equation (A13.1).

singularity. Thus, from the residue theorem,

$$\oint_C dz \exp(iz^2) = 0$$

$$= \lim_{\substack{R \to \infty \\ \alpha \to 0^+}} \left\{ \int_{z=\alpha(\cot\phi+i)}^{z=(R+i\alpha)} dz \exp(iz^2) \right.$$

$$+ \int_{r=R}^{r=(\alpha/\sin\varphi)} dr \, e^{i\varphi} \exp[ir^2(\cos 2\varphi + i\sin 2\varphi)]$$

$$\left. + \int_0^\varphi d\theta(iR\,e^{i\theta})\exp[iR^2(\cos 2\theta + i\sin 2\theta)] \right\}. \quad (A13.8)$$

In equation (A13.8), φ must be determined in such a way that we can use this equation to evaluate the real integrals in equation (A13.1). The last integral in equation (A13.8) is zero, due to the factor $\exp(-R^2\sin 2\theta)$ in the integrand, taken in the limit $R \to \infty$, provided $\varphi < \pi/2$. Consider the second integral,

$$I = e^{i\varphi} \int_\infty^0 dr \exp[ir^2(\cos 2\varphi + i\sin 2\varphi)]. \quad (A13.9)$$

Here, the integral becomes real, and can be evaluated, if $\varphi = \pi/4$, so that $\cos(2\varphi) = \cos(\pi/2) = 0$. We then have

$$I = e^{i\pi/4}\left(-\frac{1}{2}\right)\int_{-\infty}^\infty dr\, e^{-r^2} = -\frac{1}{2}e^{i\pi/4}(\pi)^{1/2}. \quad (A13.10)$$

Returning to equation (A13.8), we now have

$$0 = \int_0^\infty dx\, e^{ix^2} - \frac{\pi^{1/2}}{2} e^{i\pi/4},$$

or

$$\int_0^\infty dx \cos(x^2) + i \int_0^\infty dx \sin(x^2) = \frac{\pi^{1/2}}{2}\left\{\cos\left(\frac{\pi}{4}\right) + i \sin\left(\frac{\pi}{4}\right)\right\}$$

$$= \frac{1}{2}\left(\frac{\pi}{2}\right)^{1/2}(1+i). \tag{A13.11}$$

Equating real and imaginary parts of equation (A13.11), we have

$$\int_0^\infty dx \, \frac{\cos}{\sin}(x^2) = \frac{1}{2}\left(\frac{\pi}{2}\right)^{1/2}, \qquad \text{QED.} \tag{A13.12}$$

Chapter 14

Charge density waves in solids

14.1 Introduction

The periodicity of the electronic density in crystals is a consequence of the translationally invariant symmetry of the atomic ordering. It is manifest in the solution, in terms of Bloch functions, for the many-electron problem in the presence of static nuclei. We shall see that nuclear vibrations, or phonons, introduce a further effective potential into the problem, and that this potential may tend to induce a periodicity into the electronic distribution that is distinct from that of the static-crystal symmetry. This result was first articulated by Fröhlich (1954). It had been presaged in work by Slater (1951) and was described qualitatively by Peierls in 1955. The approach of this chapter is based on a study by the present author [Vail (1964)] that examined the relationship between the Fermi level and charge-density-wave periodicity on the one hand, and stability of the charge density wave on the other.

As we shall see, the charge density wave ground state of an interacting electron gas, when it exists, is fundamentally distinct from the more common uniform state. Both theoretically and experimentally, such a system provides a fertile situation for the study of fundamental mechanisms in the solid state. The thermodynamic transition from the normal to the charge density wave state is evidenced in the specific heat, in the Kohn anomaly [Kohn (1959)] of the phonon spectrum, in commensurate or incommensurate perturbation of the crystal lattice, in quasiparticle excitation processes, and in electronic transport characteristics. Many special aspects of charge density wave systems are also relevant to the fields of superconductivity and spin density waves. There is extensive literature, dating back many years, on experimental studies of charge density waves in solids [see for example the review by Wilson and Yoffe (1969)]. The current archival literature in condensed matter physics presents frequent examples of new studies. An up-to-date pedagogical treatment of the subject is given by Grüner (1994). An earlier work that gives a wide-ranging view of research topics is Hutiray and Sólyom (1985). A more recent work discusses charge density waves in nanoscale systems [Kim *et al.* (2001)].

    In the present work, we discuss the charge density wave ground state of a solid using a simple mathematical approximation to the one-dimensional model. In section 14.2 we show how electron–phonon interaction can modify the effective electron–electron interaction. In section 14.3 we apply the Hartree approximation to determine the uniform-density solution and to illustrate the periodic (charge density wave) case. In section 14.4 we solve the Hartree equation in the form of the Mathieu equation for the periodic case, using perturbation theory. In section 14.5 we discuss the analytical features of the result in the context of perturbation theory, emphasizing the importance of a physically correct unperturbed state as a starting point.

14.2   Effective electron–electron interaction

In section 7.2 we laid out the hamiltonian of a solid in terms of nuclei and electrons, equations (7.2)–(7.6). We then averaged this hamiltonian over the electronic state, defined in terms of the average field version of the adiabatic approximation, section 7.3, equation (7.10) with equation (7.8), to obtain the effective hamiltonian for the nuclei, equations (7.11) and (7.12). Suppose we have a metallic system such that, for example in Hartree–Fock approximation, equations (12.93) and (12.94), the single-particle Fock eigenvalues for occupied states consist of one or more filled valence bands, and a partially filled conduction band, all represented schematically in figure 14.1, in a one-dimensional model. Now we want to concentrate on the conduction electrons, and think of the valence electrons as bound with

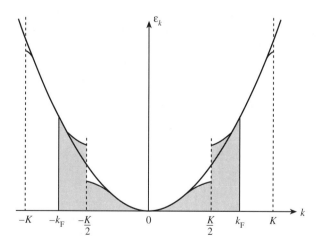

Figure 14.1. Valence and conduction bands of a one-dimensional 'metal', showing electron single-particle energies ε_k, band gaps at $k = \pm K/2$, and the Fermi level at $k = \pm k_\mathrm{F}$. Shaded area represents total energy.

the nuclei to constitute atomic cores. It is a tricky theoretical and computational problem to do this with a realistic treatment of the electrons. If we can do it in some acceptable approximation, however, we can then consider a system of n conduction electrons interacting with atomic cores, rather than with nuclei. Equations (7.2)–(7.6) will then be replaced by an effective hamiltonian H:

$$H = (H_c + H_e + H_{ce}), \tag{14.1}$$

$$H_c = \left\{ \sum_{J=1}^{N_1} \left(-\frac{\hbar^2 \nabla_J^2}{2M_J} \right) + W(\underline{R}) \right\}, \tag{14.2}$$

$$H_e = \left\{ \sum_{j=1}^{n} \left(-\frac{\hbar^2 \nabla_j^2}{2m} \right) + \frac{e^2}{4\pi\varepsilon_0} \frac{1}{2} \sum_{j,j'}^{(n)}{}' |\vec{r}_j - \vec{r}_{j'}|^{-1} \right\}, \tag{14.3}$$

$$H_{ce} = V(\underline{r}, \underline{R}). \tag{14.4}$$

In equation (14.1), we have contributions H_c for cores, H_e for conduction electrons, and H_{ce} for core–electron interaction. In equation (14.2), $W(\underline{R})$ is the core–core interaction energy, where the cores are assumed to be rigid, and M_J is the mass of the core centered on nuclear or core position $\vec{R}_J$.

We now consider small oscillations of core positions $\underline{R}$, relative to reference positions $\underline{R}_0$, by small displacements $\underline{u}$. Then in equation (14.2) we have, in harmonic approximation,

$$W(\underline{R}) \approx \{ W_0 + \underline{W}_1^{\mathrm{T}} \cdot \underline{u} + \tfrac{1}{2} \underline{u}^{\mathrm{T}} \cdot \underline{\underline{W}}_2 \cdot \underline{u} \} \tag{14.5}$$

where

$$W_0 = W(\underline{R}_0), \tag{14.6}$$

$$\underline{W}_1(\underline{R}_0) = \left(\frac{\partial W}{\partial \underline{R}} \right)_{\underline{R}=\underline{R}_0}, \tag{14.7}$$

$$\underline{\underline{W}}_2(\underline{R}_0) = \left(\frac{\partial^2 W}{\partial \underline{R} \, \partial \underline{R}} \right)_{\underline{R}=\underline{R}_0}. \tag{14.8}$$

The matrix notation of equation (14.5) is the same as that of section 7.4. Similarly, in equation (14.4), we have

$$V(\underline{r}, \underline{R}) \approx \{ V_0 + \underline{V}_1^{\mathrm{T}} \cdot \underline{u} + \tfrac{1}{2} \underline{u}^{\mathrm{T}} \cdot \underline{\underline{V}}_2 \cdot \underline{u} \}, \tag{14.9}$$

where

$$V_0(\underline{r}, \underline{R}_0) = V(\underline{r}, \underline{R}_0), \tag{14.10}$$

$$\underline{V}_1(\underline{r}, \underline{R}_0) = \left(\frac{\partial V}{\partial \underline{R}} \right)_{\underline{R}=\underline{R}_0}, \tag{14.11}$$

$$\underline{\underline{V}}_2(\underline{r}, \underline{R}_0) = \left(\frac{\partial^2 V}{\partial \underline{R} \, \partial \underline{R}} \right)_{\underline{R}=\underline{R}_0}. \tag{14.12}$$

We now simplify the effective hamiltonian given by equations (14.1)–(14.4) with equations (14.5)–(14.12). First, in equation (14.10), $V_0(\underline{r}, \underline{R}_0)$ represents the interaction of conduction electrons with atomic cores in the periodic crystal lattice. Let us represent this simply in terms of the band mass m_b for the conduction electrons [see any textbook of solid state physics]. Then, from equations (14.3), (14.4) and (14.10),

$$\left\{ -\frac{\hbar^2}{2m} \sum_j \nabla_j^2 + V_0(\underline{r}, \underline{R}_0) \right\} \rightarrow \left\{ -\frac{\hbar^2}{2m_b} \sum_j \nabla_j^2 \right\}. \tag{14.13}$$

Without loss of generality, let $W(\underline{R}_0)$ be the reference value for energy. Then $W_0 = 0$. Then from equations (14.5) and (14.9), introduce $\underline{V}_1'$ and $\underline{V}_2'$, as follows:

$$\underline{V}_1'(\underline{r}, \underline{R}_0) = \{\underline{V}_1(\underline{r}, \underline{R}_0) + \underline{W}_1(\underline{R}_0)\}, \tag{14.14}$$

$$\underline{V}_2'(\underline{r}, \underline{R}_0) = \{\underline{V}_2(\underline{r}, \underline{R}_0) + \underline{W}_2(\underline{R}_0)\}. \tag{14.15}$$

At this stage, we have the following effective hamiltonian:

$$H \approx \left\{ -\frac{\hbar^2}{2m_b} \sum_j \nabla_j^2 + \frac{e^2}{8\pi\varepsilon_0} {\sum_{j,j'}}' |\vec{r}_j - \vec{r}_{j'}|^{-1} + \underline{V}_1'^{\mathrm{T}}(\underline{r}, \underline{R}_0) \cdot \underline{u} \right.$$

$$\left. + \sum_J \frac{P_J^2}{2M_J} + \frac{1}{2} \underline{u}^{\mathrm{T}} \cdot \underline{V}_2'(\underline{r}, \underline{R}_0) \cdot \underline{u} \right\}, \tag{14.16}$$

where we have introduced the momentum operator $\vec{P}_J$ for cores,

$$\vec{P}_J = (-i\hbar\vec{\nabla}_J). \tag{14.17}$$

We can see that the core reference positions $\underline{R}_0$ are not determined in any meaningful way from equation (14.16) by the usual prescription,

$$\left(\frac{\partial H}{\partial \underline{u}} \right)_{\underline{u}=0} = \underline{V}_1'(\underline{r}, \underline{R}_0) = 0, \tag{14.18}$$

because of the $\underline{r}$-dependence from the conduction electrons. We take $\underline{R}_0$ to be the average core positions at finite temperature, in the presence of the electrons. With this prescription, the linear term in $\underline{u}$ remains non-zero in equation (14.16), indicating that the small oscillations of core positions in the presence of conduction electrons are not centered on $\underline{R}_0$ in any particular electronic configuration.

We can eliminate the linear term in $\underline{u}$ in equation (14.16), therefore, by a formal transformation to new variables $\underline{u}'$ that are translated from the original variables $\underline{u}$, as

$$\underline{u}' = (\underline{u} + \underline{\varphi}). \tag{14.19}$$

We substitute for $\underline{u}$ in terms of $\underline{u}'$, equation (14.19), into equation (14.16). The linear terms in $\underline{u}'$ are

$$(\underline{\underline{V}}_1'^{\mathrm{T}} - \underline{\varphi}^{\mathrm{T}} \cdot \underline{\underline{V}}_2') \cdot \underline{u}'. \tag{14.20}$$

Setting the coefficient of $\underline{u}'$ equal to zero in equation (14.20) gives

$$\underline{\varphi} = (\underline{\underline{V}}_2')^{-1} \cdot \underline{\underline{V}}_1'. \tag{14.21}$$

In obtaining equation (14.21), we have used

$$\underline{\underline{V}}_2'^{\mathrm{T}} = \underline{\underline{V}}_2'; \tag{14.22}$$

see equations (14.15), (14.12) and (14.8). Also note, from equations (14.21), (14.15), and (14.14), that φ, and therefore $\underline{u}'$, equation (14.19), are functions of $\underline{r}$ and $\underline{R}_0$:

$$\underline{u}'(\underline{r}, \underline{R}_0) = \{\underline{u} + \underline{\varphi}(\underline{r}, \underline{R}_0)\}$$
$$= \{\underline{u} + [\underline{\underline{V}}_2'(\underline{r}, \underline{R}_0)]^{-1} \cdot \underline{\underline{V}}_1'(\underline{r}, \underline{R}_0)\}. \tag{14.23}$$

The substitution for $\underline{u}$ from equation (14.19) into equation (14.16) also introduces a term

$$\{\tfrac{1}{2}\underline{\varphi}^{\mathrm{T}} \cdot \underline{\underline{V}}_2' \cdot \underline{\varphi} - \underline{\underline{V}}_1'^{\mathrm{T}} \cdot \underline{\varphi}\} \tag{14.24}$$

so that the effective hamiltonian, equation (14.16) with equations (14.19) and (14.21), becomes

$$H \approx \left\{ -\frac{\hbar^2}{2m_{\mathrm{b}}} \sum_j \nabla_j^2 + \frac{e^2}{8\pi\varepsilon_0} \sum_{j,j'}{}' |\vec{r}_j - \vec{r}_{j'}|^{-1} \right. \tag{14.25}$$

$$-\frac{1}{2}\underline{\underline{V}}_1'^{\mathrm{T}} \cdot [\underline{\underline{V}}_2']^{-1} \cdot \underline{\underline{V}}_1' \tag{14.26}$$

$$\left. + \sum_j \frac{P_j^2}{2M_J} + \frac{1}{2}\underline{u}'^{\mathrm{T}} \cdot \underline{\underline{V}}_2' \cdot \underline{u}' \right\}. \tag{14.27}$$

In equation (14.27), the fact that the modified phonon variable $\underline{u}'$, equation (14.23), and the force constant matrix $\underline{\underline{V}}_2'$, equation (14.15) are dependent on conduction electron coordinates $\underline{r}$ represents the renormalization of the phonon variables $\underline{u}$ and the force constant matrix $\underline{\underline{W}}_2$ by the conduction electrons.

We now make some major approximations, in order to simplify the mathematical form of the problem while retaining the physical basis of charge density waves. We treat the renormalized phonons as classical phonons. Then at zero temperature (Kelvin), we have $\underline{P} = \dot{\underline{u}}' = 0$. The effective hamiltonian, equations (14.25)–(14.27), reduces to an electronic hamiltonian:

$$H \approx \left\{ -\frac{\hbar^2}{2m_{\mathrm{b}}} \sum_j \nabla_j^2 + \frac{e^2}{8\pi\varepsilon_0} \sum_{j,j'}{}' |\vec{r}_j - r_{j'}|^{-1} - v'(\underline{r}) \right\}. \tag{14.28}$$

In equation (12.28)

$$v'(\underline{r}) = \{ \tfrac{1}{2} \underline{V}_1'^{\mathrm{T}} \cdot [\underline{V}_2']^{-1} \cdot \underline{V}_1' \}, \tag{14.29}$$

which is also a function of $\underline{R}_0$, the equilibrium configuration for the cores alone. The potential energy $v'(\underline{r})$, equation (14.29), is an effective many-electron potential energy arising from dynamical electron–phonon interaction: see equation (14.14), equation (14.15) and equation (14.4). We can imagine fitting $v'(\underline{r})$ as a sum of single-particle, two-particle, ... n-particle terms. We then assume that the two-particle term, which we denote v'', is dominant, and we neglect all others. We further assume that the two-body terms are functions of $|\vec{r}_j - \vec{r}_{j'}|$, i.e. two-body distances only. Then

$$v'(\underline{r}) \approx \frac{1}{2} \sum_{j,j'}{}' v''(|\vec{r}_j - \vec{r}_{j'}|). \tag{14.30}$$

When this is incorporated in equation (14.28), we have

$$H \approx \left\{ -\frac{\hbar^2}{2m_{\mathrm{b}}} \sum_j \nabla_j^2 - \frac{1}{2} \sum_{j,j'}{}' v(|\vec{r}_j - r_{j'}|) \right\}, \tag{14.31}$$

where

$$v(r) = \left\{ v''(r) - \frac{e^2}{4\pi\varepsilon_0 r} \right\}. \tag{14.32}$$

The above discussion shows how electron–phonon interaction can modify the effective pairwise interaction among the electrons from pure Coulomb repulsion to an effective interaction $v(r)$, as in equation (14.32). We reiterate that this depends implicitly on the core-position configuration $\underline{R}_0$.

14.3 The Hartree equation: uniform and periodic cases

14.3.1 The Hartree approximation

The Hartree approximation represents an intuitive simplification of the Hartree–Fock approximation, which we have discussed in sections 12.2 and 12.3. In the Hartree approximation, the many-electron wave function

$\psi(\underline{r})$ is approximated by a single product,

$$\psi(\underline{r}) \approx \prod_{j=1}^{n} \varphi_j(\underline{r}_j), \tag{14.33}$$

where $\underline{r}_j = (\vec{r}_j, s_j)$, as at the beginning of section 12.2.1. Note that the difference between equation (14.33) and the Hartree–Fock approximation is that here the product is not antisymmetrized: see equations (12.25) and (12.26). Strictly speaking, the particles are distinguishable in a wave function like equation (14.33), where the jth particle is in the jth orbital φ_j. This is contrary to the eminently logical quantum-mechanical principle that identical particles are indistinguishable. Nevertheless, we persist with it. In order to satisfy the Pauli principle approximately, we require the set of orbitals $(\varphi_1, \varphi_2, \dots, \varphi_n)$ to be distinct: we choose them to be an orthonormal set. The derivation carried out in section 12.3.1 to determine the optimal manifold $\{\varphi_j\}$, in terms of total energy minimization, can be adapted to the Hartree approximation, equation (14.33). The only difference is that the exchange term does not come into the two-body part of the energy, equation (12.62). The result, from the effective hamiltonian equation (14.31), is the Hartree equation:

$$\left\{ -\frac{\hbar^2}{2m_b} \nabla^2 - \sideset{}{'}\sum_{j'} \langle j'|v|j'\rangle \right\} |j\rangle = \varepsilon_j |j\rangle. \tag{14.34}$$

In equation (14.34) we have introduced Dirac notation:

$$\varphi_j(\vec{r}) \rightarrow |j\rangle. \tag{14.35}$$

The summation in equation (14.34) is restricted to $j' \neq j$ by the prime on $\sum$. If $v(r=0)$, equation (14.32), were finite, we should be able to include the term $j' = j$ in the effective hamiltonian, equation (14.31), and measure total energy relative to the value $[nv(0)/2]$ in which case, the self-consistent field in the Hartree equation (14.34) would include the term $j' = j$, and be the same for all j. We assume that this is the case. Then the total energy E is

$$E \approx \langle\psi|H|\psi\rangle = \left\{ \sum_j \varepsilon_j + \frac{1}{2} \sum_{j,j'} \langle jj'|v|jj'\rangle \right\}. \tag{14.36}$$

We can recast the Hartree equation (14.34) and the Hartree energy E, equation (14.36) in terms of the particle density, as defined in equation (12.44),

$$\rho_{op}(\vec{r}) = \sum_j \delta(\vec{r} - \vec{r}_j). \tag{14.37}$$

Let us Fourier analyse the potential v,

$$v(r) = \sum_{\vec{k}} v_{\vec{k}} \, e^{i\vec{k} \cdot \vec{r}}, \tag{14.38}$$

where

$$v_{\vec{k}} = \frac{1}{\Omega} \int d^3 r' \, v(r') \, e^{-i\vec{k} \cdot \vec{r}'}, \tag{14.39}$$

where Ω is the Born–von Karmann volume: see section 7.5.1, equations (7.22)–(7.31). If we Fourier analyse $\rho_{\mathrm{op}}(\vec{r})$ in the same way, we have

$$\rho_{\mathrm{op}} = \sum_{\vec{k}} \rho_{\vec{k}} \, e^{i\vec{k} \cdot \vec{r}}. \tag{14.40}$$

From the general form of Fourier's theorem, equations (14.38) and (14.39), we have the well-known representation of the delta function,

$$\frac{1}{\Omega} \sum_{\vec{k}} e^{i\vec{k} \cdot (\vec{r} - \vec{r}')} = \delta(\vec{r} - \vec{r}'). \tag{14.41}$$

Then, from equation (14.37) with equation (14.41),

$$\rho_{\mathrm{op}}(\vec{r}) = \sum_{\vec{k}} \left\{ \frac{1}{\Omega} \sum_{j} e^{-i\vec{k} \cdot \vec{r}_j} \right\} e^{i\vec{k} \cdot \vec{r}}. \tag{14.42}$$

Comparing equations (14.40) and (14.42), we conclude that

$$\rho_{\vec{k}} = \frac{1}{\Omega} \sum_{j} e^{-i\vec{k} \cdot \vec{r}_j}. \tag{14.43}$$

Now the second term in the total energy, equation (14.36), is

$$\frac{1}{2} \sum_{j,j'} \langle \psi | v(|\vec{r}_j - \vec{r}_{j'}|) | \psi \rangle = \frac{1}{2} \sum_{j,j'} \langle \psi | \sum_{\vec{k}} v_{\vec{k}} \, e^{i\vec{k} \cdot (\vec{r}_j - \vec{r}_{j'})} | \psi \rangle$$

$$= \frac{1}{2} \sum_{\vec{k}} v_{\vec{k}} \left\{ \sum_{j} \int d^3 r_j \, \varphi_j^*(\vec{r}_j) \, e^{i\vec{k} \cdot \vec{r}_j} \, \varphi_j(\vec{r}_j) \right\}$$

$$\cdot \left\{ \sum_{j'} \int d^3 r_{j'} \, \varphi_{j'}^*(\vec{r}_{j'}) \, e^{-i\vec{k} \cdot \vec{r}_{j'}} \, \varphi_{j'}(\vec{r}_{j'}) \right\} \tag{14.44}$$

where we have used equation (14.39) and equations (14.35) and (14.33). Now consider equation (14.43). We find

$$\langle \psi | \rho_{\vec{k}} | \psi \rangle = \langle \psi | \frac{1}{\Omega} \sum_{j} e^{-i\vec{k} \cdot \vec{r}_j} | \psi \rangle$$

$$= \frac{1}{\Omega} \sum_{j} \int d^3 r_j \, \varphi_j^*(\vec{r}_j) \, e^{-i\vec{k} \cdot \vec{r}_j} \, \varphi_j(\vec{r}_j). \tag{14.45}$$

Thus, we identify the quantities in { } brackets in equation (14.44) as

$$\frac{1}{2} \sum_{j,j'} \langle \psi | v(|\vec{r}_j - \vec{r}_{j'}|) | \psi \rangle = \frac{\Omega^2}{2} \sum_{\vec{k}} v_{\vec{k}} |\langle \rho_{\vec{k}} \rangle|^2 \qquad (14.46)$$

where

$$\langle \rho_{\vec{k}} \rangle \equiv \langle \psi | \rho_{\vec{k}} | \psi \rangle, \qquad (14.47)$$

and we have used equation (14.45), plus the fact that, from equation (14.43),

$$\rho_{\vec{k}} = \rho^*_{-\vec{k}}. \qquad (14.48)$$

Combining equations (14.46) and (14.36), we obtain the total energy algorithm for the Hartree approximation,

$$E = \left\{ \sum_j \varepsilon_j + \frac{n^2}{2\rho_0^2} \sum_{\vec{k}} v_{\vec{k}} |\langle \rho_{\vec{k}} \rangle|^2 \right\}, \qquad (14.49)$$

where we have introduced the mean density ρ_0, in terms of the number of electrons n and the volume Ω,

$$\rho_0 = \frac{n}{\Omega}. \qquad (14.50)$$

We can similarly rewrite the self-consistent field term in the Hartree equation (14.34) as

$$\sum_{j'} \langle j' | v | j' \rangle = \sum_{j'} \int \mathrm{d}^3 r_{j'}\, \varphi^*_{j'}(r_{j'}) v(|\vec{r}_j - \vec{r}_{j'}|) \varphi_{j'}(\vec{r}_{j'})$$

$$= \Omega \sum_{\vec{k}} v_{\vec{k}} \langle \rho_{-\vec{k}} \rangle\, \mathrm{e}^{\mathrm{i}\vec{k}\cdot\vec{r}_j}, \qquad (14.51)$$

having used equations (14.38), (14.45), and (14.47). Then equation (14.34) with equation (14.50) becomes

$$\left\{ -\frac{\hbar^2}{2m_\mathrm{b}} \nabla^2 - \frac{n}{\rho_0} \sum_{\vec{k}} v_{\vec{k}} \langle \rho_{-\vec{k}} \rangle\, \mathrm{e}^{\mathrm{i}\vec{k}\cdot\vec{r}} \right\} | j \rangle = \varepsilon_j | j \rangle. \qquad (14.52)$$

14.3.2   The uniform solution

In the absence of electron–electron interaction $v(r)$, equation (14.32), the Hartree equation (14.52) becomes

$$-\frac{\hbar^2}{2m_\mathrm{b}} \nabla^2 | j \rangle = \varepsilon_j | j \rangle. \qquad (14.53)$$

We now consider the one-dimensional case, involving the x-axis. Then, in position representation,

$$\varphi_j(\vec{r}) \equiv \langle \vec{r} | j \rangle, \qquad (14.54)$$

we have

$$-\frac{\hbar^2}{2m_b}\frac{d^2}{dx^2}\varphi_j = \varepsilon_j\varphi_j. \tag{14.55}$$

This has solutions

$$\varphi_j(x) \sim e^{ik_j x}, \tag{14.56}$$

with

$$\varepsilon_j = \frac{\hbar^2 k_j^2}{2m_b}. \tag{14.57}$$

Suppose there are n electrons in a *length* Ω: see equation (14.50). We apply periodic boundary conditions,

$$\varphi_j(x+\Omega) = \varphi_j(x). \tag{14.58}$$

Substituting equation (14.56) into (14.58) we have

$$e^{ik_j\Omega} = 1, \tag{14.59}$$

or

$$k_j = \frac{2\pi}{\Omega}n_j, \tag{14.60}$$

where n_j is an integer: positive, negative or zero. The ground state of the n-electron system, the state of lowest total energy, is found from equation (14.36), here with $v = 0$. It consists of particles in the n states $\varphi_j(x)$, equation (14.56) with equation (14.60), with the smallest values of ε_j, equation (14.57). To be a bit more realistic, we let two particles, of spins $\pm\hbar/2$, occupy each state φ_j, equation (14.56). The single-particle energy spectrum, equation (14.57), as a function of the quasi-continuous variable k (for large Ω: see equation (14.60)), is parabolic. The ground state encompasses k-values symmetrical about $k = 0$. There are two states of opposite spins in a k-space interval $(2\pi/\Omega)$: see equation (14.60). The density of states $d(k)$ in k-space is therefore

$$d(k) = 2\left(\frac{2\pi}{\Omega}\right)^{-1}. \tag{14.61}$$

There must be n states in a region $(-k_F \leq k \leq k_F)$, where k_F, the *Fermi wave number*, is the highest-energy occupied single-particle state for the ground state of the n-electron system. Thus

$$n = \int_{-k_F}^{k_F} dk\, d(k) = 2\left(\frac{\Omega}{2\pi}\right)(2k_F) = \frac{2\Omega k_F}{\pi}, \tag{14.62}$$

or, using equation (14.50),

$$k_F = \frac{\pi}{2}\rho_0. \tag{14.63}$$

The ground state energy E is

$$E = \sum_{j=1}^{n} \varepsilon_j = \int_{-k_F}^{k_F} dk \, d(k) \frac{\hbar^2 k^2}{2m_b} = \frac{\Omega}{\pi} \frac{\hbar^2}{2m_b} \frac{2k_F^3}{3},$$

having used equation (14.36) with $v = 0$, and equations (14.57) and (14.51). If we define the Fermi energy ε_F by

$$\varepsilon_F = \frac{\hbar^2 k_F^2}{2m_b}, \tag{14.64}$$

then the total energy can be written in the alternative forms

$$E = \frac{n}{3} \varepsilon_F = \frac{n(\pi\hbar\rho_0)^2}{24m_b}. \tag{14.65}$$

The first form shows that the total energy is *extensive*, that is, proportional to the size of the system, if one notes than $n = \rho_0 \Omega$.

It is trivial to show that the normalized single-particle Hartree eigenfunctions φ_j, equation (14.56), are

$$\varphi_j(x) = \Omega^{-1/2} e^{ik_j x}. \tag{14.66}$$

Then from equation (14.33), the many-electron wave function is

$$\psi(\underline{x}) = \Omega^{-n/2} \prod_{j=1}^{n} e^{ik_j x} \tag{14.67}$$

and the electronic probability density $|\psi|^2$ is

$$|\psi|^2 = \Omega^{-n}. \tag{14.68}$$

In equation (14.68) we see explicitly that the probability density for particles to be at $\vec{r}_1, \ldots, \vec{r}_n$ is uniform; that is, independent of all the electronic coordinates. This has the consequence, of course, that the expectation value for the particle density, proportional to the charge density, from equations (14.37) and (14.67), is

$$\langle \psi | \rho_{op} | \psi \rangle = \frac{1}{\Omega^n} \sum_{j=1}^{n} \int dx_1, \ldots, dx_n \, \delta(x - x_j)$$

$$= \Omega^{-n} \Omega^{n-1} \sum_{j=1}^{n} \int_0^{\Omega} dx_j \, \delta(x - x_j)$$

$$= \frac{n}{\Omega} = \rho_0. \tag{14.69}$$

The result of the last integration in equation (14.69) is valid, of course, only in the limit $\Omega \to \infty$. The fact that the expectation value is uniform in equation

(14.69), taken together with the Fourier expansion equation (14.40), shows that

$$\langle \rho_k \rangle = 0 \qquad \text{for } k \neq 0 \qquad (14.70\text{a})$$

$$= \rho_0 \qquad \text{for } k = 0. \qquad (14.70\text{b})$$

Equation (14.70a) follows directly from the definition of ρ_k, equation (14.43) as well, when taken with equation (14.67).

Equations (14.70) have an important consequence for the Hartree equation with non-zero interactions, equation (14.52), which in one dimension is

$$\left\{ -\frac{\hbar^2}{2m_b} \frac{d^2}{dx^2} - \frac{n}{\rho_0} \sum_k v_k \langle \rho_{-k} \rangle e^{ikx} \right\} |j\rangle = \varepsilon_j |j\rangle. \qquad (14.71)$$

We see that, if we substitute the uniform solution into the self-consistent field terms in $\langle \rho_{-k} \rangle$, then equation (14.71) becomes

$$-\frac{\hbar^2}{2m_b} \frac{d^2}{dx^2} \varphi_j = (\varepsilon_j + nv_0)\varphi_j, \qquad (14.72)$$

so that the uniform solution equation (14.56) is a solution of the Hartree equation even with non-zero electron interaction, where now, however, the eigenvalue ε_j and eigenfunction φ_j are

$$\varepsilon_j = \left(\frac{\hbar^2 k_j^2}{2m_b} - nv_0 \right), \qquad \varphi_j = \Omega^{-1/2} e^{ik_j x}. \qquad (14.73)$$

In equations (14.72) and (14.73) v_0 is

$$v_0 = (v_k)_{k=0} = \frac{1}{\Omega} \int_0^\Omega dx \, v(x), \qquad (14.74)$$

from equation (14.39), adapted to one dimension. Thus v_0 is simply the spatial average of the pairwise electronic interaction. The only difference between the total energy in the uniform state with arbitrary pairwise electronic interaction, and that with none, is at most an additive constant: see equations (14.74), (14.63) and (14.49). The uniform state, equation (14.67), is therefore a solution in the Hartree approximation for the pairwise interacting electron gas.

14.3.3 The periodic solution

Referring to equation (14.71), suppose that there were a self-consistent solution such that, for a particular value of k, say $k = K$, the coefficient $v_K \langle \rho_{-K} \rangle$ is dominant in comparison with other k values, and that the self-consistent field is symmetric in $\vec{r}$, as v is: see equation (14.38). Then

$$v_{-K} \langle \rho_K \rangle = v_K \langle \rho_{-K} \rangle, \qquad (14.75)$$

and the Hartree equation reduces to

$$\left\{ -\frac{\hbar^2}{2m_b}\frac{d^2}{dx^2} - \frac{2n}{\rho_0}v_K\langle\rho_{-K}\rangle\cos(Kx) \right\}\varphi_j = \tilde{\varepsilon}_j\varphi_j, \qquad (14.76)$$

where

$$\tilde{\varepsilon}_j = (\varepsilon_j + nv_0). \qquad (14.77)$$

Equation (14.76) is a form of the Bloch equation for electrons in a periodic potential. From elementary solid state theory, and to be shown in the next section, the Bloch eigenstates are of the form

$$\varphi_j(x) = e^{ik_j x}u_j(x) \qquad (14.78)$$

where $u_j(x)$ is periodic with wavelength

$$\lambda = \frac{2\pi}{K}, \qquad (14.79)$$

which is the periodicity of the potential: see equation (14.76). It follows from equation (14.78) with equation (14.33) that, similarly to equation (14.69), the charge density is

$$\langle\psi|\rho_{op}|\psi\rangle = \Omega^{-n}\,\Omega^{n-1}\sum_{j=1}^{n}\int_0^\Omega dx_j\,\delta(x - x_j)|e^{ik_j x_j}\,u_j(x_j)|^2$$

$$= \frac{1}{\Omega}\sum_{j=1}^{n}|u_j(x)|^2. \qquad (14.80)$$

which is periodic in x because $u_j(x)$ are. Let us at this point compare the results of the uniform solution with the postulated periodic case. In the uniform case we have

$$\tilde{\varepsilon}_k = \left(\frac{\hbar^2 k^2}{2m_b}\right), \qquad \varphi_k = \Omega^{-1/2}\,e^{ikx}. \qquad (14.81)$$

from equation (14.73) with (14.77) where here and henceforth we shall label Hartree eigenstates and eigenvalues by k rather than j. In the periodic case we have $\tilde{\varepsilon}_k$, an eigenvalue of the Bloch equation (14.76), with eigenfunction of the form

$$\varphi_k(x) = e^{ikx}\,u_k(x). \qquad (14.82)$$

We know, and shall later show, that in the periodic case, the Bloch eigenvalue spectrum will have gaps at $k = (\pm mK/2)$, with m an integer and K the primitive translation vector of the reciprocal lattice of the periodic potential, as illustrated in figure 14.1. Below the gap, introduction of a periodic potential lowers the single-particle energy relative to the parabola of the uniform case, equation (14.81); just above the gap the energy is raised. We therefore arrive at the following conclusion. If the interacting one-dimensional electron gas

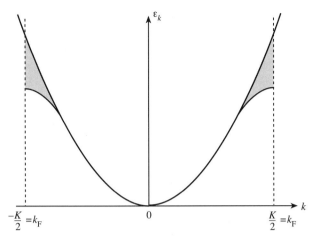

Figure 14.2. Lowering of total electronic energy (shaded area), relative to the uniform solution, of a periodic solution with first band gap at $k = K/2 = k_F$, in one dimension.

has a periodic state such that the first gap occurs at $|k| = k_F$, then the total energy will be lower than that of the uniform solution. The difference in energy will be given by the shaded area in figure 14.2. In that case, the ground state will be this particular periodic state: that is why we have limited the previous statement to the first gap, although the statement may still be correct for higher gaps.

We note that the periodicity associated with $K/2 = k_F$ has a wavelength λ in x-space of

$$\lambda = \frac{2\pi}{K} = \frac{\pi}{k_F}. \tag{14.83}$$

In general, this has no simple relationship with the periodicity of the static crystal lattice, i.e. the atomic ordering, which has been removed from the problem by introduction of the band mass m_b, in order to illustrate the charge density wave effect most simply. However, if we consider a monatomic one-dimensional 'crystal' with one atom in the primitive unit cell, and one conduction electron per atom, the average distance between atomic cores in the crystal is

$$\Lambda = \frac{\Omega}{n}. \tag{14.84}$$

From equation (14.83) with equation (14.62), we have for the wavelength of the electronic density distribution in the periodic case,

$$\lambda = \frac{\pi}{k_F} = \frac{2\Omega}{n}, \tag{14.85}$$

i.e. twice the average distance in the atomic ordering. For this 'crystal' as a whole, therefore, electrons plus atomic cores, the periodicity will be *twice* the average interatomic distance, even though the atomic cores are identical. In equilibrium, the cores will not be evenly spaced with interatomic distances Λ, but rather will be spaced according to a pattern $(\Lambda + \alpha)$, $(\Lambda - \alpha)$, $(\Lambda + \alpha)$, etc. In experimentally observed charge density wave materials, this perturbation of the crystalline order can be observed upon transition to the charge density wave state. Equation (14.85) represents the simplest case of a *commensurate charge density wave*, where the wavelength is a rational multiple of the average atomic spacing. It may also happen that atomic ordering, expressed more accurately than simply through the band mass, prevents formation of a self-consistent field whose periodicity places a band gap at $k = k_F$. The total energy in a charge density wave state may still be lower in such a system than that of the 'uniform' state, however, provided that, in the one-dimensional illustration, only a small number of states are occupied above the gap, so that the total area under the curve in figure 14.2 up to $k_F \gtrsim K/2$ is still less than that under the parabola for $k \leq k_F$.

In two and three dimensions, a new consideration enters the problem. In three dimensions, with a periodic potential, the regions of forbidden energy (the band gaps) occur for $\vec{k}$-values that define a set of planes in k-space (reciprocal space) as orthogonal bisectors of the reciprocal lattice vectors. These planes constitute *Brillouin zone boundaries*. Working outward from the origin of k-space one first encounters the first Brillouin zone boundary. Continuing outward, one encounters the second Brillouin zone boundary, and so on. The Brillouin zones are therefore *volumes* in the three-dimensional case, bounded by *surfaces*. These matters are covered in standard discussions of crystal structure and band theory: see for example Grosso and Pastori Parravicini (2000). In a two-dimensional system, the Brillouin zones are areas, bounded by straight-line segments, and in one dimension the Brillouin zones are segments of the one-dimensional line, bounded by points $k = \pm(mK/2)$, as we have said. In figure 14.3 we show the first Brillouin zone boundary for a square Bravais crystal lattice in two dimensions. The first Brillouin zone is also square. (For a simple cubic crystal lattice the first Brillouin zone is a cube.) Thus in two dimensions, there is a band gap everywhere on the square boundary of the first Brillouin zone. For a weak periodic potential, the one-electron energies are approximately proportional to $|\vec{k}|^2$, and the band gaps are narrow. In such a case it commonly occurs that the energy in the corners of the first Brillouin zone is higher than that above the band gap at the center of a side of the zone: points A and B respectively in figure 14.3. Thus generally the question whether or not there will be a charge density wave ground state depends on details of the crystal structure and of the pairwise interaction for electrons. What is clear, however, is that if the electronic distribution spills over the band gap at some points without filling the Brillouin zone, then the material will not be an insulator: infinitely small energy excitations will be possible into the corners of the Brillouin zone, and to states adjacent (in k-space) to those that are occupied just

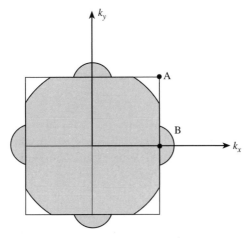

Figure 14.3. Illustration of how the Fermi region (shaded) in a two-dimensional charge density wave solution may fail to have an energy gap everywhere at its boundary, even though the first Brillouin zone (the square) contains exactly one state per electron.

above the gap, as in figure 14.3. Exactly similar remarks apply to the three-dimensional case, but with even greater likelihood of the ground state not being an insulator. We emphasize the possible origin of an insulator ground state associated with a charge density wave electronic configuration. It is an insulator state not due to the atomic ordering in the crystal, but rather due to an ordering induced by the characteristics of the effective pairwise interaction among conduction electrons.

14.4 Charge density waves: the Mathieu equation

14.4.1 The Mathieu equation

We now return to the one-dimensional example with a single dominant sinusoidal term in the periodic self-consistent field in the Hartree approximation, equation (14.76). This is in fact a form of the Mathieu equation (Mathieu, 1868). Apart from the self-consistency feature, our treatment follows that of Slater (1952). For the wave number K, we choose the value that puts the first band gap at k_F, namely $K = 2k_F$, equation (14.83). We introduce dimensionless variables and parameters as follows:

$$\eta = k_F x, \qquad q = \frac{k}{k_F}, \qquad \Gamma = \frac{v_K}{\varepsilon_F}, \qquad \varepsilon' = \frac{\tilde{\varepsilon}}{\varepsilon_F}$$

$$s = 4n \frac{v_K \langle \rho_{-K} \rangle}{\rho_0 \varepsilon_F}, \qquad (14.86)$$

where k_F and ε_F are given in equations (14.63) and (14.64), and $\tilde{\varepsilon}$ is given in equation (14.77). From equation (14.77) we should have

$$\tilde{\varepsilon}_k = (\varepsilon_k + nv_0), \tag{14.87}$$

labelling Hartree eigenstates by k rather than j, and therefore

$$\varepsilon' \equiv \varepsilon'_q. \tag{14.88}$$

In equations (14.86), Γ is a dimensionless form of the pairwise electron–electron coupling constant, and s is a dimensionless amplitude of the periodic potential. With these notations we have, from equation (14.76),

$$\left\{ -\frac{d^2}{d\eta^2} - \frac{1}{2}s\cos(2\eta) - \varepsilon'_q \right\}\varphi_q(\eta) = 0. \tag{14.89}$$

We now derive the properties of $\varphi_q(\eta)$, equation (14.89), that follow from the periodic potential. We introduce a complete orthonormal set of basis functions for the region $0 \le x \le \Omega$ with periodic boundary conditions. Consider the function $\xi_q(\eta)$,

$$\xi_q(\eta) \sim e^{iq\eta}, \tag{14.90}$$

where $q = k/k_F$; $\eta = k_F x$. Periodic boundary conditions require

$$qk_F\Omega = 2\pi m, \qquad m = 0, \pm 1, \pm 2, \ldots \tag{14.91}$$

Using equation (14.62) for k_F, this becomes

$$q = \frac{4m}{n}, \tag{14.92}$$

where n is the number of particles. Since we have $\eta = n\pi/2$ when $x = \Omega$, for orthonormality we consider

$$\int_0^{n\pi/2} d\eta\, e^{-iq\eta}\, e^{iq'\eta} = \int_0^{n\pi/2} d\eta \exp\left[\frac{4i}{n}(m' - m)\eta\right]$$

$$= \frac{n\pi}{2}\frac{e^{i\pi(m'-m)}\sin[(m'-m)\pi]}{[(m'-m)\pi]}. \tag{14.93}$$

In equation (14.93), if $m \ne m'$, the result is zero, from the sine. For the case $m = m'$ we consider

$$\lim_{x \to 0}\left(\frac{\sin x}{x}\right) = 1, \tag{14.94}$$

and conclude

$$\int_0^{n\pi/2} d\eta\, e^{-iq\eta}\, e^{iq'\eta} = \frac{n\pi}{2}\delta_{q,q'} \tag{14.95}$$

whence the normalized basis functions

$$\xi_q(\eta) = \left(\frac{2}{n\pi}\right)^{1/2} e^{iq\eta}. \tag{14.96}$$

From equation (14.83), the wavelength of the periodic potential in x is (π/k_F), and in η it is π (since $\eta = k_F x$). The elements of the Bravais lattice in η are therefore $(\mu\pi)$, $\mu = 0, \pm 1, \pm 2, \ldots$, and the elements of the reciprocal lattice are

$$Q_\nu = \frac{2\pi}{(\pi)} \nu = 2\nu, \qquad \nu = 0, \pm 1, \pm 2, \ldots \tag{14.97}$$

Now Bloch's theorem [see for example Ashcroft and Mermin (1976, Chapter 8)] requires the eigenfunction $\varphi_q(\eta)$, equation (14.89), to be of the form given in equation (14.82), namely,

$$\varphi_q(\eta) = \left(\frac{2}{n\pi}\right)^{1/2} e^{iq\eta} u_q(\eta), \tag{14.98}$$

where $u_q(\eta)$ has periodicity of π in η, in the present case. Thus $u_q(\eta)$ has the form

$$u_q(\eta) = \sum_\lambda a_\lambda(q) e^{2i\lambda\eta}, \tag{14.99}$$

where $\lambda = 0, \pm 1, \pm 2, \ldots$. Consider now the normalization condition for the coefficients a_λ in equation (14.99),

$$
\begin{aligned}
1 &= \int_0^{n\pi/2} d\eta \, |\varphi_q(\eta)|^2 \\
&= \left(\frac{2}{n\pi}\right) \sum_{\lambda,\lambda'} a_\lambda^*(q) a_{\lambda'}(q) \int_0^{n\pi/2} d\eta \, e^{2i(\lambda'-\lambda)\eta} \\
&= \sum_{\lambda,\lambda'} a_\lambda^*(q) a_{\lambda'}(q) e^{i(\lambda'-\lambda)n\pi/2} \frac{\sin[(\lambda'-\lambda)n\pi/2]}{[(\lambda'-\lambda)n\pi/2]}.
\end{aligned}
\tag{14.100}
$$

Now we have an even number $n = 2n'$ of electrons, each occupied spatial Hartree eigenfunction $\varphi_q(\eta)$ being found in both spin-up and spin-down manifolds. Thus in equation (14.100) we have

$$
\begin{aligned}
\frac{\sin[(\lambda'-\lambda)n'\pi]}{[(\lambda'-\lambda)n'\pi]} &= 0 \qquad \text{for } \lambda' \neq \lambda, \\
&= 1 \qquad \text{for } \lambda' = \lambda,
\end{aligned}
\tag{14.101}
$$

see equation (14.94). Thus equation (14.100) reduces to

$$1 = \sum_{\lambda} |a_\lambda|^2. \tag{14.102}$$

We now substitute the Bloch wave, equation (14.98) with equation (14.99), into the Mathieu equation (14.89), writing the cosine in terms of complex exponentials,

$$\sum_{\lambda=-\infty}^{\infty} \{[(q+2\lambda)^2 - \varepsilon_q'] e^{i(q+2\lambda)\eta} - \tfrac{1}{4}s[e^{i[q+2(\lambda+1)]\eta} + e^{i[q+2(\lambda-1)]\eta}]\} a_\lambda(q) = 0. \tag{14.103}$$

Thus if we multiply equation (14.103) by $\exp[-i(q+2\lambda')\eta]$ and integrate over $(0, n\pi/2)$, we obtain

$$0 = \{[(q+2\lambda')^2 - \varepsilon_q']a_{\lambda'}(q) - \tfrac{1}{4}s[a_{\lambda'-1}(q) + a_{\lambda'+1}(q)]\}. \tag{14.104}$$

This is an alternative form of the Mathieu equation (14.89): it gives the relationship that must be satisfied by the Fourier coefficients $a_\lambda(q)$ of the Bloch-type solution, equations (14.98) and (14.99).

14.4.2   Solution away from the band gap

We now apply perturbation theory to solve equation (14.104), based on the assumed weakness of the self-consistent field periodic potential: $s \ll 1$ in equations (14.89) and (14.104): see also equation (14.86) for the definition of s. In the relationship

$$s = \frac{4nv_K\langle\rho_{-K}\rangle}{\rho_0\varepsilon_F}, \tag{14.105}$$

$\langle\rho_{-K}\rangle$ will be a function of s, as is the wave function used in the expectation value. Equation (14.105) will therefore determine the functional relationship between s and v_K, which are both being assumed small. The specific relationship that will be obtained enables us to discuss, in section 14.5, the non-analytical relationship between the uniform and charge density wave solutions. Such a non-analytical relationship occurs in a variety of quantum-mechanical systems.

If the Bloch electrons are to form a sinusoidal charge density wave, of wave number $K = 2k_F$, so that $Kx = 2\eta$, then we should expect the periodic part of the Bloch wave, u_q, equation (14.99), to be dominated by the terms in $\exp(\pm 2i\eta)$, i.e. the terms in $\lambda = \pm 1$. We assume that these terms have coefficients $a_{\pm 1}$ of the order of s, with higher harmonics, $|\lambda| > 1$, having coefficients $\sim s^{|\lambda|}$. We shall find that such a solution exists. On this basis, we shall apply perturbation theory to the Mathieu equation (14.104), in order to obtain a solution to second order. This entails terms up to

$|\lambda'| = 2$, namely

$$\lambda' = 0: \qquad (q^2 - \varepsilon_q')a_0 = \tfrac{1}{4}s(a_{-1} + a_1), \tag{14.106}$$

$$\lambda' = 1: \qquad [(q+2)^2 - \varepsilon_q']a_1 = \tfrac{1}{4}s(a_0 + a_2), \tag{14.107}$$

$$\lambda' = -1: \qquad [(q-2)^2 - \varepsilon_q']a_{-1} = \tfrac{1}{4}s(a_{-2} + a_0), \tag{14.108}$$

$$\lambda' = 2: \qquad [(q+4)^2 - \varepsilon_q']a_2 \approx \tfrac{1}{4}s(a_1), \tag{14.109}$$

$$\lambda' = -2: \qquad [(q-4)^2 - \varepsilon_q']a_{-2} \approx \tfrac{1}{4}s(a_{-1}), \tag{14.110}$$

We also write ε_q' as

$$\varepsilon_q' \approx [\varepsilon_q^{(0)} + s\varepsilon_q^{(1)} + s^2\varepsilon_q^{(2)}]. \tag{14.111}$$

We now build up the solution in successive orders of s. In zeroth order, from equations (14.111) and (14.106),

$$\varepsilon_q^{(0)} = q^2. \tag{14.112}$$

In first order, from equations (14.106), (14.111) and (14.112),

$$\varepsilon_q^{(1)} = 0, \tag{14.113}$$

and, from equations (14.107) and (14.108),

$$[(q+2)^2 - q^2]a_1 = \frac{1}{4}sa_0 \rightarrow a_1 = \frac{s}{16}\frac{a_0}{(q+1)}, \tag{14.114}$$

$$[(q-2)^2 - q^2]a_{-1} = \frac{1}{4}sa_0 \rightarrow a_{-1} = \frac{s}{16}\frac{a_0}{(-q+1)}. \tag{14.115}$$

In second order, we have, from equations (14.106) and (14.112)–(14.115),

$$-s^2\varepsilon_q^{(2)}a_0 = \frac{1}{4}\frac{s^2}{16}\left[\frac{1}{(q+1)} + \frac{1}{(-q+1)}\right]a_0$$

or

$$\varepsilon_q^{(2)} = -\frac{1}{32}\frac{1}{(1-q^2)}, \tag{14.116}$$

and, from equations (14.109) and (14.110),

$$[(q+4)^2 - q^2]a_2 = \frac{1}{4}\frac{s^2}{16}\frac{a_0}{(q+1)}, \tag{14.117}$$

$$[(q-4)^2 - q^2]a_{-2} = \frac{1}{4}\frac{s^2}{16}\frac{a_0}{(1-q)} \tag{14.118}$$

or

$$a_{\pm 2} = \frac{1}{2}\left(\frac{s}{16}\right)^2 \frac{a_0}{(q \pm 1)(q \pm 2)}. \tag{14.119}$$

Normalization of the eigenfunctions, equation (14.98) with equation (14.102) determines a_0,

$$1 = (|a_0|^2 + |a_1|^2 + |a_{-1}|^2), \tag{14.120}$$

to second order in s. From equations (14.114) and (14.115), equation (14.120) becomes

$$1 = a_0^2 \left\{ 1 + \left(\frac{s}{16}\right)^2 \frac{2(1+q^2)}{(1-q^2)^2} \right\},$$

or, still to second order,

$$a_0^2 = \left\{ 1 + \left(\frac{s}{16}\right)^2 \frac{2(1+q^2)}{(1-q^2)^2} \right\}^{-1} \approx \left\{ 1 - \left(\frac{s}{16}\right)^2 \frac{2(1+q^2)}{(1-q^2)^2} \right\}. \tag{14.121}$$

14.4.3   Solution near the band gap

Examining equations (14.114), (14.115) and (14.119), we see that $a_{\pm 1}$ and $a_{\pm 2}$ diverge as $q \to \pm 1$, and therefore, since $q = k/k_F$, and the first band gap is to be at $|k| = K/2 = k_F$, i.e. at $|q| = 1$, the solutions obtained so far are not valid near the gap. As is well known [see for example Mott and Jones (1936, Chapter II, section 4)] the method of degenerate perturbation theory must be applied in this case. Near $q = +1$ ($k = +k_F$), we see from equation (14.114) that a_{-1} becomes comparable in magnitude with a_0, and therefore cannot be considered to be first order in s. The zeroth-order problem from the Mathieu equation (14.104) then is

$$\lambda' = 0: \quad (q^2 - \varepsilon_q^{(0)})a_0 = \tfrac{1}{4}sa_{-1} \tag{14.122}$$

$$\lambda' = -1: \quad [(q-2)^2 - \varepsilon_q^{(0)}]a_{-1} = \tfrac{1}{4}sa_0. \tag{14.123}$$

The necessary and sufficient condition for a solution to exist for this system of homogeneous linear equations is that the determinant of coefficients of a_0 and a_{-1} should vanish,

$$\left\{ (q^2 - \varepsilon_q^{(0)})[(q-2)^2 - \varepsilon_q^{(0)}] - \left(\frac{s}{4}\right)^2 \right\} = 0, \tag{14.124}$$

whose solution is

$$\varepsilon_q^{(0)} = \left\{ q^2 + 2(1-q) - \left[4(q-1)^2 + \left(\frac{s}{4}\right)^2 \right]^{1/2} \right\}. \tag{14.125}$$

In equation (14.125) we have chosen the negative square root, so that $\varepsilon_q^{(0)}$ will be less than q^2 for $0 < q \lesssim 1$. Substituting equation (14.125) into equation (14.122) we obtain

$$\left\{ -2(1-q) + \left[4(q-1)^2 + \left(\frac{s}{4} \right)^2 \right]^{1/2} \right\} a_0 = \frac{1}{4} s a_{-1}. \tag{14.126}$$

We substitute the zeroth order solution $\varepsilon_q^{(0)}$, equation (14.125), into equations (14.107) and (14.110), noting that now, for $0 < q \lesssim 1$, a_{-2} will be *first* order small

$$[(q+2)^2 - \varepsilon_q^{(0)}] a_1 = \frac{1}{4} s a_0, \tag{14.127}$$

$$[(q-4)^2 - \varepsilon_q^{(0)}] a_{-2} = \frac{1}{4} s a_{-1}. \tag{14.128}$$

We substitute for a_1 and a_{-2} into equations (14.106) and (14.108), using equations (14.127) and (14.128) to express a_1 and a_{-2} in terms of the dominant coefficients a_0 and a_{-1}. The results are

$$a_1 = [(q+2)^2 - \varepsilon_q^{(0)}]^{-1} \left(\frac{s}{4} \right) a_0, \tag{14.129}$$

$$a_{-2} = [(q-4)^2 - \varepsilon_q^{(0)}]^{-1} \left(\frac{s}{4} \right) a_{-1}, \tag{14.130}$$

whence, from equations (14.106) and (14.108),

$$(q^2 - \varepsilon_q') a_0 = \left(\frac{s}{4} \right) \left\{ a_{-1} + [(q+2)^2 - \varepsilon_q^{(0)}]^{-1} \left(\frac{s}{4} \right) a_0 \right\} \tag{14.131}$$

$$[(q-2)^2 - \varepsilon_q'] a_{-1} = \left(\frac{s}{4} \right) \left\{ [(q-4)^2 - \varepsilon_q^{(0)}]^{-1} \left(\frac{s}{4} \right) a_{-1} + a_0 \right\}. \tag{14.132}$$

We rewrite equations (14.131) and (14.132) in standard form as

$$\left\{ q^2 - \varepsilon_q' - \left(\frac{s}{4} \right)^2 [(q+2)^2 - \varepsilon_q^{(0)}]^{-1} \right\} a_0 - \left(\frac{s}{4} \right) a_{-1} = 0 \tag{14.133}$$

$$\left(\frac{s}{4} \right) a_0 - \left\{ (q-2)^2 - \varepsilon_q' - \left(\frac{s}{4} \right)^2 [(q-4)^2 - \varepsilon_q^{(0)}]^{-1} \right\} a_{-1} = 0. \tag{14.134}$$

In the above equations (12.127)–(12.134), $\varepsilon_q^{(0)}$ is given by equation (14.125). The necessary and sufficient condition that equations (14.133) and (14.134) have a solution is that the determinant of the coefficients for a_0 and a_{-1} be zero, determining ε_q' in second order. We note that equation (14.125) for $\varepsilon_q^{(0)}$ contains a term of order s. We assume that the region near the band gap at $q = k_F$ is defined by

$$(1-q) \leq s. \tag{14.135}$$

Let us now obtain the correction of order s^2 to $\varepsilon_q^{(0)}$, equation (14.125). Write

$$\varepsilon_q' \approx (\varepsilon_q^{(0)} + s^2 \varepsilon_q^{(2)}). \tag{14.136}$$

The determinantal condition applied to equations (14.133) and (14.134) gives

$$\left\{ (q^2 - \varepsilon_q^{(0)})[(q-2)^2 - \varepsilon_q^{(0)}] - \left(\frac{s}{4}\right)^2 \right\}$$

$$+ (q^2 - \varepsilon_q^{(0)}) \left[-s^2 \varepsilon_q^{(2)} - \left(\frac{s}{4}\right)^2 \gamma_2 \right]$$

$$+ [(q-2)^2 - \varepsilon_q^{(0)}] \left[-s^2 \varepsilon_q^{(2)} - \left(\frac{s}{4}\right)^2 \gamma_1 \right]$$

$$+ \left[-s^2 \varepsilon_q^{(2)} - \left(\frac{s}{4}\right)^2 \gamma_1 \right] \left[-s^2 \varepsilon_q^{(2)} - \left(\frac{s}{4}\right)^2 \gamma_2 \right] = 0. \tag{14.137}$$

In equation (14.137), we have defined

$$\gamma_1 = [(q+2)^2 - \varepsilon_q^{(0)}]^{-1}, \tag{14.138}$$

$$\gamma_2 = [(q-4)^2 - \varepsilon_q^{(0)}]^{-1}. \tag{14.139}$$

In equation (14.137), note that the first line is zero: see equation (14.124). We note that γ_1 and γ_2 are of leading order s^0. Expanding out equation (14.137), we now have

$$(s^2 \varepsilon_q^{(2)})^2 + \left(\frac{s}{4}\right)^2 (\gamma_1 + \gamma_2) s^2 \varepsilon_q^{(2)} + \left(\frac{s}{4}\right)^4 \gamma_1 \gamma_2$$

$$- [(q^2 - \varepsilon_q^{(0)}) + (q-2)^2 - \varepsilon_q^{(0)}] s^2 \varepsilon_q^{(2)}$$

$$- \left(\frac{s}{4}\right)^2 \{(q^2 - \varepsilon_q^{(0)})\gamma_2 + [(q-2)^2 - \varepsilon_q^{(0)}]\gamma_1\} = 0. \tag{14.140}$$

Now from equation (14.125) for $\varepsilon_q^{(0)}$, we find

$$(q^2 - \varepsilon_q^{(0)}) = -2(1-q) + \left[4(1-q)^2 + \left(\frac{s}{4}\right)^2 \right]^{1/2}, \tag{14.141}$$

and

$$[(q-2)^2 - \varepsilon_q^{(0)}] = 2(1-q) + \left[4(1-q)^2 + \left(\frac{s}{4}\right)^2 \right]^{1/2}. \tag{14.142}$$

We therefore see from equation (14.135) that both equations (14.141) and (14.142) are of order s. Thus the first line of equation (14.140) is of order s^4, while the second and third lines are of order s^3. Neglecting the first line, we have

$$-2\left[4(1-q)^2 + \left(\frac{s}{4}\right)^2\right]^{1/2} s^2\varepsilon_q^{(2)} = \left(\frac{s}{4}\right)^2 \{(q^2 - \varepsilon_q^{(0)})\gamma_2 + [(q-2)^2 - \varepsilon_q^{(0)}]\gamma_1\}.$$
$$(14.143)$$

We now want γ_1 and γ_2 to order s^0. From equations (14.138),

$$\gamma_1 = [q^2 + 4(q+1) - \varepsilon_q^{(0)}]^{-1}$$
$$= [4(q-1) + 8 + q^2 - \varepsilon_q^{(0)}]^{-1}$$
$$= \left\{8 - 4(1-q) - 2(1-q) + \left[4(1-q)^2 + \left(\frac{s}{4}\right)^2\right]^{1/2}\right\}^{-1}$$
$$\approx \frac{1}{8}.$$
$$(14.144)$$

Similarly from equation (14.139),

$$\gamma_2 \approx \frac{1}{8}.$$
$$(14.145)$$

Now, using equations (14.141), (14.142), (14.144) and (14.145), equation (14.143) becomes

$$-2\left[4(1-q)^2 + \left(\frac{s}{4}\right)^2\right]^{1/2} s^2\varepsilon_q^{(2)} \approx \left(\frac{s}{4}\right)^2 \frac{1}{8} 2\left[4(1-q)^2 + \left(\frac{s}{4}\right)^2\right]^{1/2}$$

or

$$s^2\varepsilon_q^{(2)} \approx -\frac{1}{8}\left(\frac{s}{4}\right)^2.$$
$$(14.146)$$

Thus, from equations (14.136), (14.125) and (14.146), we have the single-particle energy to order s^2 for $(1-q) \leq s$, near the gap,

$$\varepsilon_q' = \left\{q^2 + 2(1-q) - \left[4(1-q)^2 + \left(\frac{s}{4}\right)^2\right]^{1/2} - \frac{1}{8}\left(\frac{s}{4}\right)^2\right\}.$$
$$(14.147)$$

We now return to equations (14.133) and (14.134) to determine a_{-1} in terms of a_0, to order s^2,

$$\left(\frac{s}{4}\right)a_{-1} = \left\{-2(1-q) + \left[4(1-q)^2 + \left(\frac{s}{4}\right)^2\right]^{1/2}\right\}a_0,$$
$$(14.148)$$

$$\left(\frac{s}{4}\right)a_0 = \left\{+2(1-q) + \left[4(1-q)^2 + \left(\frac{s}{4}\right)^2\right]^{1/2}\right\}a_{-1},$$
$$(14.149)$$

having used equation (14.147) along with equations (14.138), (14.139), (14.144) and (14.145). Thus, from equations (14.148) and (14.149),

$$\left(\frac{a_{-1}}{a_0}\right)^2 = \frac{\left\{-2(1-q) + \left[4(1-q)^2 + \left(\frac{s}{4}\right)^2\right]^{1/2}\right\}}{\left\{+2(1-q) + \left[4(1-q)^2 + \left(\frac{s}{4}\right)^2\right]^{1/2}\right\}}. \tag{14.150}$$

Normalization to order s^2 can be achieved now by using equations (14.150), (14.129) and (14.130),

$$1 = \{|a_0|^2 + |a_{-1}|^2 + |a_1|^2 + |a_{-2}|^2\}. \tag{14.151}$$

Now, to order s, from equations (14.129), (14.130), (14.144) and (14.145),

$$a_1 = -\frac{s}{32} a_0, \tag{14.152}$$

$$a_{-2} = -\frac{s}{32} a_{-1}, \tag{14.153}$$

so, from equation (14.151),

$$1 = \left\{\left[1 + \left(\frac{s}{32}\right)^2\right](|a_0|^2 + |a_{-1}|^2)\right\}, \tag{14.154}$$

or

$$(|a_0|^2 + |a_{-1}|^2) \approx \left[1 - \left(\frac{s}{32}\right)^2\right] = \frac{2|a_0|^2}{\left\{\dfrac{2(1-q)}{\left[4(1-q)^2 + \left(\dfrac{s}{4}\right)^2\right]^{1/2}} + 1\right\}}. \tag{14.155}$$

The last line of equation (14.155) follows from equation (14.150). Solving equation (14.155) we have

$$|a_0|^2 = \frac{1}{2}\left\{1 + \frac{2(1-q)}{\left[4(1-q)^2 + \left(\dfrac{s}{4}\right)^2\right]^{1/2}}\right\}\left[1 - \left(\frac{s}{32}\right)^2\right]. \tag{14.156}$$

14.4.4   The self-consistency condition

We are now in a position to establish self consistency for the Hartree wave functions by solving equation (14.105) for s as a function of $\Gamma = (v_K/\varepsilon_F)$, thereby obtaining the wave functions, equations (14.82) or (14.98) and (14.99), directly in terms of the physical parameters of the crystal. From equation (14.105), we must evaluate $\langle \rho_{-K} \rangle$. From equations (14.47) and

(14.45), with equation (14.33), we have

$$\langle \rho_{-K} \rangle = \frac{1}{\Omega} \sum_k \int dx\, \varphi_k^*(x)\, e^{iKx}\, \varphi_k(x). \qquad (14.157)$$

In equation (14.157) we now introduce $\eta = k_F x$, equation (14.86) and equation (14.90), along with $q = k/k_F$, equation (14.86). In the present case we are considering $K = 2k_F$, so we have $Kx = 2\eta$. From equations (14.98) and (14.99), the integrals in equation (14.157) become

$$\int_0^{\Omega k_F} d\eta |\varphi_q(\eta)|^2\, e^{2i\eta} = \left(\frac{2}{n\pi}\right) \sum_{\lambda,\lambda'} a_\lambda^*(q) a_{\lambda'}(q) \int_0^{n\pi/2} d\eta\, e^{2i(\lambda' - \lambda + 1)\eta}$$

$$= \left(\frac{2}{n\pi}\right) \sum_{\lambda,\lambda'} a_\lambda^*(q) a_{\lambda'}(q)\, \frac{n\pi}{2}\, \delta_{\lambda',\lambda - 1} \qquad (14.158)$$

whence

$$\int_0^{\Omega k_F} d\eta |\varphi_q(\eta)|^2\, e^{2i\eta} = \sum_\lambda a_\lambda^*(q) a_{\lambda - 1}(q). \qquad (14.159)$$

In equation (14.158), the Kronecker delta comes as a trivial adaptation of equations (14.100) and (14.101). Equations (14.157) and (14.159) now give us

$$\langle \rho_{-K} \rangle = \frac{1}{\Omega} \sum_q \sum_\lambda a_\lambda^*(q) a_{\lambda - 1}(q). \qquad (14.160)$$

We can now address the self-consistency condition, equation (14.105) which, with equations (14.160) and (14.86) becomes

$$s = \frac{4n}{\rho_0} \frac{\Gamma}{\Omega} \sum_q \sum_\lambda a_\lambda^*(q) a_{\lambda - 1}(q). \qquad (14.161)$$

Now from $q = k/k_F$, the density of points in q space is

$$d(q) = \frac{2}{\left(\dfrac{2\pi}{\Omega k_F}\right)} = \frac{\Omega k_F}{\pi} = \frac{n}{2}. \qquad (14.162)$$

Thus, converting the sum over k in equation (14.161) to an integral, we have

$$s = 2n\Gamma \sum_\lambda \int_{-1}^{1} dq\, a_\lambda^*(q) a_{\lambda - 1}(q), \qquad (14.163)$$

having used $\rho_0 = n/\Omega$, and the fact that the range of k in the first Brillouin zone for this charge density wave case is $(-k_F \le k \le k_F)$, corresponding to $(-1 \le q \le 1)$. In fact, from the symmetry of $u_k(x)$ about $k = 0$, we can write equation (14.163) as

$$s = 4n\Gamma \sum_\lambda \int_0^1 dq\, a_\lambda^*(q) a_{\lambda-1}(q).$$ (14.164)

We now have the problem of substituting into equation (14.164) expressions for $a_\lambda(q)$, equations (14.114), (14.115), (14.119) with (14.121) for $0 \le q \le (1-s)$, and equations (14.129) and (14.130) with equations (14.150) and (14.156) for $(1-s) \le q \le 1$, and then evaluating the integrals. We first collect the above expressions.

For $0 \le q \le (1-s)$,

$$a_{\pm1} = \frac{s}{16} \frac{a_0}{(1 \pm q)},$$ (14.165)

$$a_{\pm2} = \frac{1}{2} \left(\frac{s}{16}\right)^2 \frac{a_0}{(1 \pm q)(2 \pm q)},$$ (14.166)

$$a_0^2 = \left\{1 - \left(\frac{s}{16}\right)^2 \frac{2(1+q^2)}{(1-q^2)^2}\right\}.$$ (14.167)

For $(1-s) \le q \le 1$,

$$a_1 = [(q+2)^2 - \varepsilon_q^{(0)}]^{-1} \left(\frac{s}{4}\right) a_0,$$ (14.168)

$$a_{-2} = [(q-4)^2 - \varepsilon_q^{(0)}]^{-1} \left(\frac{s}{4}\right) a_{-1}.$$ (14.169)

From equations (14.138), (14.139), (14.144) and (14.145), we have, to lowest order,

$$a_1 = \frac{s}{32} a_0,$$ (14.170)

$$a_{-2} = \frac{s}{32} a_{-1}.$$ (14.171)

In addition, equations (14.156) and (14.155) give

$$|a_0|^2 = \frac{1}{2} \left\{1 + \frac{2(1-q)}{\left[4(1-q)^2 + \left(\frac{s}{4}\right)^2\right]^{1/2}}\right\} \left[1 - \left(\frac{s}{32}\right)^2\right],$$ (14.172)

and

$$|a_{-1}|^2 = \frac{1}{2} \left\{1 - \frac{2(1-q)}{\left[4(1-q)^2 + \left(\frac{s}{4}\right)^2\right]^{1/2}}\right\} \left[1 - \left(\frac{s}{32}\right)^2\right].$$ (14.173)

Returning to equation (14.164), the lowest order terms are:

For $0 \leq q \leq (1 - s)$, away from the band gap, from equation (14.165),

$$(a_0 a_{-1} + a_1 a_0) \approx \frac{s}{16} \frac{2}{(1 - q^2)}. \tag{14.174}$$

For $(1 - s) \leq q \leq 1$, near the band gap, from equations (14.170) and (14.171),

$$[a_0 a_{-1} + (a_1 a_0 + a_{-1} a_{-2})] = \left\{ a_0 a_{-1} + (|a_0|^2 + |a_{-1}|^2) \left(\frac{s}{32} \right) \right\}. \tag{14.175}$$

From equations (14.172) and (14.173), to lowest order,

$$|a_0|^2 = \tfrac{1}{2}(1 + \gamma), \tag{14.176}$$

$$|a_{-1}|^2 = \tfrac{1}{2}(1 - \gamma), \tag{14.177}$$

where

$$\gamma = \frac{2(1 - q)}{\left[4(1 - q)^2 + \left(\frac{s}{4} \right)^2 \right]^{1/2}}. \tag{14.178}$$

For evaluating $(a_0 a_{-1})$ in equation (14.175), we choose the positive square root for a_0,

$$a_0 = 2^{-1/2}(1 + \gamma)^{1/2}. \tag{14.179}$$

For a_{-1}, we must choose the sign that matches a_{-1}, away from the band gap, i.e. from the region $0 \leq q \leq (1 - s)$ at $q = (1 - s)$. In this region, from equation (14.165),

$$a_{-1}(q \to (1 - s)^-) = -\frac{s}{16} \frac{a_0}{(1 - 1 + s)} = \frac{a_0}{16}. \tag{14.180}$$

In the same region, from equation (14.167),

$$|a_0(q \to (1 - s)^-)|^2 = \left[1 - \frac{2}{16^2} \right]. \tag{14.181a}$$

Thus

$$a_0(q \to (1 - s)^-) \approx \left(1 - \frac{1}{16^2} \right) \approx 1, \tag{14.181b}$$

so equation (14.180) gives

$$a_{-1}(q \to (1 - s)^- \approx \frac{1}{16}. \tag{14.182}$$

From equation (14.177), for $(1 - s) \leq q \leq 1$, near the band gap,

$$a_{-1} = \pm 2^{-1/2}(1 - \gamma)^{1/2}. \tag{14.183}$$

From equation (14.178),

$$\gamma(q \to (1 - s)^+) = \frac{2s}{\left[4s^2 + \left(\frac{s}{4}\right)^2\right]^{1/2}} = \frac{1}{\left[1 + \frac{1}{64}\right]^{1/2}} \approx \left(1 - \frac{1}{128}\right). \tag{14.184}$$

Returning to equation (14.183) with equation (14.184),

$$a_{-1}(q \to (1 - s)^+) = \pm 2^{-1/2}\left[1 - \left(1 - \frac{1}{128}\right)\right]^{1/2}$$

$$= \pm 2^{-1/2}(128)^{-1/2} = \pm \frac{1}{16}. \tag{14.185}$$

Comparing equations (14.182) and (14.185), we conclude that

$$a_{-1} = 2^{-1/2}(1 - \gamma)^{1/2}, \qquad (1 - s) \leq q \leq 1. \tag{14.186}$$

Returning now to equation (14.175), using equations (14.179), (14.186) and (14.155), we have, near the band gap,

$$[a_0 a_{-1} + (a_1 a_0 + a_{-1} a_{-2})] = \left\{\frac{1}{2}(1 - \gamma^2)^{1/2} + \left(\frac{s}{32}\right)\left[1 - \left(\frac{s}{32}\right)^2\right]\right\}. \tag{14.187}$$

From equation (14.178), this reduces to

$$\left(\frac{s}{8}\right)\left\{\frac{1}{\left[4(1 - q)^2 + \left(\frac{s}{4}\right)^2\right]^{1/2}} + \frac{1}{4}\right\} \tag{14.188}$$

to order s.

For the self-consistency condition, equation (14.164), we have the following integrals, from equations (14.174), away from the band gap, and (14.188), near the band gap,

$$\sum_\lambda \int_0^1 dq\, a_\lambda^*(q)a_{\lambda-1}(q) \approx \frac{s}{8}\int_0^{(1-s)} dq\, \frac{1}{(1 - q^2)}$$

$$+ \frac{s}{8}\int_{(1-s)}^1 dq\, \left[\frac{1}{\left[4(1 - q)^2 + \left(\frac{s}{4}\right)^2\right]^{1/2}} + \frac{1}{4}\right]. \tag{14.189}$$

Now

$$\int_0^{(1-s)} dq \frac{1}{(1-q^2)} = \int_0^{(1-s)} dq \frac{1}{2} \left\{ \frac{1}{(1-q)} + \frac{1}{(1-q)} \right\}$$

$$= \frac{1}{2} \ln \left(\frac{1+q}{1-q} \right) \Big|_0^{(1-s)}$$

$$= \frac{1}{2} \{ -\ln s + \ln(2-s) \}$$

$$= \frac{1}{2} \{ -\ln s + \ln 2 \} \qquad (14.190)$$

for small s. Also in equation (14.189)

$$\frac{1}{4} \int_{(1-s)}^1 dq = \frac{s}{4}, \qquad (14.191)$$

which gives a term $\sim s^2$ in equation (14.189), which we are neglecting. Furthermore,

$$\int_{(1-s)}^1 dq \frac{1}{\left[4(1-q)^2 + \left(\frac{s}{4} \right)^2 \right]^{1/2}}, \qquad u = (1-q)$$

$$= \int_0^s du \frac{1}{\left[4u^2 + \left(\frac{s}{4} \right)^2 \right]^{1/2}} = \frac{1}{2} \int_0^s du \frac{1}{\left[u^2 + \left(\frac{s}{8} \right)^2 \right]^{1/2}}$$

$$= \frac{1}{2} \ln \left(u + \left[u^2 + \left(\frac{s}{8} \right)^2 \right]^{1/2} \right) \Big|_0^s$$

$$= \frac{1}{2} \ln \left(s + \left[s^2 + \left(\frac{s}{8} \right)^2 \right]^{1/2} \right) - \frac{1}{2} \ln \left(\frac{s}{8} \right)$$

$$= \frac{1}{2} \ln \left\{ s \left(1 + \left[1 + \frac{1}{64} \right]^{1/2} \right) \right\} - \frac{1}{2} \ln \left(\frac{s}{8} \right)$$

$$= \frac{1}{2} \ln \left\{ \left(1 + \frac{(65)^{1/2}}{8} \right) \cdot 8 \right\}$$

$$\approx \frac{1}{2} \ln(16) = 2 \ln 2. \qquad (14.192)$$

Thus, collecting results from equations (14.189) to (14.192) we have the self-consistency condition equation (14.164),

$$s = 4n\Gamma \left\{ \frac{s}{8} \left[\frac{1}{2} (-\ln s + \ln 2) + 2 \ln 2 \right] \right\}. \qquad (14.193)$$

Thus

$$1 = -4n\Gamma \frac{1}{16} (\ln s - 5 \ln 2),$$

whence

$$\frac{1}{16} \ln\left(\frac{s}{32}\right) = \frac{-1}{4n\Gamma},$$

or finally,

$$s = 32 \exp[-4/(n\Gamma)]. \qquad (14.194)$$

Now our perturbative treatment requires $s \ll 1$, which can only be true in equation (14.194) if Γ is small and positive. From the definition of Γ (equation (14.86)), this requires not only that $v_K \langle \rho_{-K} \rangle$ be dominant in the scf in equation (14.71) for $K = 2k_F$, but also

$$v_K = v_{2k_F} = \Gamma k_F > 0, \qquad (14.195)$$

that is, that the effective pairwise potential in equation (14.32) should be attractive at Fourier component $k = K = 2k_F$. For this to be the case, the corresponding Fourier component of the effective pairwise potential due to electron–phonon interaction, $[-v'(r)]$ must be attractive also, with sufficient strength to exceed in magnitude the corresponding Fourier component of the electron–electron Coulomb repulsion: see equations (14.28)–(14.32).

14.4.5　The total energy

We can now compare the total energy of the charge density wave solution, in Hartree approximation, with that of the uniform solution. From equation (14.73), we see that in the presence of the periodic potential, as in equation (14.76), the uniform solution gives

$$\varepsilon_j = \frac{\hbar^2 k_j^2}{2m_b}, \qquad (14.196)$$

since $v_0 = 0$ for a sinusoidal potential. The total energy E_0 for the uniform case is therefore given by equation (14.49),

$$E_0 = \sum_j \varepsilon_j, \qquad (14.197)$$

since $\langle \rho_k \rangle = 0$ for $k \neq 0$, equation (14.70a), and $v_k = 0$ for $k = 0$, as mentioned above. Thus, from equation (14.197) with equation (14.86) for $\varepsilon(q)$,

$$\frac{E_0}{\varepsilon_F} = \int_{-1}^{1} \mathrm{d}q\, \mathrm{d}(q)q^2 = 2 \int_{0}^{1} \mathrm{d}q\, \mathrm{d}(q)q^2 \qquad (14.198)$$

where the density of states $d(q)$ of q-space points is given, in equation (14.162) to be $(n/2)$. Thus, equation (14.198) gives

$$\frac{E_0}{\varepsilon_F} = 2\frac{n}{2}\int_0^1 dq\, q^2 = \frac{n}{3}. \qquad (14.199)$$

Now consider the total energy of the periodic solution given in the previous section. Let us first consider the term from the self-consistent field (scf) in the energy expression (14.49); denote it E_1:

$$\frac{E_1}{\varepsilon_F} = \frac{n^2}{2\rho_0^2\varepsilon_F}\{v_K|\langle\rho_{-K}\rangle|^2 + v_{-K}|\langle\rho_K\rangle|^2\}$$

$$= \frac{n^2}{\rho_0^2\varepsilon_F}v_K|\langle\rho_K\rangle|^2, \qquad (14.200)$$

the last line coming from equation (14.75) plus the fact that $v_K = v_{-K}$. In our dimensionless units, equations (14.86), this becomes

$$\frac{E_1}{\varepsilon_F} = s^2\frac{1}{16}\frac{\varepsilon_F}{v_K} = s^2\frac{1}{16}\frac{1}{\Gamma}. \qquad (14.201)$$

Now from our scf solution, equation (14.194), this is

$$\frac{E_1}{\varepsilon_F} = -\frac{s^2}{16}\frac{n}{4}\ln\left(\frac{s}{32}\right). \qquad (14.202)$$

The rest of the energy, E_2, comes from the Hartree eigenvalues ε_j, or $\varepsilon(q)$,

$$\frac{E_2}{\varepsilon_F} = \sum_j\frac{\varepsilon_j}{\varepsilon_F} = 2\int_0^1 dq\,\frac{n}{2}\varepsilon(q) \qquad (14.203)$$

where $\varepsilon(q)$ is as follows:

For $0 \le q \le (1-s)$,

$$\varepsilon(q) = \left\{q^2 - \frac{s^2}{32}\frac{1}{(1-q^2)}\right\} \qquad (14.204)$$

from equations (14.111)–(14.113) and (14.116);

for $(1-s) \le q \le 1$,

$$\varepsilon(q) = \left\{q^2 + 2(1-q) - \left[4(1-q)^2 + \left(\frac{s}{4}\right)^2\right]^{1/2} - \frac{1}{8}\left(\frac{s}{4}\right)^2\right\} \qquad (14.205)$$

from equation (14.147). The total energy E in the periodic case, is

$$E = (E_1 + E_2). \qquad (14.206)$$

Thus, the difference in total energy between uniform and periodic solutions is

$$\frac{(E_0 - E)}{\varepsilon_F} = n \frac{s^2}{64} \ln\left(\frac{s}{32}\right) - n \left\{ -\frac{s^2}{32} \int_0^{(1-s)} dq \frac{1}{(1-q^2)} + 2 \int_{(1-s)}^1 dq \, (1-q) \right.$$

$$- \int_{(1-s)}^1 dq \left[4(1-q)^2 + \left(\frac{s}{4}\right)^2 \right]^{1/2}$$

$$\left. - \frac{1}{8} \left(\frac{s}{4}\right)^2 \int_{(1-s)}^1 dq \right\}. \tag{14.207}$$

Note that the terms in q^2 in the integrand are completely cancelled. The various integrals in equation (14.207) are

$$\int_0^{(1-s)} dq \frac{1}{(1-q^2)} = \frac{1}{2} \int_0^{(1-s)} dq \left\{ \frac{1}{(1-q)} + \frac{1}{(1+q)} \right\}$$

$$= \frac{1}{2} \{ -\ln s + \ln(2-s) \}; \tag{14.208}$$

$$\int_{(1-s)}^1 dq \, (1-q) = \left(q - \frac{q^2}{2} \right)_{(1-s)}^1 = \frac{s^2}{2}; \tag{14.209}$$

$$\int_{(1-s)}^1 dq \, \frac{1}{8} \left(\frac{s}{4}\right)^2 = \frac{s^3}{128}; \tag{14.210}$$

$$\int_{(1-s)}^1 dq \left[4(1-q)^2 + \left(\frac{s}{4}\right)^2 \right]^{1/2}, \qquad u = (1-q)$$

$$= 2 \int_0^s du [u^2 + a^2]^{1/2}, \qquad a = \frac{s}{8}$$

$$= \{ u(u^2 + a^2)^{1/2} + a^2 \ln(u + [u^2 + a^2]^{1/2}) \}_0^s$$

$$= \left\{ s^2 \left(1 + \frac{1}{8^2} \right)^{1/2} + \left(\frac{s}{8}\right)^2 \ln\left[s + s\left(1 + \frac{1}{8^2} \right)^{1/2} \right] - \left(\frac{s}{8}\right)^2 \ln\left(\frac{s}{8}\right) \right\}$$

$$= s^2 \left\{ \left(1 + \frac{1}{8^2} \right)^{1/2} + \frac{1}{8^2} \ln\left[1 + \left(1 + \frac{1}{8^2} \right)^{1/2} \right] + \frac{1}{8^2} \ln 8 \right\}. \tag{14.211}$$

Using these results, equation (14.207) becomes

$$\frac{(E_0 - E)}{\varepsilon_F} = ns^2 \left\{ \frac{1}{64} (\ln s - \ln 32) + \frac{1}{32} \frac{1}{2} [-\ln s + \ln(2-s)] - 1 \right.$$

$$+ \left[\left(1 + \frac{1}{8^2} \right)^{1/2} + \frac{1}{8^2} \ln\left[1 + \left(1 + \frac{1}{8^2} \right)^{1/2} \right] + \frac{1}{8^2} \ln 8 \right]$$

$$\left. + \frac{1}{8} \left(\frac{s}{4^2}\right) \right\}. \tag{14.212}$$

Note that the terms in ($s^2 \ln s$) cancel, and that $\ln(2 - s) \approx \ln 2$ for $s \ll 1$. We neglect the term in s^3, and are left with

$$\frac{(E_0 - E)}{\varepsilon_F} = ns^2 \left\{ -\frac{1}{64} \ln 32 + \frac{1}{64} \ln 2 - 1 + \left(1 + \frac{1}{8^2}\right)^{1/2} \right.$$

$$\left. + \frac{1}{8^2} \ln\left[1 + \left(1 + \frac{1}{8^2}\right)^{1/2}\right] + \frac{1}{8^2} \ln 8 \right\}. \tag{14.213}$$

We expand

$$\left(1 + \frac{1}{8^2}\right)^{1/2} \approx \left(1 + \frac{1}{128}\right) \tag{14.214}$$

and

$$\ln\left[1 + \left(1 + \frac{1}{8^2}\right)^{1/2}\right] = \ln\left[1 + \left(1 + \frac{1}{128}\right)\right]$$

$$= \ln\left[2\left(1 + \frac{1}{256}\right)\right] \approx \left(\ln 2 + \frac{1}{256}\right). \tag{14.215}$$

Combining equations (14.213) to (14.215), the final result is

$$\frac{(E_0 - E)}{\varepsilon_F} = ns^2 \left\{ \frac{1}{128} + \frac{1}{8^2}\frac{1}{256} \right\} \approx \frac{ns^2}{128}. \tag{14.216}$$

The conclusions from equation (14.216) are that the periodic solution has energy lower than that of the uniform solution, proportional to the number of electrons n and of order s^2. Harking back to equation (14.194) for s, we can express equation (14.216) as

$$\frac{(E_0 - E)}{\varepsilon_F} = 8n \exp[-8/(n\Gamma)]. \tag{14.217}$$

14.5 Discussion

We have been considering a one-dimensional interacting electron gas in Hartree approximation. The interactions include Coulomb forces, and other effective pairwise forces arising from interaction of the electrons with an external field, such as the phonon field of a solid. We have found that in such a system, where the Hartree equation takes the form of equation (14.71), the spatially uniform state, equation (14.67), is always an exact solution. We have then raised the question whether the uniform solution is the ground state. We have found that if the net effective pairwise interaction, Coulomb plus externally mediated, is attractive, and is dominated by a particular Fourier component, then a spatially periodic solution (a Bloch state) exists, determined by a Mathieu-type Hartree equation (14.76). If the wavenumber K of the periodic self-consistent field has the value $K = 2k_F$, equation (14.83), then the first band-gap of the Bloch wave spectrum occurs at the Fermi level, $k = k_F$, and the total energy may be expected to be lower than that of the uniform state. In the approximation of weak net effective pairwise interaction,

$$\frac{v_K}{\varepsilon_F} = \Gamma \ll 1, \tag{14.218}$$

a self-consistent solution of the Hartree equation exists only for $\Gamma > 0$. This means that the net effective pairwise interaction is an attractive force. Since the Coulomb force between electrons is repulsive, it means that the externally (phonon) mediated effective force must itself be attractive, and large enough to overcome Coulomb repulsion at the given wavenumber $k = 2k_F$.

Since the uniform solution is an exact solution of the Hartree equation for non-interacting electrons equations (14.55), (14.59) and (14.60), we might think of trying to get to the periodic solution in the presence of weak interactions by a perturbative expansion based on power series in Γ for the Hartree eigenfunctions and eigenvalues, of the form

$$f(\Gamma) = \{f_0 + f_1\Gamma + f_2\Gamma^2 + \cdots\}. \tag{14.219}$$

Now the effect of the perturbation due to weak interaction is borne by the strength s of the periodic self-consistent field in the Hartree equation, expressed in dimensionless form in equation (14.89) where, from equations (14.86),

$$s = 4n\Gamma\frac{\langle\rho_{-K}\rangle}{\rho_0}. \tag{14.220}$$

Applying perturbation theory to equation (14.89), the power series expansions for Hartree eigenfunctions and eigenvalues are of the form

$$F(s) = \{F_0 + F_1s + F_2s^2 + \cdots\}. \tag{14.221}$$

Now the dependence of s on Γ is not simply linear, as equation (14.220) may seem to imply, because $\langle\rho_{-K}\rangle$ depends on Γ through the Hartree eigenfunctions (see equation (14.45)), and the eigenfunctions depend on Γ. If the dependence of s on Γ could be expressed in terms of powers of Γ, then equation (14.221) could be rewritten in the form of equation (14.219), and the interacting periodic state would be obtainable in terms of small perturbative corrections to the uniform non-interacting state. In fact, we have found the explicit relationship between s and Γ to be of the form

$$s \sim \exp(-\alpha/\Gamma), \qquad \Gamma > 0; \tag{14.222}$$

see equation (14.195). Write

$$\Gamma = \alpha g, \tag{14.223}$$

so that equation (14.222) becomes

$$s \sim e^{-1/g}. \tag{14.224}$$

Clearly, from equation (14.224), s is small when g, or Γ, is small. But consider the possibility of expanding s as a Taylor series in g, about the value $g = 0$. For that, we need to know the values of the derivatives of s with respect to g, evaluated at $g = 0$. Now,

$$\frac{ds}{dg} \sim \frac{1}{g^2} e^{-1/g}. \tag{14.225}$$

The limit $(g \to 0^+)$ is indeterminate, and L'Hôpital's rule does not help, for differentiating numerator and denominator of equation (14.225) gives

$$\frac{e^{-1/g}}{(2g^3)}. \tag{14.226}$$

The result, as $g \to 0^+$, is still indeterminate as are all higher order derivatives. We therefore have non-analytic behavior for s as a function of g, or Γ, near $g = 0$, although the curve approaches $s = 0$ smoothly as $g \to 0^+$. Consequently, the properties of the system cannot be described in terms of relationships of the form of equation (14.219). From the second derivative,

$$\frac{d^2 s}{dg^2} \sim \left(-\frac{2}{g^3} + \frac{1}{g^4} \right) e^{-1/g}, \tag{14.227}$$

we see that $s(g)$ has a point of inflection, and a horizontal asymptote as $g \to \infty$. Our derivation, however, is only valid near $g = 0$, $s = 0$.

In summary, we have the situation that a weak perturbation, the specific effective pairwise potential described above, with strength Γ, when introduced to the non-interacting electron gas, creates a periodic ground state that is not accessible in any order of perturbation theory based on the small parameter Γ, by starting from the non-interacting uniform ground state. The reason is because the perturbed ground state behaves non-analytically as a function of Γ near $\Gamma = 0$. This then amounts to a cautionary note, that the application of perturbation theory must be limited to cases where the unperturbed starting point is qualitatively like the perturbed state, in the sense that the latter is analytically related to the former. The situation, illustrated here by charge density waves in the electron gas, also occurs in some other vitally interesting quantum-mechanical systems, among them the BCS theory of superconductivity [Bardeen *et al.* (1957)].

References

Agullo-Lopez F, Catlow C R A and Townsend P 1988 *Point Defects in Materials* (New York: Academic)

Appel J 1968 'Polarons' in *Solid State Physics* **21** eds F Seitz, D Turnbull and H Ehrenreich (New York: Academic) p 193

Ashcroft N W and Mermin N D 1976 *Solid State Physics* (Philadelphia: Saunders)

Bachelet G B, Hamann D R and Schlüter M 1982 *Phys. Rev. B* **26** 4199

Baldacchini G 1989 *Cryst. Latt. Def. and Amorph. Mat.* **18** 43

Baldacchini G and Mollenauer L F 1973 *J. Phys. Colloq.* **9** Suppl. 11–12, 141

Bardeen J, Cooper L N and Schrieffer J R 1957 *Phys. Rev.* **108** 1175

Baughman R H, Shacklette J M, Zakhidov A A and Stafström S 1998 *Nature* **392** 362

Baym G 1969 *Lectures on Quantum Mechanics* (New York: Benjamin)

Berg J M and McClure D S 1989 *J. Chem. Phys.* **90** 3915

Bhatia A B and Singh R N 1986 *Mechanics of Deformable Media* (Bristol: Hilger)

Bhyrappa P, Vaijayanthinala G and Suslick K S 1999 *J. Am. Chem. Soc.* **121** 262

Bilz H and Kress W 1979 *Phonon Dispersion Relations in Insulators* (Berlin: Springer-Verlag)

Bilz H 1985 *Cryst. Latt. Def. Amorph. Mater.* **12** 31

Birringer R 1989 *Mat. Sci. and Eng. A* **117** 33

Borg R J and Dienes G J 1988 *An Introduction to Solid State Diffusion* (New York: Academic)

Born M 1952 *Atomic Physics* 5th edition (London: Blackie)

Born M and Huang K 1954 *Dynamical Theory of Crystal Lattices* (Oxford: Clarendon)

Born M and Mayer J E 1932 *Z. Physik* **75** 1

Bourgouin J and Lannoo M 1983 *Point Defects in Semiconductors II* (Berlin: Springer-Verlag)

Brillouin L 1933 *Actualités sci. et ind.* **71**

Brillouin L 1934 *Actualités sci. et ind.* **159**

Burns G 1985 *Solid State Physics* (Orlando, FL: Academic)

Catlow C R A and Mackrodt W C 1982 *Computer Simulation of Solids* (Berlin: Springer-Verlag)

Cavenett B C, Hayes W and Hunter I C 1967 *Solid State Commun.* **5** 653

Chu P C W 1987 *Proc. Nat. Acad. Sci.* **84** 4681

Cioslowski J 2000 *Many-Electron Densities and Reduced Density Matrices* (New York: Kluwer Academic/Plenum)

Coleman A J and Yukalov V I 2000 *Reduced Density Matrices* (Berlin: Springer-Verlag)

334

Courant R 1937 *Differential and Integral Calculus* volume 1, 2nd edition (London: Blackie)

CRYSTAL: see Pisani C, Dovesi R and Roetti C 1988 *Hartree–Fock ab initio Treatment of Crystalline Systems* (Berlin: Springer-Verlag)

Davidson E R 1976 *Reduced Density Matrices in Quantum Chemistry* (New York: Academic)

Davidson E R and Silver D W 1977 *Phys. Lett.* **52** 403

Daw M S and Baskes M I 1983 *Phys. Rev. Lett.* **50** 1285

Daw M S and Baskes M I 1984 *Phys. Rev. B* **29** 6443

Daw M S, Foiles S M and Baskes M I 1993 *Materials Science Reports* **9** 251

Dederichs P H, Zeller R and Schroeder K 1980 *Point Defects in Metals II* (Berlin: Springer-Verlag)

Dick B G and Overhauser W A 1958 *Phys. Rev.* **112** 90

Eastman D E and Freeouf J L 1975 *Phys. Rev. Lett.* **34** 395

Feher G 1959 *Phys. Rev.* **114** 1219 1245

Feynman R P 1955 *Phys. Rev.* **97** 660

Flynn C P and Stoneham A M 1970 *Phys. Rev. B* **1** 3966

Fowler W B 1964 *Phys. Rev.* **135** A1725

Fowler W B 1968 *Physics of Color Centers* (New York: Academic)

Friedel J 1964 *Dislocations* (Oxford: Pergamon)

Fröhlich H 1954 *Proc. Phys. Soc. (London) A* **223** 296

Fröhlich H 1963 in *Polarons and Excitons* eds C G Kuper and G D Whitfield (Edinburgh: Oliver and Boyd) p 1

Gale J D 1997 *J. Chem. Soc. Faraday Trans.* **93** 629

GAUSSIAN Inc, Carnegie Office Park, Bldg. 6, Carnegie PA 15160 USA, Email: info@gaussian.com

Gilbert T L 1964 'Self-consistent equations for local orbitals in polyatomic systems' in *Molecular Orbitals in Chemistry, Physics and Biology* eds P-O Löwdin and B Pullman (New York: Academic) p 405

Goldstein H 1980 *Classical Mechanics* 2nd edition (Reading, MA: Addison-Wesley)

Goswami A 1992 *Quantum Mechanics* (Dubuque, IA: Wm C Brown)

Green A E and Zerna W 1968 *Theoretical Elasticity* (Oxford: Clarendon)

Greiner W 1989 *Quantum Mechanics: An Introduction* (Berlin: Springer-Verlag)

Grosso G and Pastori Parravicini G 2000 *Solid State Physics* (New York: Academic)

Grüner G 1994 *Density Waves in Solids* (Reading MA: Addison-Wesley)

Harding J H 1982 Harwell Report AERE-R 10546

Harding J H, Harker A H, Keegstra P, Pandey R, Vail J M and Woodward C 1985 *Physica* **131B** 151

Henke K P, Richtering M and Ruhrberg W 1986 *Solid State Ionics* **21** 171

Hofmann D M, Lohse F, Paus H J, Smith D Y and Spaeth J-M 1985 *J. Phys. C: Solid State Phys.* **18** 443

Hohenberg P and Kohn W 1964 *Phys. Rev.* **136** B864

Huang K 1967 *Statistical Mechanics* (New York: Wiley)

Hull D and Bacon D J 1995 *Introduction to Dislocations* 3rd edition (Woburn, MA: Butterworth-Heinemann)

Husimi K 1940 *Proc. Math. Phys. Soc. Japan* **22** 264

Hutiray Gy and Sólyom J eds 1985 *Charge Density Waves in Solids* (Berlin: Springer-Verlag)

Huzinaga S ed 1984 *Gaussian Basis Sets for Molecular Calculations* (New York: Elsevier)

Itoh N and Stoneham A M 2000 *Materials Modification by Electronic Excitation* (Cambridge: Cambridge University Press)

Jackson J D 1962 *Classical Electrodynamics* (New York: Wiley)

Jacobs P W M, Corish J and Catlow C R A 1980 *J. Phys. C: Solid State Physics* **13** 1977

Jahnke E and Emde F 1945 *Tables of Functions* 4th edition (New York: Dover)

Kadono R, Kiefl R F, Ausaldo E J, Brewer J H, Celio M, Kreitzman S R and Luke G M 1990 *Phys. Rev. Lett.* **64** 665

Kagan Yu and Klinger M I 1974 *J. Phys. C* **7** 2791

Kagan Yu and Prokofev N V 1990 *Phys. Lett.* **150** 320

Kiefl R F, Kadono R, Brewer J H, Luke G M, Yen H K, Celio M and Ansaldo E J 1989 *Phys. Rev. Lett.* **62** 792

Kim P, Zhang J and Lieber C M 2001 *Solid State Physics* **55** 120 ed H Ehrenreich and F Spaenen (New York: Academic)

Kittel C 1953 *Introduction to Solid State Physics* 3rd edition (New York: Wiley)

Kohn W 1959 *Phys. Rev. Letters* **2** 393

Kuiper P, Kruizinga G, Ghijsen J, Sawatzky G H and Verweij H 1989 *Phys. Rev. Lett.* **62** 221

Kunz A B 1982 *Phys. Rev. B* **26** 2056

Kunz A B and Klein D L 1978 *Phys. Rev. B* **17** 4614

Lakes R 1987 *Science* **235** 1038

Lamb W E and Retherford R C 1947 *Phys. Rev.* **72** 241

Landau L 1933 *Phys. Zeit. Sowjetunion* **3** 664

Landau L D and Lifshitz E M 1970 *Theory of Elasticity* (Oxford: Pergamon)

Lannoo M and Bourgouin J 1981 *Point Defects in Semiconductors I* (Berlin: Springer-Verlag)

Leibfried G and Breuer N 1978 *Point Defects in Metals I* (Berlin: Springer-Verlag)

Leslie M 1981 Daresbury Laboratory Report No. DL-SCI-TM31T, UK Science and Engineering Research Council

Lidiard A B and Norgett R M J 1972 in *Computational Solid State Physics* eds F Herman, N W Dalton and T R Koehler (New York: Plenum) p 385

Low W and Offenbacher E L 1965 in *Solid State Physics*, eds F Seitz and D Turnbull (New York: Academic) volume 17, section IV and table AI, p 135

Madelung O 1978 *Introduction to Solid State Theory* (Berlin: Springer-Verlag)

Majlis N 2000 *The Quantum Theory of Magnetism* (Singapore: World Scientific)

Maradudin A A 1963 'Topics in the theory of vibrations of perfect crystals' in *Astrophysics and the Many-Body Problem* 1962 Brandeis Lectures, volume 2 (New York: Benjamin) p 107

Marder M P 2000 *Condensed Matter Physics* (New York: Wiley)

Mathieu E 1868 *J. de Math.* (2) **XIII** 137

Mattis D C 1988 *The Theory of Magnetism I. Statics and Dynamics* (Berlin: Springer-Verlag)

McMullen T and Bergersen B 1978 *Solid State Commun.* **28** 31

McMullen T, Meng J, Vail J M and Jena P 1995 *Phys. Rev. B* **51** 15879

Meng J, Jena P and Vail J M 1990b *J. Phys.: Condens. Matter* **2** 10371

Meng J, Kunz A B and Woodward C 1988 *Phys. Rev. B* **38** 10870

Meng J, Pandey R and Vail J 1989 *J. Phys.: Condens. Matter* **1** 6049

Meng J, Vail J M, Stoneham A M and Jena P 1990a *Phys. Rev. B* **42** 1156

Mollenauer L F and Baldacchini G 1972 *Phys. Rev. Lett.* **29** 465

Morrish A H 1965 *The Physical Principles of Magnetism* (New York: Wiley)

Morse P M and Feshbach H 1953 *Methods of Theoretical Physics* (2 volumes) (New York: McGraw-Hill)

Mott N F and Gurney R W 1948 *Electronic Processes in Ionic Crystals*, 2nd edition (Oxford: Oxford University Press)

Mott N F and Jones H 1936 *The Theory of the Properties of Metals and Alloys* (Oxford: Clarendon Press)

Müller E W and Tsong T T 1969 *Field Ion Microsopy* (New York: Elsevier)

Müller K A and Bednorz J G 1987 *Proc. Nat. Acad. Sci.* **84** 4678

Nabarro F R N 1967 *Theory of Crystal Dislocations* (Oxford: Clarendon)

Nabarro F R N 1979–1996 Series Editor *Dislocations in Solids* vols 1–10 (North-Holland: Elsevier)

Nadgorny E 1988 *Dislocation Dynamics and Mechanical Properties of Crystals* (Oxford: Pergamon)

Norgett M J 1974 *Harwell Rep. No.* AERE-R.7650

Nye J F 1957 *Physical Properties of Crystals* (Oxford: Clarendon)

Pandey R and Kunz A B 1990 *J. Phys. Chem. Solids* **51** 929

Pandey R, Zuo J and Kunz A B 1989 *Phys. Rev. B* **39** 12565

Parr R G and Yang W 1989 *Density Functional Theory of Atoms and Molecules* (New York: Oxford)

Payne S A, Goldberg A B and McClure D S 1984 *J. Chem. Phys.* **81** 1529

Peierls R 1955 *Quantum Theory of Solids* (Oxford: Clarendon)

Pershits Ya N and Kallenikova T A 1981 *Sov. Phys. Solid State* **23** 1497

Pines D 1963 in *Polarons and Excitons* eds C G Kuper and G D Whitfield (Edinburgh: Oliver and Boyd) p 33

Pippard A B 1962 'The dynamics of conduction electrons' in *Low Temperature Physics*, Les Houches Summer School of Theoretical Physics, Grenoble, 1961, eds C DeWitt, B Dreyfus and P G de Gennes (New York: Gordon and Breach) p 4

Pippard A B 1964 *The Elements of Classical Thermodynamics* (Cambridge: Cambridge University Press)

Pryce M H L 1961 'The theory of transition-element ions' Lectures presented at the 1961 Seminar of the Theoretical Physics Division of the Canadian Association of Physicists, at the University of Montreal, August 14 to September 9, unpublished

Puls M P, Woo C H and Norgett M J 1977 *Philos. Mag.* **36** 1457

Rabier J and Puls M P 1987 *Cryst. Latt. Defects and Amorph. Mat.* **16** 131

Rao S I, Parthasarathy T A and Woodward C 1999 *Philos. Mag. A* **79** 1167

Rayleigh Lord 1885 *Proc. London Math. Soc.* **17** 4

Reitz J R 1955 'Methods of the one-electron theory of solids' in *Solid State Physics* **1** eds F Seitz and D Turnbull (New York: Academic) p 1

Reitz J R, Milford F J and Christy R W 1979 *Foundations of Electromagnetic Theory* 3rd edition (Reading, MA: Addison-Wesley)

Roessler D M and Walker W C 1966 *Phys. Rev. Lett.* **17** 319

Roessler D M and Walker W C 1967 *Phys. Rev.* **159** 733

Sakurai J J 1967 *Advanced Quantum Mechanics* (Reading, MA: Addison-Wesley)

Schafroth M R 1958 *Phys. Rev.* **111** 72

Schafroth M R 1960 in *Solid State Physics* **10** eds F Seitz and D Turnbull (New York: Academic) p 295

Schrödinger E 1957 *Statistical Thermodynamics* 2nd edition (Cambridge: Cambridge University Press)

Schultz T D 1963 in *Polarons and Excitons* eds C G Kuper and G D Whitfield (Edinburgh: Oliver and Boyd) pp 72–73

Seidel H and Wolf H C 1968 in *Physics of Color Centers* ed W B Fowler (New York: Academic) ch 8

Simmons J P, Rao S I and Dimiduk D M 1997 *Philos. Mag. A* **75** 1299

Slater J C 1951 *Phys. Rev.* **84** 179

Slater J C 1952 *Phys. Rev.* **87** 807

Slater J C 1963 *Quantum Theory of Molecules and Solids* volume 1 *Electronic Structure of Molecules* (New York: McGraw-Hill)

Sokolnikoff I S 1956 *Mathematical Theory of Elasticity* (New York: McGraw-Hill)

Stoneham A M 1975 *Theory of Defects in Solids* (Oxford: Clarendon)

Stoneham A M 1981 *Handbook of Interatomic Potentials: I—Ionic Crystals* Harwell Report AERE-R 9598

Sushko P V, Shluger A L and Catlow C R A 2000 *Surface Science* **450** 153

Szabo A and Ostlund N S 1989 *Modern Quantum Chemistry* (New York: McGraw-Hill)

Tessman J, Kahn A and Shockley W 1953 *Phys. Rev.* **92** 890

Thouless D J 1961 *The Quantum Mechanics of Many-Body Systems* (New York: Academic)

Topiol S, Moskowitz J W and Melius C F 1978 *J. Chem. Phys.* **68** 3264

Tosi M P 1968 'Cohesion of ionic solids in the Born model' in *Solid State Physics* **21** eds F Seitz, D Turnbull and H Ehrenreich p 1

Vail J M 1964 *Can. J. Phys.* **42** 329

Vail J M 2001 *Radiation Effects and Defects in Solids* **154** 211

Vail J M, Bromirski M, Emberley E, Lu T, Yang Z and Pandey R 1998b *Radiation Effects and Defects in Solids* **145** 29

Vail J M, Coish W A, He H and Yang A 2002 *Phys. Rev. B* **66** 014109

Vail J M, Emberly E, Lu T, Gu M and Pandey R 1998a *Phys. Rev. B* **57** 764

Vail J M, Pandey R and Kunz A B 1991 *Revs. Solid State Science* **5** 241

Vail J M and Yang Z 1993 *J. Phys.: Condens. Matter* **5** 7649

van Hove L 1953 *Phys. Rev.* **89** 1189

van Vleck J H 1932 *The Theory of Electric and Magnetic Susceptibilities* (Oxford: Oxford University Press)

Verwey E J W, Haaijman P H, Romeijn F C and Van Oosterhout G W 1950 *Philips Res. Rep.* **5** 173, cited in Henderson B and Wertz J E 1977 *Defects in Alkaline Earth Oxides* (London: Taylor and Francis) p 43

Vineyard G H 1957 *J. Phys. Chem. Sol.* **3** 121

Wannier G H 1959 *Elements of Solid State Theory* (Cambridge: Cambridge University Press)

Weber M J, Lecoq P, Ruchti R C, Woody C, Yen W M and Zhu R Y eds 1994 *Scintillators and Phosphor Materials* MRS Symposia Proc No. 348 (Pittsburgh, PA: Materials Research Society)

Weertman J and Weertman J R 1964 *Elementary Dislocation Theory* (New York: Macmillan)

White R M 1983 *Quantum Theory of Magnetism* (Berlin: Springer-Verlag)

Wilson J A and Yoffe A D 1969 *Adv. Phys.* **18** 193

Woods A D B, Cochran W and Brockhouse B N 1960 *Phys. Rev.* **119** 980

Woody C L, Levy P W and Kierstead J A 1989 *IEEE Trans. Nucl. Sci.* **36** 536

Wyckoff R W G 1963 *Crystal Structures* 2nd edition (New York: Interscience)

Zhu R Y 1994 *Nucl. Instrum. Methods Phys. Res. A* **340** 442

Exercises

Chapter 1

1.1 Consider a deformation field $\vec{u}(\vec{r})$ in a homogeneous isotropic solid material, given by

$$\vec{u}(\vec{r}) = \sum_{\alpha=1}^{3} u_\alpha(\vec{r}) \hat{\varepsilon}_\alpha$$

in SI units, where $\hat{\varepsilon}_\alpha$ is a unit vector in the direction of the αth cartesian coordinate axis, and

$$\vec{r} = \sum_{\alpha=1}^{3} x_\alpha \hat{\varepsilon}_\alpha$$

is the position vector in the material. Suppose

$$u_1(\vec{r}) = 0.2x_1 - 5.6x_2 + 2.1x_3,$$
$$u_2(\vec{r}) = 0.2x_1 - 5.6x_2 + 2.1x_3,$$
$$u_3(\vec{r}) = 3.1x_3,$$

in units of 10^{-5} m.

(a) Evaluate the fractional volume change of the sample.
(b) Evaluate the shear angle of the material in the (x_1, x_2), (x_2, x_3) and (x_3, x_1) planes, and sketch the shear deformation of a rectangle in each of these planes.
(c) Evaluate the rotation of the sample, giving magnitude and direction of the rotation. For the direction of rotation, give the cosines of the angles between the axis of rotation and the x_1, x_2 and x_3 axes.

1.2 Consider a cubic sample of a homogeneous isotropic solid material of volume $1\,\mathrm{cm}^3$. It is subject to static stresses described by the stress tensor $\underline{\underline{\sigma}}$:

$$\underline{\underline{\sigma}} = \begin{pmatrix} 2.3 & -3.1 & 0.8 \\ -3.1 & -0.4 & 2.2 \\ 0.8 & 2.2 & 4.6 \end{pmatrix}$$

in units of $10^6 \, \mathrm{N \, m^{-2}}$.

Evaluate the force (vector) that must be applied uniformly to each of the six faces of the cube to produce this stress tensor. Show directly that the resultant torque from all these forces is zero.

1.3 The stresses described in problem 1.2 produce the following strain tensor $\underline{\underline{\varepsilon}}$:

$$\underline{\underline{\varepsilon}} = \begin{pmatrix} 0.6831 & -3.2529 & 0.8395 \\ -3.2529 & 2.1500 & 2.3085 \\ 0.8395 & 2.3085 & 3.0965 \end{pmatrix}$$

in units of 10^{-5}. What are the Voigt elastic constants c_{11} and c_{12}, accurate to two significant figures, for the material?

1.4 Using only equations (1.67), (1.68) and (1.77), derive the expression for Poisson's ratio in terms of Young's modulus and the bulk modulus.

Chapter 2

2.1 From the definitions of grad, div and curl, derive equations (2.1), (A2.2) and (A2.12).

2.2 Equation (2.33) is consistent to first-order small quantities, for the dynamics of an isotropic solid continuum. Carry out the derivation of this chapter to second order, beginning with equation (2.26), to obtain the corrections to the quantity $(\rho \vec{a})$, equation (2.18), right-hand side. Can the new expression be equated to the force per unit volume, equation (2.16)? Explain your answer.

2.3 Suppose that, in equation (2.33), ρ_0 is replaced by

$$\rho_0 (1 - \vec{\nabla} \cdot \vec{u})$$

introducing a second-order correction.

(a) Do shear waves propagate undeformed?
(b) If so, are plane shear waves transverse, and what is their speed of propagation?
(c) Do dilatational waves propagate undeformed?
(d) If so, are plane dilatational waves longitudinal, and what is their speed of propagation?

In each part, explain your answer.

Chapter 3

3.1 From equation (3.47), evaluate the phase difference (in radians) between the damped plane wave and its associated thermal distribution wave in an isotropic continuous solid medium.

3.2 Derive equations (3.63) and (3.64) from equations (3.61) and (3.62).

3.3 Derive equations (3.68) and (3.69) from equations (3.66) and (3.67).

Chapter 4

4.1 Using computer software, plot the curve γ versus ν, for $-1 \leq \nu \leq \frac{1}{2}$, equation (4.22). Then plot the family of cubics in (ξ^2), the left-hand side of equation (4.21), for a selection of values of γ within the appropriate range. From the resulting plot, verify the qualitative assertions that are deduced between equations (4.24) and (4.26).

4.2 Derive equation (4.34) from equation (4.33). Then derive equation (4.35). For $z = 0$, i.e. on the surface, use computer software to plot R versus ν in the physical range, and thereby verify the qualitative assertions that are made between equations (4.35) and (4.36). Plot R, equation (4.35), versus z for $c_T = 2.31 \times 10^3$ m/s and $c_L = 4.69 \times 10^3$ m/s, at a frequency of 10^6 s^{-1} and thereby verify the qualitative assertions that are made in the last two sentences of Chapter 4.

4.3 Derive an expression in terms of c_T, ξ and angular frequency ω for the depth $|z|$ at which a Rayleigh wave has no x-component: see equation (4.35) etc.

Chapter 5

5.1 Instead of our equations (5.38) for the deformation field of an edge dislocation, the Weertmans (1964) give the following solution:

$$u_1 = -\frac{b}{2\pi}\left\{ \tan^{-1}\left(\frac{y}{x}\right) + \frac{(\lambda+\mu)}{(\lambda+2\mu)}\frac{xy}{(x^2+y^2)}\right\}$$

$$u_2 = -\frac{b}{2\pi}\left\{ -\frac{\mu}{2(\lambda+2\mu)}\ln\left(\frac{x^2+y^2}{R^2}\right) + \frac{(\lambda+\mu)}{(\lambda+2\mu)}\frac{y^2}{(x^2+y^2)}\right\}.$$

(a) Derive the strain tensor for this case.

(b) Derive the stress tensor. Note that σ_{33} is not zero in this case.

(c) Derive the force per unit area (P_1, P_2) on the surface of a cylinder of radius R with axis coinciding with the dislocation. Show that in cylindrical coordinates, $P_1 \sim \sin(2\theta)$ and $P_2 \sim \cos(2\theta)$.

(d) Prove that the net force on the surface at $r = R$ is zero.
(e) Show that the given deformation field is a function of only one material constant, ν, Poisson's ratio.

Chapter 6

6.1 (a) Evaluate the polaron fractional mass renormalization

$$\left(\frac{m^{*\prime}}{m} - 1\right) = \frac{n_0}{3\,\mu m}\left(\frac{e}{\varepsilon_0\omega^2}\right)^2$$

[taken from equation (6.71)], for KCl. In order to do so, use Fröhlich's formula, adjusted for SI units

$$\varepsilon_0\mu\omega^2 = \overline{K}, \qquad \overline{K} = (K_\infty^{-1} - K_0^{-1})^{-1}$$

where K_0 and K_∞ are low- and high-frequency dielectric constants respectively. Also use the relationship

$$\omega^2 = \left(\frac{K_0}{K_\infty}\right)\omega_t^2.$$

To evaluate n_0, the number of particles per unit volume, use the estimate given by

$$n_0 = \frac{\rho_0}{\frac{1}{2}(m_K + m_{Cl})}$$

where m_K and m_{Cl} are the atomic masses of K and Cl respectively, and ρ_0 is the mass density. For KCl, use the following data:

$$\rho_0 = 1.99\,\mathrm{g\,cm^{-3}}$$
$$m_K = 64.9 \times 10^{-27}\,\mathrm{kg}$$
$$m_{Cl} = 58.9 \times 10^{-27}\,\mathrm{kg}$$
$$K_0 = 4.68$$
$$K_\infty = 2.13$$
$$\omega_t = 2.71 \times 10^{13}\,\mathrm{s^{-1}}$$

(b) Evaluate the polaron self energy E_s', equation (6.69), for KCl.
6.2 Referring to section 6.4, evaluate:

$$\left(\pi_\alpha\pi_\alpha\pi_\beta\pi_\beta - \pi_\alpha\pi_\beta\pi_\beta\pi_\alpha\right),$$

in terms of the applied $\vec{B}$ field or its source current density.

Chapter 7

7.1 Consider a form of the adiabatic approximation, equation (7.8), in which the electronic part of the wave function, taken to be normalized, depends linearly on the nuclear displacements $\underline{u}$, equation (7.14), as follows:

$$\psi_\lambda(\underline{r}, \underline{R}) = \{\psi_\lambda^{(0)}(\underline{r}, \underline{R}_0) + \underline{\psi}_\lambda^{(1)\mathrm{T}} \cdot \underline{u}\}$$

$$\underline{\psi}_\lambda^{(1)}(\underline{r}, \underline{R}_0) = (\nabla_J \psi_\lambda)_{\underline{R} = \underline{R}_0}.$$

(a) Determine the new effective 'hamiltonian' for nuclear motion in the harmonic approximation. We have placed 'hamiltonian' in quotation marks because the kinetic energy term now involves the variables $\underline{u}$.

(b) Ignoring this complication, determine the modified equilibrium condition and force constant matrix.

Note: If we do not ignore the complication in the kinetic energy term, we must treat the $\underline{u}$-dependent parts as momentum-dependent contributions to the potential energy, or as constituting a mass renormalization in the kinetic energy.

7.2 Determine the analytical expression for the coefficient of thermal expansion at constant pressure for the crystalline solid in terms its phonon frequencies.

Chapter 8

8.1 Consider the monatomic linear chain of section 8.2 where, however, the atomic motions are constrained to one transverse direction (say the y direction). Assume that atomic displacements are all small compared with the interatomic spacing.

(a) Assuming that the interatomic 'springs' are not under compression or tension when the atoms are undisplaced, derive the equations of motion for the atoms. Is the motion harmonic? Explain your answer.

(b) Suppose the 'springs' are all under tension, corresponding to elongation x_0, when the atoms are undisplaced. Derive the equations of motions for the atoms. Is the motion harmonic? Explain your answer.

(c) What is the situation if the 'springs' are under compression when the atoms are undisplaced?

Chapter 9

9.1 Consider a monatomic linear chain, as in section 8.2, consisting of shell-model type atoms, with uncharged cores and shells. Let the core and shell masses be M and m respectively, and let shell–shell and shell–core force constants be K and k respectively.

(a) Derive the equations of motion for the system (in longitudinal vibration).

(b) In general, such a system will have two branches to its phonon spectrum, as in section 8.3, even though there is only one atom per primitive unit cell. This will not be the case, experimentally. This problem is commonly dealt with in shell-model simulations by taking $m = 0$, which may seem reasonable if the shells represent the polarizable part of the electron cloud. Making the approximation $m = 0$, evaluate the dispersion relation.

Compare this dispersion relation with that obtained for rigid atoms, equation (8.22).

Chapter 10

10.1 Derive equation (10.64), using equations (10.61)–(10.63).

10.2 Consider a large (essentially infinite) sample of a binary alloy of atomic species X and Y. Initially, to the left of a planar interface, the concentration of X is C_1, and to the right it is C_2, with $C_1 > C_2$. The concentration difference, $C_0 = (C_1 - C_2)$ is small, so that the diffusion constant is essentially the same for species X in both regions. Solve equation (10.19), i.e. evaluate equation (10.43), to determine the concentration difference in the region to the right of the interface as a function of position and time.

Chapter 12

12.1 Carry out the demonstration mentioned in the last sentence of section 12.2.5, expressing the two-particle part of the total energy in Hartree–Fock approximation as a single linear operator acting on the single-particle density matrix.

12.2 Derive equation (12.195) from equation (12.193).

12.3 Explain in words the meaning of

$$\sum_{\substack{a<b(\text{occ}) \\ c<d(\text{virt})}}$$

in equation (12.211).

12.4 Refer to the two-particle density formulation of the many-body problem, equations (12.243) and (12.251).

    (a) Express total spin in terms of ρ_2.
    (b) Express total angular momentum in terms of ρ_2.

12.5 Refer to equation (12.110). We want to analyse, at the Hartree–Fock level, a hydrogen diatomic molecule, using an atomic orbital basis consisting of an s-type ($l = 0$) and a p-type ($l = 1$) orbital centred on each of the nuclei (protons). Write out explicit formulae for each of these four basis functions, normalized. Give a diagram on which your notation is based.

Chapter 13

13.1 The low-temperature paramagnetic susceptibility of an electron–gas system is 1.36×10^{-5}. What is the electron density of the system? Assume free-electron mass for the electrons.
13.2 From equation (13.130), prove equation (13.132).
13.3 Evaluate the integral of equation (A13.1) by considering a contour C in the *third* quadrant of the complex-number plane; i.e. construct the argument corresponding to equations (A.13.8)–(A.13.12).

Chapter 14

14.1 Prove the assertion following equation (14.35): 'If $v(r = 0)$ ... same for all j.'
14.2 From equation (14.194) and the discussion following it, show that this theory is valid only for $s < 32$: see also equation (14.202).
14.3 Show, both analytically and graphically, that at $g = 0$

$$\frac{\mathrm{d}}{\mathrm{d}g}(\mathrm{e}^{-1/g}) = 0.$$

See equation (14.217), and the discussion following equation (14.225).

Answers

Chapter 1

1.1   (a) $\Delta V/V = -2.3 \times 10^{-5}$.

     (b) $\gamma_{12} = -2.7 \times 10^{-5}$ radians, <0.
         $\gamma_{23} = 1.05 \times 10^{-5}$ radians, >0.
         $\gamma_{31} = 1.05 \times 10^{-5}$ radians, >0.

     (c) $R = 3.3 \times 10^{-5}$ radians: $\cos\theta_1 = -1.05 \times 10^{-5}$,
         $\cos\theta_2 = 1.05 \times 10^{-5}$, $\cos\theta_3 = 2.9 \times 10^{-5}$.

1.2   Unit outward normal vectors $\hat{n}_j, j = 1, 2, \ldots, 6$ on the faces of the cube:

$$\hat{n}_1 = (-1, 0, 0), \vec{F}_1 = (-2.3, 3.1, -0.8);$$
$$\hat{n}_2 = (1, 0, 0), \vec{F}_2 = (2.3, -3.1, 0.8);$$
$$\hat{n}_3 = (0, -1, 0), \vec{F}_3 = (3.1, 0.4, -2.2);$$
$$\hat{n}_4 = (0, 1, 0), \vec{F}_4 = (-3.1, -0.4, 2.2);$$
$$\hat{n}_5 = (0, 0, -1), \vec{F}_5 = (-0.8, -2.2, -4.6);$$
$$\hat{n}_6 = (0, 0, 1), \vec{F}_6 = (0.8, 2.2, 4.6);$$

all in units ($\times 10^2$ N).

       Resultant torque $\vec{\tau}$:

$$\vec{\tau} = \sum_j (\vec{r}_j \times \vec{F}_j),$$

$\vec{r}_j = (0.5 \times 10^{-2})\hat{n}_j N$, whence $\vec{\tau} = 0$.

1.3 $c_{11} = 2.0 \times 10^{11}\,\mathrm{N\,m^{-2}}$,
     $c_{12} = 1.0 \times 10^{11}\,\mathrm{N\,m^{-2}}$.

1.4 $\nu = \dfrac{1}{2}\left(1 - \dfrac{E}{3B}\right)$.

Chapter 2

2.2   Correction to $(\rho\vec{a})$ is

$$\rho_0\{-(\vec{\nabla}\cdot\vec{u})\partial_t^2 + \partial_t(\vec{u}\cdot\vec{\nabla})\partial_t + \partial_t(\partial_t\vec{u})\cdot\vec{\nabla}\}\vec{u}.$$

The force per unit volume, equation (2.16), has not been derived to second order, so the new expression for $(\rho\vec{a})$ cannot be used in this way.

2.3 (a) Yes, because for shear waves $(\vec{\nabla}\cdot\vec{u}) = 0$, so ρ is unchanged.

(b) Shear waves are still transverse with speed v_T because the wave equation is unchanged.

(c) No. The equation for dilatational waves is now

$$\{(\lambda + 2\mu)\nabla^2\vec{u} - \rho_0\partial_t^2\vec{u}\} = -\rho_0(\vec{\nabla}\cdot\vec{u})\partial_t^2\vec{u}.$$

A travelling wave like equation (2.45) does not satisfy this equation. The effective speed of propagation is

$$\left[\frac{(\lambda + 2\mu)}{\rho_0(1 - \vec{\nabla}\cdot\vec{u})}\right]^{1/2},$$

which depends on $(\vec{r}, t)$.

Chapter 3

3.1 $\theta = (\pi + \theta')$; $\theta' = \tan^{-1}\left(\dfrac{\omega c_v}{k^2\chi}\right)$.

Chapter 4

4.3 $|z| = \dfrac{c_T\xi}{\omega}\left|\dfrac{1}{[(1 - \xi^2\gamma)^{1/2} - (1 - \xi^2)^{1/2}]}\ln\left\{\dfrac{(1 - \xi^2)^{1/2}(1 - \xi^2\gamma)^{1/2}}{(1 - \xi^2/2)}\right\}\right|.$

Chapter 5

5.1 (a) $\varepsilon_{11} = \dfrac{by}{2\pi}\left\{\dfrac{(2\lambda + 3\mu)x^2 + \mu y^2}{(\lambda + 2\mu)(x^2 + y^2)^2}\right\},$

$\varepsilon_{22} = -\dfrac{by}{2\pi}\left\{\dfrac{(2\lambda + \mu)x^2 - \mu y^2}{(\lambda + 2\mu)(x^2 + y^2)^2}\right\},$

$\varepsilon_{12} = -\dfrac{1}{2}\dfrac{by}{2\pi}\dfrac{2(\lambda + \mu)}{(\lambda + 2\mu)}\dfrac{(x^2 - y^2)}{(x^2 + y^2)^2}.$

Note the factor $\frac{1}{2}$ in ε_{12}, omitted in Weertmans. Note:

$$\frac{2(\lambda + \mu)}{(\lambda + 2\mu)} = \frac{1}{(1 - \nu)}.$$

(b) $\sigma_{11} = \dfrac{by}{2\pi}\dfrac{\mu}{(1-\nu)}\dfrac{(3x^2+y^2)}{(x^2+y^2)^2}$,

$\sigma_{22} = -\dfrac{by}{2\pi}\dfrac{\mu}{(1-\nu)}\dfrac{(x^2-y^2)}{(x^2+y^2)^2}$,

$\sigma_{33} = \dfrac{by}{2\pi}\dfrac{\nu\mu}{(1-\nu)}\dfrac{1}{(x^2+y^2)}$,

$\sigma_{12} = -\dfrac{bx}{2\pi}\dfrac{\mu}{(1-\nu)}\dfrac{(x^2-y^2)}{(x^2+y^2)^2}$.

(c) $P_1 = \dfrac{b\mu\sin(2\theta)}{2\pi(1-\nu)r}$,

$P_2 = \dfrac{-b\mu\cos(2\theta)}{2\pi(1-\nu)r}$.

(d) $\displaystyle\int_{\theta=0}^{2\pi} d\theta\, RP_j = 0,\qquad j=1,2.$

(e)   From $\dfrac{2(\lambda+\mu)}{(\lambda+2\mu)} = \dfrac{1}{(1-\nu)}, \nu = f\left(\dfrac{\mu}{\lambda}\right) \rightarrow \left(\dfrac{\mu}{\lambda}\right) = g(\nu).$

Chapter 6

6.1 (a) $\left(\dfrac{m^{*\prime}}{m}-1\right) = 5.4\times10^3.$

(b) $E_s' = 1.5\,\text{eV}.$

6.2 $(\pi_\alpha\pi_\alpha\pi_\beta\pi_\beta - \pi_\alpha\pi_\beta\pi_\beta\pi_\alpha) = -e\hbar^2\vec{\pi}\cdot(\vec{\nabla}\times\vec{B}) = -e\hbar^2\mu\vec{\pi}\cdot\vec{J}.$

Chapter 7

7.1 Omitting the index λ, and replacing bracketed superscripts by subscripts, write

$$\psi = (\psi_0 + \underline{\psi}_1^{\text{T}}\cdot\underline{u})$$

or, in Dirac notation,

$$|\psi\rangle = (|0\rangle + |\underline{1}\rangle^{\text{T}}\cdot\underline{u}),$$

where column matrices $|\underline{1}\rangle$ and $\underline{u}$ have the dimensionality of $\underline{R}_0$, section 7.3.

(a) Then the effective hamiltonian H_{eff} is

$$H_{\text{eff}} = \langle \psi | H | \psi \rangle$$

$$\approx \{ \langle 0|0 \rangle T_n + \underline{u}^{\text{T}} \cdot (T_n \langle \underline{1}|0 \rangle) + (T_n \langle 0|\underline{1} \rangle^{\text{T}} \cdot \underline{u})$$

$$+ (\underline{u}^{\text{T}} \cdot \langle \underline{1}|)(|\underline{1} \rangle^{\text{T}} T_n \cdot \underline{u}) \} + \langle 0|0 \rangle W_0$$

$$+ \{ \langle 0|\underline{\tilde{W}}_1^{\text{T}}|0 \rangle + 2 W_0 \langle 0|\underline{1} \rangle^{\text{T}} \} \cdot \underline{u}$$

$$+ \tfrac{1}{2} \underline{u}^{\text{T}} \cdot \{ \langle 0|\underline{\tilde{W}}_2|0 \rangle + 2 \langle 0|\underline{\tilde{W}}_1|\underline{1} \rangle^{\text{T}}$$

$$+ 2 \langle \underline{1}|\underline{\tilde{W}}_1^{\text{T}}|0 \rangle + 2 W_0 \langle \underline{1}|\underline{1} \rangle^{\text{T}} \} \cdot \underline{u}.$$

(b) Equilibrium condition:

$$\{ \langle 0|\underline{\tilde{W}}_1^{\text{T}}|0 \rangle + 2 W_0 \langle 0|\underline{1} \rangle^{\text{T}} \} = 0.$$

Force constant matrix:

$$\{ \langle 0|\underline{\tilde{W}}_2|0 \rangle + 2 \langle 0|\underline{\tilde{W}}_1(|\underline{1} \rangle^{\text{T}}) + 2 \langle \underline{1}|\underline{\tilde{W}}_1^{\text{T}}|0 \rangle + 2 W_0 \langle \underline{1}|\underline{1} \rangle^{\text{T}} \}.$$

Note outer products of the form $\underline{\tilde{W}}_1(|\underline{1} \rangle^{\text{T}})$, giving square matrices. We have introduced the notation:

$$\underline{\tilde{W}}_1 = \left(\frac{\partial \tilde{W}}{\partial \underline{R}} \right)_{\underline{R} = \underline{R}_0}; \qquad \underline{\tilde{W}}_2 = \left(\frac{\partial^2 \tilde{W}}{\partial \underline{R} \, \partial \underline{R}} \right)_{\underline{R} = \underline{R}_0},$$

$$\tilde{W} = \{ V_n(\underline{R}) + H_e(\underline{r}) + V_{ne}(\underline{R}, \underline{r}) \},$$

see equations (7.3)–(7.6).

7.2

$$\frac{1}{V} \left(\frac{\partial V}{\partial T} \right)_{\text{p}} = \frac{-\dfrac{1}{V} \left(\dfrac{\partial p}{\partial T} \right)_{\text{V}}}{\left(\dfrac{\partial p}{\partial V} \right)_{\text{T}}}$$

$$= \frac{\dfrac{1}{V} \sum_j \left\{ \dfrac{\hbar \omega_j}{2 k_B T^2} \left(\dfrac{d\omega_j}{\partial V} \right) \text{csch}^2 \left(\dfrac{\hbar \omega_j}{2 k_B T} \right) \right\}}{\sum_j \left\{ \dfrac{\hbar}{2 k_B T} \left(\dfrac{d\omega_j}{\partial V} \right)^2 \text{csch}^2 \left(\dfrac{\hbar \omega_j}{2 k_B T} \right) - \left(\dfrac{d^2 \omega_j}{dV^2} \right) \coth \left(\dfrac{\hbar \omega_j}{2 k_B T} \right) \right\}}.$$

Chapter 8

8.1 (a) $\dfrac{M \, d^2 y_j}{dt^2} = -\dfrac{K}{2 a^2} \{ (y_j - y_{j-1})^3 + (y_j - y_{j+1})^3 \},$

cubic, not linear in small displacements: anharmonic.

(b) $$\frac{M\,\mathrm{d}^2 y_j}{\mathrm{d}t^2} = -\frac{K}{2a^2}\{(y_j - y_{j-1})^3 + (y_j - y_{j+1})^3\}$$

$$-Kx_0\left\{\frac{(y_j - y_{j-1})}{[a^2 + (y_j - y_{j-1})^2]^{1/2}} + \frac{(y_j - y_{j+1})}{[a^2 + (y_j - y_{j+1})^2]^{1/2}}\right\}$$

$$\approx -\frac{Kx_0}{a}\{(y_j - y_{j-1}) + (y_j - y_{j+1})\},$$

to first order in small displacements: harmonic.

(c)   Under compression, the straight-line configuration is in unstable equilibrium, so the small oscillations take place relative to bow-shaped linear configurations above or below the x-axis.

Chapter 9

9.1   (a) $$\frac{M\,\mathrm{d}^2 U_j}{\mathrm{d}t^2} = -k(U_j - u_j),$$

$$\frac{m\,\mathrm{d}^2 u_j}{\mathrm{d}t^2} = -K[2u_j - (u_{j-1} + u_{j+1})] - k(u_j - U_j).$$

(b) $$\omega_n = 2\left(\frac{K}{M}\right)\left|\sin\left(\frac{n\pi}{N}\right)\right|\cdot\left[1 + \frac{4K}{k}\sin^2\left(\frac{n\pi}{N}\right)\right]^{1/2}.$$

Chapter 10

10.2 $$C(x,t) = \frac{C_0}{2}\left\{1 - \mathrm{erf}\left[\frac{x}{2(Dt)^{1/2}}\right]\right\}, \quad \text{for } x \geq 0,$$

where

$$\mathrm{erf}(y) = \int_0^y \mathrm{d}y'\exp(-y'^2).$$

Chapter 12

12.1 $$E_2 = \frac{1}{2}\int \mathrm{d}r\,\mathrm{d}r'\,|\vec{r} - \vec{r}'|^{-1}\int \mathrm{d}\mu\,\mathrm{d}\mu'\,\delta(\underline{\mu} - \underline{r})\delta(\underline{\mu}' - \underline{r}')$$

$$\times [1 - P(\underline{\mu}, \underline{\mu}')]\rho_1(\underline{r}, \underline{\mu})\rho_1(\underline{r}', \underline{\mu}'),$$

where $P(\underline{\mu}, \underline{\mu}')$ is the pairwise interchange operator.

12.3 We have Fock eigenstates labelled by $a, b, c, d, \ldots$, ordered in increasing energy $\varepsilon_a < \varepsilon_b < \varepsilon_c < \varepsilon_d \ldots$, divided into occupied and virtual (unoccupied) manifolds, with $\varepsilon_a < \varepsilon_b$ if a is occupied and b unoccupied. In equation (12.211), we have sums over pairs of occupied states and over pairs of unoccupied states:

$$\frac{1}{4} \underset{\substack{a,b \\ (\text{occ})}}{\sum}{}' \; \underset{\substack{c,d \\ (\text{virt})}}{\sum}{}',$$

the factor $(1/4) = (1/2)^2$ eliminating double counting, and primes avoiding self-interaction in the summand. An alternative way of doing such a double sum is:

$$\frac{1}{2} \sum_{a,b}{}' = \sum_{a<b},$$

where a and b are positive integer indices $1, 2, \ldots$

In $\sum_{a<b}$, the sum over b for a given value of a is limited to $b > a$.

12.4 (a) Total spin $\vec{S}$:

$$\vec{S} = \sum_{j=1}^{N} \vec{S}_1(r_j)$$

where $\vec{S}_1(r_j)$ is the single-electron spin operator, acting on the spin variable dependence of particle j. Then

$$\langle \psi | \vec{S} | \psi \rangle = \int d\underline{r}_1, \ldots, d\underline{r}_N \, \psi(\underline{r}_1, \ldots, \underline{r}_N) \sum_{j=1}^{N} \vec{S}_1(r_j) \psi(\underline{r}_1, \ldots, \underline{r}_N)$$

$$\equiv N \int d\underline{r} \, d\underline{r}' \int d\underline{q} \, \delta(\underline{r} - \underline{q}) \vec{S}_1(\underline{q}) \rho_2(\underline{r}, \underline{r}'; \underline{q}, \underline{r}').$$

(b) Total angular momentum $\vec{J}$:

$$\vec{J} = \sum_{j=1}^{N} J_1(r_j),$$

whence

$$J = N \int d\underline{r} \, d\underline{r}' \int d\underline{q} \, \delta(\underline{r} - \underline{q}) J_1(\underline{q}) \rho_2(\underline{r}, \underline{r}'; \underline{q}, \underline{r}').$$

12.5

$$\langle \underline{r}|1 \rangle = n_1 [\exp(-\alpha_{s1}|\vec{r} - \vec{R}_1|^2)],$$

$$\langle \underline{r}|2 \rangle = n_2 [\exp(-\alpha_{p1}|\vec{r} - \vec{R}_1|^2)] \cos \theta_1,$$

$$\langle \underline{r}|3 \rangle = n_3 [\exp(-\alpha_{s2}|\vec{r} - \vec{R}_2|^2)],$$

$$\langle \underline{r}|4 \rangle = n_4 [\exp(-\alpha_{p2}|\vec{r} - \vec{R}_2|^2)] \cos \theta_2.$$

θ_1 is the angle between the z axis and the vector $(\vec{r} - \vec{R}_1)$; θ_2 is the corresponding angle with $(\vec{r} - \vec{R}_2)$. $\vec{R}_j$ are the proton positions, $j = 1, 2$. By symmetry, we should take $\alpha_{s1} = \alpha_{s2}$; $\alpha_{p1} = \alpha_{p2}$. Then

$$n_1 = n_3 = \left(\frac{2\alpha_{s1}}{\pi}\right)^{3/4}, \qquad n_2 = n_4 = 3^{1/2}\left(\frac{2\alpha_{p1}}{\pi}\right)^{3/4}.$$

Chapter 13

13.1 $4.0 \times 10^{29}\,\mathrm{m}^{-3}$.

13.3 Choose $\varphi = 5\pi/4$. Then:

$$0 = \oint_c dz\, e^{iz^2}$$

$$= \lim_{\substack{R \to \infty \\ \alpha \to 0^+}} \left\{ \int_{z=-\alpha(1+i)}^{z=-(R+i\alpha)} dz\, e^{iz^2} + \int_{r=R}^{r=\sqrt{2}\cdot\alpha} dr\, e^{i\cdot 5\pi/4} \exp(-r^2) \right.$$

$$\left. + \int_{\pi}^{5\pi/4} d\theta(iR\,e^{i\theta}) \exp[iR^2(\cos 2\theta + i\sin 2\theta)] \right\}$$

$$= \left\{ -\int_0^\infty dx\, e^{ix^2} + \frac{1}{2}\left(\frac{\pi}{2}\right)^{1/2}(1+i) \right\}$$

Chapter 14

14.1

$$H = \left\{ -\frac{\hbar^2}{2m}\sum_j \nabla_j - \frac{1}{2}\sum_{j,j'} v(|\vec{r}_j - \vec{r}_{j'}|) + \frac{N}{2}v(0) \right\}.$$

14.2 From equation (14.194), $s > 0$. This requires, in equation (14.202), that $\ln(s/32)$ be negative, whence $s < 32$.

14.3

$$\lim_{g \to 0}\frac{d}{dg}(e^{-1/g}) = \lim_{g \to 0}\left(\frac{e^{-1/g}}{g^2}\right), \quad \text{let } x = g^{-1}$$

$$= \lim_{x \to \infty}(x^2 e^{-x}) = \lim_{x \to \infty}\left(\frac{x^2}{e^x}\right)$$

$$= \lim_{x \to \infty}\left(\frac{2x}{e^x}\right) = \lim_{x \to \infty}\left(\frac{2}{e^x}\right) = 0,$$

having used l'Hôpital's rule twice.

Author index

Agullo-Lopez, F., 334, 154
Ansaldo, E. J., *see* Kadono
   see Kiefl
Appel, J., 334, 73
Ashcroft, N. W., 334, 109, 114, 163,
   187, 269, 315

Bachelet, G. B., 334, 194
Bacon, D. J., *see* Hull
Baldacchini, G., 334, 176, 186
   see Mollenauer
Bardeen, J., 334, 333
Baskes, M. I., *see* Daw (1983), (1984),
   (1993)
Baughman, R. H., 334, 19
Baym, G., 334, 73
Bednorz, J. G., *see* Müller, K. A.
Berg, J. M., 334, 179
Bergersen, B., *see* McMullen
   (1978)
Bhatia, A. B., 334, 2, 22, 34, 37
Bhyrappa, P., 334, 197
Bilz, H., 334, 129
Birringer, R., 334, 163
Borg, R. J., 334, 140, 141
Born, M., 334, 91, 109, 128, 166
Bourgouin, J., 334, 168
   see Lannoo
Breuer, N., *see* Leibfried
Brewer, J. H., *see* Kadono
   see Kiefl
Brillouin, L., 334, 235
Brockhouse, B. N., *see* Woods
Bromirski, M., *see* Vail (1998b)
Burns, G., 334, 164

Catlow, C. R. A., 334, 129
   see Agullo-Lopez
   see Jacobs
   see Sushko
Cavenett, B. C., 334, 181, 186
Celio, M., *see* Kadono
   see Kiefl
Christy, R. W., *see* Reitz (1979)
Chu, P. C. W., 334, 183
Cioslowski, J., 334, 248
Cochran, W., *see* Woods
Coish, W. A., 45, 46
   see Vail (2002)
Coleman, A. J., 334, 248
Cooper, L. N., *see* Bardeen
Corish, J., *see* Jacobs
Courant, R., 334, 161

Davidson, E. R., 335, 231, 248
Daw, M. S., 335, 72
Dederichs, P. H., 335, 168
Dick, B. G., 335, 128
Dienes, G. J., *see* Borg
Dimiduk, D. M., *see* Simmons
Dovesi, R., *see* Pisani

Eastman, D. E., 335, 183
Emberly, E., *see* Vail (1998a), (1998b)
Emde, F., *see* Jahnke

Feher, G., 335, 181
Feshbach, H., *see* Morse
Feynman, R. P., 335, 85, 187
Flynn, C. P., 335, 187
Foiles, S. M., *see* Daw (1993)

Fowler, W. B., 335, 73, 168
Freeouf, J. L., *see* Eastman
Friedel, J., 335, 71
Fröhlich, H., 335, 74, 85, 186, 298

Gale, J. D., 335, 129
Ghijsen, J., *see* Kuiper
Gilbert, T. L., 335, 219, 221
Goldberg, A. B., *see* Payne
Goldstein, H., 335, 74, 87, 251
Goswami, A., 335, 164
Green, A. E., 335, 21
Greiner, W., 335, 251
Grosso, G., 335, 251, 312
Grüner, G., 335, 298
Gu, M., *see* Vail (1998a)
Gurney, R. W., *see* Mott (1948)

Haaijman, P. H., *see* Verwey, E. J. W.
Hamann, D. R., *see* Bachelet
Harding, J. H., 335, 126, 167
Harker, A. H., *see* Harding (1985)
Hayes, W., *see* Cavenett
He, H., *see* Vail (2002)
Henke, K. P., 335, 169
Hofmann, D. M., 335, 173
Hohenberg, P., 335, 239, 248, 249
Huang, K., 335, 107, 250, 252, 255, 257, 259
Huang, K., *see* Born (1954)
Hull, D., 335, 71
Hunter, I. C., *see* Cavenett
Husimi, K, 335, 244
Hutiray, Gy, 335, 298
Huzinaga, S., 335, 174

Itoh, N., 335, 178

Jackson, J. D., 335, 284
Jacobs, P. W. M., 335, 128
Jahnke, E., 335, 261
Jena, P., *see* McMullen (1995)
  see Meng (1990a), (1990b)
Jones, H., *see* Mott (1936)

Kadono, R., 336, 184
  see Kiefl
Kagan, Yu, 336, 187, 188
Kahn, A., *see* Tessman

Kallenikova, T. A., *see* Pershits
Keegstra, P., *see* Harding (1985)
Kiefl, R. F., 336, 187
  see Kadono
Kierstead, J. A., *see* Woody, C. L.
Kim, P., 336, 298
Kittel, C., 336, 125
Klein, D. L., *see* Kunz (1978)
Klinger, M. I., *see* Kagan (1974)
Kohn, W., 336, 241, 298
  see Hohenberg
Kreitzman, S. R., *see* Kadono
Kress, W., *see* Bilz (1979)
Kruizinga, G., *see* Kuiper
Kuiper, P., 336, 183
Kunz, A. B., 336, 103, 194
  see Meng (1988)
  see Pandey (1989), (1990)
  see Vail (1991)

Lakes, R., 336, 19
Lamb, W. E., 336, 73
Landau, L., 336, 2, 16, 19, 22, 34, 48, 84
Lannoo, M., 336, 168
  see Bourgouin
Lecoq, P., *see* Weber
Leibfried, G., 336, 168
Leslie, M., 336, 128
Levy, P. W., *see* Woody, C. L.
Lidiard, A. B., 336, 128
Lieber, C. M., *see* Kim
Lifshitz, E. M., *see* Landau (1970)
Lohse, F., *see* Hofmann
Low, W., 336, 177
Lu, T., *see* Vail (1998a), (1998b)
Luke, G. M., *see* Kadono
  see Kiefl

Mackrodt, W. C., *see* Catlow
Madelung, O., 336, 269, 276
Majlis, N., 336, 250
Maradudin, A. A., 336, 94, 125
Marder, M. P., 336, 84, 85
Mathieu, E., 336, 313
Mattis, D. C., 336, 250
Mayer, J. E., *see* Born (1932)
McClure, D. S., *see* Berg
  see Payne

McMullen, T., 336, 187
Melius, C. F., *see* Topiol
Meng, J., 336, 168, 176, 179, 183
    see McMullen (1995)
Mermin, N. D., *see* Ashcroft
Milford, F. J., *see* Reitz (1979)
Mollenauer, L. F., 336, 186
    see Baldacchini (1973)
Morrish, A. H., 336, 250
Morse, P. M., 336, 24
Moskowitz, J. W., *see* Topiol
Mott, N. F., 337, 84, 318
Müller, E. W., 337, 216
Müller, K. A., 337, 183

Nabarro, F. R. N., 337, 69, 71
Nadgorny, E., 337, 71
Norgett, M. J., 337, 193
    see Lidiard
    see Puls
Nye, J. F., 337, 136

Offenbacher, E. L., *see* Low
Ostlund, N. S., *see* Szabo
Overhauser, W. A., *see* Dick

Pandey, R., 337, 178, 179
    see Harding (1985)
    see Meng (1989)
    see Vail (1998a), (1998b), (1991)
Parr, R. G., 337, 241
Parthasarathy, T. A., *see* Rao
Pastori Parravicini, G., *see* Grosso
Paus, H. J., *see* Hofmann
Payne, S. A., 337, 179
Peierls, R., 337, 94, 298
Pershits, Ya. N., 337, 169
Pines, D., 337, 85
Pippard, A. B., 337, 11, 259, 269, 274
Pisani, C., 334 (*see* CRYSTAL), 103, 218
Prokofev, N. V., *see* Kagan (1990)
Pryce, M. H. L., 337, 209
Puls, M. P., 337, 72
    see Rabier

Rabier, J., 337, 72
Rao, S. I., 337, 72
    see Simmons

Rayleigh, Lord, 337, 48
Reitz, J. R., 337, 133, 197, 254
Retherford, R. C., *see* Lamb
Richtering, M., *see* Henke
Roessler, D. M., 337, 178
Roetti, C., *see* Pisani
Romeijn, F. C., *see* Verwey, E. J. W.
Ruchti, R. C., *see* Weber
Ruhrberg, W., *see* Henke

Sakurai, J. J., 337, 73, 251
Sawatzky, G. H., *see* Kuiper
Schafroth, M. R., 337, 87, 89
Schlüter, M., *see* Bachelet
Schrieffer, J. R., *see* Bardeen
Schrödinger, E., 337, 104, 155
Schroeder, K., *see* Dederichs
Schultz, T. D., 337, 78
Seidel, H., 338, 181
Shacklette, J. M., *see* Baughman
Shluger, A. L., *see* Sushko
Shockley, W., *see* Tessman
Silver, D. W., *see* Davidson (1977)
Simmons, J. P., 338, 72
Singh, R. N., *see* Bhatia
Slater, J. C., 338, 197, 235, 298,
    313
Smith, D. Y., *see* Hofmann
Sokolnikoff, I. S., 338, 21
Sólyom, J., *see* Hutiray
Spaeth, J.-M., *see* Hofmann
Stafström, S., *see* Baughman
Stoneham, A. M., 338, 91, 129, 168, 181
    see Flynn
    see Itoh
    see Meng (1990a)
Sushko, P. V., 338, 196
Suslick, K. S., *see* Bhyrappa
Szabo, A., 338, 197, 217, 218, 235

Tessman, J., 338, 128
Thouless, D. J., 338, 231, 238
Topiol, S., 338, 194
Tosi, M. P., 338, 128
Townsend, P., *see* Agullo-Lopez
Tsong, T. T., *see* Müller, E. W.

Vaijayanthinala, G., *see* Bhyrappa

Vail, J. M., 338, 91, 172, 174, 177, 181,
    182, 186, 193, 229, 298
    see Harding (1985)
    see McMullen (1995)
    see Meng (1989), (1990a), (1990b)
van Hove, L., 338, 114
van Vleck, J. H., 338, 250
Van Oosterhout, G. W., *see* Verwey,
    E. J. W.
Verweij, H., *see* Kuiper
Verwey, E. J. W., 338, 177
Vineyard, G. H., 338, 161

Walker W. C., *see* Roessler (1966),
    (1967)
Wannier, G. H., 338, 114
Weber, M. J., 338, 171
Weertman, J., 338, 57, 61, 69, 71
Weertman, J. R., *see* Weertman, J.
White, R. M., 338, 250
Wilson, J. A., 338, 298
Wolf, H. C., *see* Seidel
Woo, C. H., *see* Puls

Woods, A. D. B., 338, 128
Woodward, C., *see* Harding (1985)
Woodward, C., *see* Meng (1988)
Woodward, C., *see* Rao
Woody, C. L., 338, 181
Woody, C., *see* Weber
Wyckoff, R. W. G., 338, 165

Yang, A., *see* Vail (2002)
Yang, W., *see* Parr
Yang, Z., *see* Vail (1998b), (1993)
Yen, H. K., *see* Kiefl
Yen, W. M., *see* Weber
Yoffe, A. D., *see* Wilson
Yukalov, V. I., *see* Coleman

Zakhidov, A. A., *see* Baughman
Zeller, R., *see* Dederichs
Zerna, W., *see* Green
Zhang, J., *see* Kim
Zhu, R. Y., 338, 180
    see Weber
Zuo, J., *see* Pandey (1989)

Subject index

Acoustic branch, phonons, 118
Activation energies, Cu and Ag in alkali halides, 168–171, *see also* Diffusion, Dissociation
Adiabatic approximation, 91–93, 126, 299
  average field approximation, 126
Aggregation, point defects, 140
Alkali halides, 128, 164, 168
American Physical Society, 197
Amorphous solids, 163
Angular momentum, many-body system, 245
Anharmonic crystal, 93
Anisotropic materials, 19
Antisymmetrizing operator, 200
Arrhenius relation, 161
Atomic, core, 127, 300
  diameters, 216
  dipole moment, 127
  multipole moments, 127
  orbitals, 216–218
    basis sets, 194, 217, 245
    primitive, 217
  quadrupole moments, 127
  units, Bohr–Hartree, 198
Atomicity of matter, 216
Atomistic modelling, crystals, classical, 126–139
Attenuation, wave, by thermal conduction, 41–47
Attenuation length, 42, 46–47
Auxetic materials, 19
Average field approximation, 91, 126, 299

BaF_2, 166, 190, 191, 192, 193, 238
  conduction band edge, defect spin splitting, chemical character, 184–185
  F center in, 86, 186
  optical absorption, 179–181
  oxygen in, 177, 193
  oxygen-vacancy defect complex in, 171–173
  radiation damage, 181
Balmer series, 166
Band mass, 85, 186, 251, 301
Band structure, electronic, 196
Band-edge modification, local, 167, 183–185
Basis of a crystal, 129, 163
Binding in a crystal, 92
Biology, molecular, 197
Bloch, equation, 310
  states, 187, 298, 310
    in quantum diffusion, 187–189
  theorem, 315
Body forces, 20
Bohr magneton, 251, 266
Bohr–Hartree atomic units, 198
Born–Mayer potential, 128
Born–von Karmann boundary conditions, 94, 95–96, 252, 305
Bose statistics, 107
Bravais lattice, of a crystal, 129, 163
  charge density waves, 315
Breakdown, 42
Breathing mode, 189
Brillouin zones, 312–313
Brillouin's theorem, 235–237

357

Brockhouse, B. N., 128
Buckingham potential, 171
  for shell model impurities, 171
  shell model, 128–129
Bulk, material, model, 94
  moduli, 17, 19, 35, 37
  modulus, adiabatic, 46
    isothermal, 46
Burgers vector, 59, 61, 63

CaF_2, 171
Calorimetric coefficient, 35
Canonical, coordinates, 98
  ensemble, 103
  momenta, 98
Cauchy, integral theorem, 295
  relation, 16
Charge density, 204
  exchange, 205
  of Bloch wave, 310
  waves, 31, 298–333
    and pertubative expansion, 332
    attractive pairwise interaction, 332
    Bravais lattice, 315
    commensurate, 312
    hamiltonian for, 301–302
    in two and three dimensions, 312
    incommensurate, 298
    non-analytic dependence on
      coupling, 333
    periodicity of, 298, 311
    reciprocal lattice, 315
    self-consistency condition for, 322
    self-consistent field, 329
    stability of, 298
    total energy, 328
Charge state stability, 167, 173, 176–177
Chemical potential, 153, 257, 264
Chlorine atom, 164
Classical, atomistic modelling, 126–139
  diffusion, 141–161, 167
Cluster-embedding interaction, 196
Cluster-localized eigenstates in
  Hartree–Fock, 223
Coherent mechanism for diffusion, 188
Cohesive energy, 129–132, 133, 134, 136
Collective coordinates, phonon modes,
  102, 109

Coloration, additive, 140
Combined first and second laws of
  thermodynamics, 104
Commensurate charge density wave, 312
Compressibility, 17, 19
Compression, 3, 6, 10, 29, 34
  hydrostatic, 18
Computational methods, 21
  modelling and simulation, 108
Concentration, fractional, 142, 147, 150
  point defects, 140, 141, 151–154
Condon approximation, 195
Conduction, band, 180, 299
  electrons, 251, 299
  ionic, 140
  modification in BaF_2 by oxygen, 173,
    184
  thermal, 34, 38, 39
Conductivity, thermal, 40
Configuration space for diffusion, 155
Continuity equations, 143–144
Continuum model, 1
  analogy, monatomic linear chain,
    112–113
Contraction, 19
  coefficients, 217, 218, 222
Convection, 39
Copper, 44, 168
  wave attenuation in, 46
  wave speed in, 45
Core, atomic, 127
  charge, shell model, 127
  –shell interaction, 130
Correlation, 166, 167, 194, 196, 229–238,
  246
  correction, optical excitations in
    $NaF:Cu$, 179
  in BaF_2, 180
  second-order, 237–238
Coulomb's law, 23
Coupling constant, electron–electron,
  314
  polaron, 85, 86
Crystal, anharmonic, 93
  atomistic modelling, classical,
    126–139
  basis, 129, 163
  functions, 94, 96–97

Bravais lattice, 129, 163
canonical coordinates and momenta, 98
cohesive energy, 129–131
dynamical matrix, 99
electronic state, 102–103
equation of state, 105–107
field splitting, Cu in NaF optical
    excitation, 179
force constant matrix, 93, 97
ideal, 164
insulating, shell model, 126–129
internal energy, 107
ionic, 164; *see also* crystal, insulating
    polarizability, 128
local mode, 109
normal modes, 100–102
nuclear hamiltonian, 98–99, 101
partition function, 103, 104
perfect, 164
phonon, mode distribution, 107
    state, 102
point defects in, 163–195
primitive translation vectors, 94
    unit cell, 129
quantum, 93
real, 164
Schrödinger equation for nuclei, 102
static lattice approximation, 103
structure, 92, 171
surface atomic relaxation, 164
surface reconstruction, 164
temperature dependence, 106
total energy, 102
translational invariance, 97
volume dependence, 106–107
with center of symmetry, 100
zero-point energy, 103
CRYSTAL program, 218
Crystalline solids, 163
    cubic, 19
CsBr, 165
CsCl, 165
CsI, 86, 165
Cu impurity in NaF, two photon
    excitations, 179
Cu^+ in NaF, 179
Cubic, crystals, 19
    equation, 53

Current density, defect, 141
Cyclotron frequency, 271

Damping coefficient, thermal, 42
Damping factors, 54
de Haas–van Alphen effect, 250, 269,
    283–291
Defect, complex, dissociation of, 166,
    168
    current density, 141
    dipolar, 171–172
    extended, 31
    in BaF_2, 171
    point, 31, *see also* Point defects
    stability, 167, 173
Deformation, 7, 22
    and polarization by a point defect,
        166
    angular, 18
    field, 2, 20, 21
        dislocation, 62–69
    plastic, 57, 59
    tensor, 3
Degenerate perturbation theory, 318
Delta function, representation of, 305
Density, functional, 238–249
    and the two-particle density, 248
    Hartree–Fock, single-particle,
        207–208
    method, single-particle, 239–241
    matrix, reduced, 239, 241
        single-particle, 207
        two-particle, 244
    of states, monatomic linear chain,
        114–115
    two-particle, 243
Diamagnetic susceptibility, low
    temperature, 293–294
Diamagnetism, at $T = 0$, 291–294
    energy considerations in, 281
    hamiltonian for, 270
    in electron gas, 250, 269–294
    strong magnetic field for, 277
    topology of the Fermi surface, 277
    total magnetization in, 287–291
Diatomic linear chain, 109, 116–121,
    see also phonons
    optical branch, 119

Dielectric constant, 135, 138
  effective, for vacancy, 186
  tensor for shell-model crystal, 138
Diffuse excited states, 186
Diffusion, activation energy, 140, 154,
      161, 168–171
  Arrhenius relation in, 161
  classical, 141–161, 167–171
  coherent mechanism for, 188
  concentration profile, 147
  configuration space for, 155
  constant, 142, 143, 150
  equation, 40, 141–145
  equipotential lines in, 154, 155
  force constant matrices in, 159
  fractional concentration, 150
  impurity in alkali halides, 166, 168
  incoherent mechanism for, 188
  interstitial mechanism, 168, 170
  kinetics, 140
  mean jump rate in, 157, 158, 160–161
  mechanisms, 188
  migration energy in, 161
  momentum space for, 155
  non-collinear mechanism, 169, 170
  normal modes in, 159, 161
  particle flux in, 157
  phase space for, 155
  planar source, 144
  probability density, 150
  quantum, 141, 167, 187–189
  random walk, 147–150
  rates, 140
  saddlepoint for, 154, 155
  temperature dependence of, 154
  vacancy mechanisms, 166, 168, 169
Dilatation, 3, 6, 10
Diltatational waves, *see* Waves,
      dilatational
Dipolar defect, 171–172
Dipole, electric, 8
  moments, atomic, 127
Dirac, delta function, 22, 305
  electron theory, 251
Dislocation, 57–72, 164
  atomistic modelling, 72
  core, 63
  deformation field, 62–69

edge, 21, 58, 59, 60
  deformation field, 65–69
  equilibrium equation, 66
  strain tensor, 68
  stress tensor, 68
  line, 60
  loop, 60
  mixed, 59
  processes, 71–72
  screw, 21, 59, 60, 63, 69
    deformation field, 63–65
    equilibrium condition, 64
    strain tensor, 64
    stress tensor, 65
  uniform motion, deformation field,
      69–71
Dispersion relation, 28, 29, 30, 41–43,
      51–54, 112–113
  diatomic linear chain, 117–118
  dielectric medium, 83
  linear elastic continuum, 113
Displacement field, 2, 76, *see also*
      distortion field
  of electron, 76–78
Disproportionation, 176
Dissipative processes, 34
Dissociation, vacancy-impurity defect
      complex, 168, 171–172
  point defect complexes, 140
Distortion field, 24, 48, 50, 93, *see also*
      displacement field
Divergence, 23
Dualism of quantum mechanics,
      particle-wave, 141
Dynamical matrix of a crystal, 97–99

Edge dislocation: *see* Dislocation, edge
Effective, electron–electron interaction,
      299–303
  pairwise interaction, 299–303, 332
  polaron mass, 81–82, 84, 85, 86
  potential energy, 127
  potential, polaron, 81–82
Einstein summation convention, 3
Elastic, constant, tensor, 12
  constants, 35, 37, 51, 139
    adiabatic, 36
    in shell-model crystal, 131

Elastic, constants (*continued*)
  isothermal, 36
  Lamé, 14, 25, 50
  scalar, 14
  moduli, 16–18
  solids, isotropic, 24
Elasticity, linear, 3, 10–20
Electric, displacement, 135, 137, 138
  field, 23
  susceptibility, 135, 138
    tensor for shell-model crystal, 138
Electrical current density, 88
Electromagnetism, 22
Electron–electron coupling constant, 314
  interaction, effective, 299–303, 332
Electron, gas, 250
    diamagnetism in, 250, 269–294
    paramagnetism in, 250, 251–268
  loss, Ni^{3+} in MgO, 177
  magnetic moment, 251
  nuclear double resonance, ENDOR, 181, 186
  phonon interaction, 92, 186, 299, 303
    in ionic crystals, 186
  trapping, Ni^{+} in MgO, 177
Electronic, band structure, 196
  localization, 185–187
  localization, F_2^{+} center in NaF, 185
  localization, hole state in NiO:Li, 185–186
  localization, in $(F_2^{+})^{*}$ center in NaF:Mg, 185
  properties, 31
    localized, 196
  state, 102
  transport, 298
Electrons, in magnetic field, 270–281
Electrostatics, 23
Electrostriction, 8, 136
Ellipse, 56
Elongation, fractional, 4
Embedded molecular cluster, 227
Embedding, in a crystal, 221–229
  potential, 226
  problem, 167, 195
  shell-model crystal, 167
ENDOR, electron–nuclear double resonance, 181, 186

Energy density, solid, 121, 131
  functional, two-particle density, 245
Entropy, 12, 104
  density, 40
Equation of state, phonons, 104–107
Equilibrium, concentration of point defects, 140, 141, 151–154
  concentration of vacancies, 151, 153
  condition, screw dislocation, 64
    nuclei in a solid, 92, 93
  equation, edge dislocation, 66
  in solid continuum, 20
  in statistical thermodynamics, 151, 254
  thermodynamic, 104, 153
Equipotential lines in diffusion, 154, 155
Euler's theorem, 13
Exchange, 204, 205
  charge density, 205
Excited states, diffuse, 186
Exciton, Frenkel, 178
Exclusion principle, Pauli, 198
Exponential, coefficients, gaussian, 217, 218, 222
  decay, 49, 124

F center, 73, 86, 140, 173, 177, 178, 179, 181, 181–182, 185, 186–187, 191, 192
F_2^{+} center in NaF, 173, 182, 185, 191, 192
(F_2^{+}) center in NaF:Mg, 173–176, 191, 192
Fluorine, impurity in MgO, 179
Factorial function, 161
Fermi, distribution, 273–281
  energy, 262, 264, 308
  surface in diamagnetism, topology of, 277
  surface of metals, 291
  wave number, 263, 264, 307
Feynman, polaron theory, 85
Fick's first law, 143
  second law, 144
Field ion microscopy, 216
Field theories, 22
Fluorite crystal structure, 171
Flux density, heat, 39
Foams, polymer, 19

Fock, equation, 194, 208–218
    modified, 220, 223, 228
    solution of, 214
    operator, 213, 230
      modified, 219, 226
    –Dirac density, 207, 219, 228
Forbidden two-photon excitations, 179
Force constants, for local modes,
    effective, 189
    in NaF:Cu, 189
    in MgO:Ni, 189
    in MgO:Cu, 189
    matrix, 93, 97
      in diffusion, 159
Forces, body, 8, 10
    compressive, 10
    dilatational, 10
    external, 10
    interatomic, 8
    internal, 8, 11
    Madelung, 8
    short-range, 8
    surface, 8, 11
Fourier, series, function of many
    variables, 198
    integral theorem, 146
    theorem, 286, 305
Free energy, Gibbs, 35, 153
    Helmholtz, 18, 35, 103, 152, 153, 255,
      256, 257
Frenkel exciton in MgO, 178
Fröhlich, polaron hamiltonian, 85

Gamma rays, 172
    detection, 171
Gauss's theorem, 8, 11, 23, 32, 39, 77
Gaussian functions, 217, 218, 222
GAUSSIAN program, 218
Geminals, 246
Generalized Hartree–Fock, 217
Gibbs, canonical ensemble, 103
    free energy, 35, 153
Gradient, thermal, 37
Grand partition function, 257
Gravity, 10
Green's function, 78

HADES program, 193

Hamiltonian, for a solid, 90–91
    for charge density waves, 300–303
    for diamagnetism, 270
    for paramagnetism, 251–252
    many-electron, 197–198
    polarization field, 75
    polaron, 85
Hard core, atomic, 128
Harmonic, approximation, 13
    for nuclei in a solid, 93
    oscillator hamiltonian, 101
Hartree, approximation, 198, 299
    for charge density waves, 303–306
    total energy algorithm for, 306
    equation, 304
Hartree–Fock
    approximation, 166, 194, 196,
      197–218, 234
    method, 238
    single-particle density functional,
      207–208
Heat, 34
    flow equation, 40, 144
    flux density, 39
Helmholtz free energy, 18, 35, 103, 152,
    153, 255, 256, 257
Hermiticity, 210, 211
Hilbert space, 200, 208, 209, 211
Hohenberg–Kohn theorem, 239, 249
Hole, capture, Ni^{3+} in MgO, 177
    loss, Ni^+ in MgO, 177
Hooke's law, 12, 50
Hydrostatic pressure, 17
Hyperfine interactions, isotropic part,
    181, 182

ICECAP, program, 179, 186, 189, 194,
    195
    method, 167, 178, 193–195
Idempotency, 209, 211
Identical particles, indistinguishability
    of, 243
Identity matrix, 4
Impurity, 164
    optical absorption, in MgO, 178, 179
    in NaF, 179
    in BaF_2, 179–181
    charge-state stability, 176

diffusion in alkali halides, 166
–vacancy dipole complex in NaF, 173
Incoherent mechanism for diffusion, 188
Indistinguishability of identical particles, 243
Insulating crystal, *see* Crystal
Insulator state of a crystal, 313
Interacting electron gas, ground state, 331
    one-dimensional, 309–331
Interatomic binding in a crystal, 92
Interfaces, 31
Intermediate coupling, polaron, 85
Internal energy, 104, 107, 153
Interstitial, 164, 165
Intrinsic luminescence of BaF_2, 180
Ionic conduction, 140
Ionic, crystal, 164
        electron–phonon interaction, 186
        modelling optical excitations in, 178
        point defect in, 193
        polarizability, insulating crystals, 128
        pseudopotentials, 167
Irreversible processes, 43
Irrotational vector, 22
Irrotational waves, *see* Waves, irrotational
Isothermal elastic constants, 35–36
Isotropic, hyperfine constants, 181, 182
    materials, 19, 20

Jump frequency, diffusion, 141
    rate, diffusion, 141

KBr, 86, 186
KCl, 86, 168, 169, 171, 187, 190
    polaron mass, 84
Kelvin temperature, 103
KI, 86, 186
Kohn, Walter, 241, 298
    anomaly, 298
Kronecker delta, 3

L'Hôpital's rule, 260, 333
Lamb, shift, 73
–Retherford effect, 73
Lamé elastic constants, 14, 18, 25, 37, 50, 68

Landau levels, 270–273
Laser, F center, 176
LCAO–MO, 218
Levi-Cività symbol, 9
Linear, approximation, 38
    combination of atomic orbitals: LCAO, 218
    dielectric, 135
    elastic continuum, dispersion relation, 113
    elasticity, 20
Local, band-edge modification, 183–185
    mode frequency, 125
        in MgO:Ni, 189
        in MgO:Cu, 189
        in NaF:Cu, 189
    modes, 109, 121–125, 167, 189
        effective force constants for, 189
    properties on surfaces, 196
Localization, electronic, 167, 185–189
    F_2^+ center in NaF, 185
    $(F_2^+)^*$ center in NaF:Mg, 185
    hole, in NiO:Li, 185–186
Localized electronic properties, 196
Localizing potential, 218–221, 223–228
Longitudinal waves, *see* Waves, longitudinal
Loop, dislocation, 60
Lorentz–Fitzgerald contraction, 70
Luminescence, intrinsic, in BaF_2, 180
Luminescent materials, 171

Madelung, field, 179, 221, 222
    force, 8
Magnetic, induction field, 87, 251
    moment, electron, 251
    permeability, 88
    susceptibility, 254
    vector potential, 87
Magnetism, 250
Magnetization, 254, 268
    diamagnetic, 287–291
    paramagnetic, high-temperature, 268
    the de Haas–van Alphen effect, 283
Many-body, perturbation theory, 231
    problem, 196
    boson wave function, 200
    electron hamiltonian, 197–198

Many-body (*continued*)
  potential, effective, 303
  fermion system, two particle density,
    245–249
  wave function, 200
Many-particle, *see* N-particle
Mass, density, 51
  renormalization, polaron, 73, 83
  effective, 82
Material time derivative, 26, 39
Mathieu equation, 299, 313–331
Maxwell, relations, 34, 35
  equation, 23, 77, 78
Mean jump rate in diffusion, 157, 158,
    160–161
MgO, 165, 176–177, 178–179, 190, 191,
    192
MgO : Cu, local mode force constants, 189
MgO : Ni, local mode force constants, 189
Migration energy in diffusion, 161
Minimum-energy variational principle,
    245
Modelling and simulation,
    computational, 108
  of point defects in ionic crystals,
    165–166
  optical excitations in ionic crystals,
    178
Modified Fock, equation, 220, 223, 228
  operator, 219, 226
Modulus, bulk, 17, 19, 35, 37
  shear, 17, 18, 19, 37, 68
  Young's, 16, 19
Molecular biology, 197
Molecular cluster computations, 196–249
  embedded, 227
  for point defect, 166, 167
Molecular orbital: MO, 218
Momentum space for diffusion, 155
Monatomic linear chain, 109, 110–115
  continuum analogy, 112–113
  density of states, 114–115
  local mode, 122–125
  see also, phonons
Multipole moments, atomic, 127
Muon, 188
Muonium, diffusion in alkali halides,
    187–189

N-body problem, 196
N-particle, density, matrix, 244
  operator, 244
  energy, two-particle density, 244
N-representability problem, 248
NaCl, 164, 165, 166, 169, 188
  polaron mass, 84
NaF, 169, 173–176, 177, 179, 181–183,
    185, 188, 190, 191, 192, 216
  F center, 181
  F_2^+ center, 173, 182
  $(F_2^+)^*$ center, 182
NaF : Cu local mode, force constants, 189
  frequencies, 189
Nanostructured solids, 163
Newton's second law, 25
Ni in MgO, 176–177
NiO, 165, 183–184, 185, 192
NiO : Li, valence band edge in, 183–184
Nobel prize, 216
  in Chemistry, 241
Non-analytic function, 333
Non-collinear mechanism, 169
Normal modes, diatomic linear chain,
    118–121
  crystal vibration (phonons), 100–102
  in diffusion, 159, 161
  monatonic linear chain, 112–115
  point defect in linear chain, 122–125
Normalization, in many-body
    perturbation theory, 232
Normalizing factor for Hartree–Fock
    wave function, 201–202
Nuclei in a solid, Schrödinger equation,
    93–94

Oblate quadrupole strain, 180
Occupation numbers, 253, 262
Occupied manifold, 218
One-dimensional interacting electron
    gas, 331
Optical branch, phonons, diatomic linear
    chain, 119
Optical emission, 178
Optical excitation, 167, 173, 177–181, 184
  forbidden, Cu in NaF, 179
  in crystals, 166, 178
  O^- in BaF$_2$, 173

point defects, 177–181
Overlap matrix, 215, 217
Oxygen in BaF$_2$, 171, 173, 177, 179–181, 193
  local perturbation of conduction band, 180
  oblate quadrupole strain in ground state, 180
  optical excitation, 180, 184
  prolate quadrupole moment of ground state, 180
  spin polarization of ground state, 180
  spin splitting of optical excitation, 180
Oxygen-vacancy defect complex in BaF$_2$, 171

Pairwise, interaction, effective, 332
  interchange of fermions, 197–198
Paramagnetic, magnetization, high-temperature, 268
  susceptibility, 293–294
    high temperature, 266–268
    low temperature, 257–267, 293–294
Paramagnetism, hamiltonian for, 251–252
  in electron gas, 250, 251–268
Particle, density, 205, 304
  flux in diffusion, 157
  –phonon coupling, 187, 188
  –wave dualism of quantum mechanics, 141
Partition function, 103, 104, 151, 254, 255
  grand, 257
Path integral method, 85
Pauli, effects, 167
  exclusion principle, 198, 246, 248
    Hartree approximation, 304
  spin matrices, 251
Periodic self-consistent field, 313
  solution for interacting electron gas, 309–331
Periodicity of charge density wave, 311
Permeability of free space, 254
Permittivity of free space, 23, 135
Perturbation theory, 333
  cautionary note, 333
  degenerate, 318
  many-body, 231

Perturbative expansion, 332
Phase space for diffusion, 155
Phonons, 94–103, 109–125
  dispersion relation, alkali halides, 128
  electron–phonon interaction, 299–303
  equation of state, 104–107
  in quantum diffusion, 187–189
  see also diatomic linear chain
    monatomic linear chain
Piezoelectric, constant, 135, 136, 138–139
  effect, converse, 136
  direct, 136, 139
Piezoelectricity, 8, 138–139
Planar source, diffusion, 144
Plane wave, 28–30, 49
Plastic deformation, 57, 59
Point defects, 121, 151–154
  aggregation of, 140
  calculations, 196
  charge, 166
  complex, dissociation of, 140
  deformation and polarization by, 166
  equilibrium concentrations of, 140, 141, 151
  in an ionic crystal, 193
    modelling, 165, 166
  in crystals, 163–195
  molecular cluster for, 166, 167
  optical excitation, 177–181
  partition function, 151
Poisson's ratio, 16, 19, 52, 53, 54, 56, 69
  negative, 19
Polarizability, ionic, 128
Polarization, 74, 137, 166
  field, 74
    canonical momenta, 75
    equation of motion, 78
    generalized coordinates, 75
    hamiltonian, 75
    interaction with electron, 76
    lagrangian density, 75
  point defect, 166
  vector, 28, 29, 135
Polaron, 73–89, 186–187
  classical, constant velocity, 78
  coupling constant, 85, 86
  effective, mass, 81–82, 85, 86

Polaron (*continued*)
    NaCl and KCl, 84
    potential, 81–82
equations of motion, 77
F center excited state, 86
hamiltonian, Fröhlich, 85
in a magnetic field, 86–89
intermediate coupling, 85
mass renormalization, 83
quasiparticle, 73
self energy, 82–83
total energy, 80
theory, Feynman, 85
velocity-dependent potential, 84
wake, 73
weak coupling, 85
Polymer foams, 19
Pople, J. A., 241
Potential, effective, 82
    energy, effective, 127
    embedding, 226
    momentum-dependent, quantization
        ambiguity, 87–89
Pressure, 10, 17
Primitive, translation vectors, 94
    unit cell, 129
Principal axes, 4, 18
Probability density for diffusion, 150
Prolate quadrupole moment, 180
Propagation vector, 28
Pseudopotentials, 167, 194

Quadrupolar strain, 180
Quadrupole moment, 180
    atomic, 127
Quantitative modelling, 189
Quantum chemistry, 166, 197
Quantum crystals, 93
Quantum diffusion, 141, 167, 187–189
Quantum-molecular cluster, 221
Quasiparticle, excitation, 298
    polaron, 73

Radiation damage, 171, 172, 181
Radiation Effects and Defects in Solids,
    journal, 168
Random walk, diffusion, 141, 147–150
Range of atomic orbital, 216

Rayleigh waves, 48, 49, 125, *see also*
    Surface waves
Rayleigh–Schrödinger many-body
    perturbation theory, 231, 234, 238
RbCl, 86, 169
RbI, 86
Reciprocal lattice, 310
    charge density waves, 315
Reduced density matrix, 239, 241, 243
    single-particle, 244
    two-particle, 244
Renormalization theory, 73
Restricted Hartree–Fock, 217
Riemann zeta function, 261
Rocksalt crystal structure, 164
Rotation, 3, 7, 28
    tensor, 3, 6
Rotational waves, *see* Waves, shear

S, impurity in MgO, 179
Saddlepoint for diffusion, 154, 155
Scalar invariants, 14
Schrödinger equation, 93
    in relation to the diffusion equation,
        144
Screw dislocation, *see* Dislocation, screw
Se, impurity in MgO, 179
Second-order correlation correction,
    237–238
Self-consistency condition for charge
    density waves, 322
Self-consistent field, 246
    charge density wave, 329
    Hartree–Fock, 214, 215
    in Hartree approximation, 304, 306,
        309, 316
    periodic, 313
Self-energy, polaron, 82–83
Self-interaction, 73
Separation of variables, 145
Shear, 3, 5, 18, 28
    angle, 6
    modulus, 17, 18, 19, 37, 68
    waves, *see* Waves, shear
Shell, charge, shell model, 127
    model, 126, 161, 165, 193
        calculations for impurities, 169–171
Shell–shell interaction, 130

Short-range forces, 8
  shell model, 128
Si, 164
Silver, 168
Simulation, computational, 108
Single-particle density, 239
  functional, Hartree–Fock, 207–208
  functional method, 196
  matrix, 207, 244
Size dependence, N-body system, 231
Slater, determinant, 201, 221
  –type exponentials, 216, 217
Slip plane, 59
Sodium, atom, 164
  chloride, 164, 165
Solid, isotropic, 13
Specific heat capacity, 35, 36, 37, 298
Spin, 180
  densities, 167, 181, 182
  density waves, 298
  matrices, Pauli, 251
  N-body system, 245
  polarization, 180
  splitting, optical excitation, 180
Spline fit, 128
Stability, defect complex, 173
  in solid continuum, 18
  of the charge density wave, 298
Static lattice approximation, 103
Statistical thermodynamics, equilibrium,
    151, 254
  of a solid, 103–108
Step frequency, diffusion, 150
Stirling's, approximation, 149
  formula, 152, 161–162
Strain, 1, 35
  in shell-model crystal, 133
  lateral, 16
  longitudinal, 16
  tensor, 3, 4–6, 50, 64
    edge dislocation, 68
Stress, 1, 7–10, 35, 71
  tensor, 8, 65
    edge dislocation, 68
Stretching, 19
Sub-nanoscale technology, 197
Substitutional impurity in crystals, 164
Summation convention, Einstein, 3

Superconductivity, 31, 92, 298, 333
  high-temperature, 183
Surface, 22, 35
  as a crystal defect, 164
  atomic relaxation in crystals, 164
  forces, 8, 11
  local properties on, 196
  reconstruction in crystals, 164
  waves, 48–56, *see also* Wave, surface
Susceptibility, electric, 135
Symmetrizing operator, 200
Symmetry, conditions for N-particle
    systems, 242
  crystalline, 163

Taylor series, 158, 267, 333
Technology, sub-nanoscale, 197
Temperature, Kelvin, 12, 103
Tension, 19
Thermal, conductivity, coefficient of, 39
  expansion, coefficient, 37
  properties, 34
Thermodynamics, classical, 11, 34, 104
  combined first and second laws, 34,
    104, 152, 153
  equation of state, 104
  equilibrium, 104, 153
Thermometric measurement, 35
Three-body forces, 128
Topology of the Fermi surface in
    diamagnetism, 277
Torque, 8
Total, angular momentum, 245
  Coulomb energy, 130
  energy, algorithm, Hartree, 306
      Hartree–Fock, 214, 227
    many-particle, 244
    charge density wave, 328
    Hartree–Fock, 202–204
    modified Hartree–Fock, 221
    solid, 102
  magnetization in diamagnetism,
    287–291
  spin, 245
Trace, 4
Transfer matrix, 187, 188, 189
Translational invariance, 97
Transport, electronic, 298

Transverse waves, *see* Waves shear
Two-particle density, 243
  atomic orbital basis set, 245
  density functional, 248
  energy functional, 245
  many-fermion system, 245–249
  matrix, 244
    reduced, 244

Uniform solution for electron gas,
    306–309
Unrestricted Hartree–Fock, 217

Vacancies, equilibrium concentration of,
    153
Vacancy, 164, 165
  and interstitial mechanisms, diffusion,
    168
  –defect complexes, 171–176
  diffusion, 154, 155, 166
    mechanism, 168–169
    non-collinear, 169, 171
  equilibrium concentration, 151
  –impurity dipole complex, 173
Valence band, 299
  edge in NiO:Li, 183
van der Waals potential, shell model, 128
van Hove singularities, 114
van Leeuwen's theorem, 269
Variational principle, minimum-energy,
    209, 245
Vector, field, 22–24, 31–33
    longitudinal, 24
    transverse, 24
  irrotational, 22
  polarization of, 28
  potential, 251
  propagation, 28
Velocity-dependent, potentials, 84
  field, 25, 26
Vineyard relation, 141, 154, 161
Virtual manifold, 218
Voigt notation, 12, 34
Volume, concentration, 142
  fractional change, 6

Wake, polaron, 73
Wave, attenuation, by thermal
    conduction, 41–47
    length, thermal, 42, 43, 44, 46–47
  charge density, 31, 298–333, *see also*
    Charge density waves
  damping coefficient, thermal, 42
  dilatational, 28–30, 41–47
  dispersion relation, dilatational, 29,
    30
    shear 28, 30
    surface wave, 51–54
    with thermal damping, 41–43
  equation, bulk media, 27
  irrotational, 28–30
  –particle dualism of quantum
    mechanics, 141
  plane, dilatational, 29
    shear, 28
    surface, 49
  polarization, 28, 29
  propagation in bulk media, 27–31
  Rayleigh, *see* Surface waves and
    Waves, surface
  rotational, 27–28
  shear, 27–28
  speed, dilatational, 29
    monatomic linear chain, 111, 113
    shear, 27–28
    with thermal damping, 42, 43,
      44–45, 46–47
  spin density, 298
  surface, 48–56
    dispersion relation, 51–54
    longitudinal component, 55–56
    plane wave, 49
    transverse component, 55–56
Weak coupling, polaron, 85
Wigner–Brillouin perturbation theory,
    233
Work, 11

Young's modulus, 16, 19

Zero-point energy, 103